CLARENDON LIBRARY OF LOGIC AND
General Editor: L. Jonathan Cohen, The Queen

A MODEL OF THE UNIVERSE

The Clarendon Library of Logic and Philosophy brings together books, by new as well as by established authors, that combine originality of theme with rigour of statement. Its aim is to encourage new research of a professional standard into problems that are of current or perennial interest.

General Editor: L. Jonathan Cohen, The Queen's College, Oxford.

Also published in this series

A MODEL OF THE UNIVERSE

Space-Time, Probability, and Decision

STORRS McCALL

CLARENDON PRESS · OXFORD

Oxford University Press, Walton Street, Oxford ox2 6dp

Oxford New York
Athens Auckland Bangkok Bombay
Calcutta Cape Town Dar es Salaam Delhi
Florence Hong Kong Istanbul Karachi
Kuala Lumpur Madras Madrid Melbourne
Mexico City Nairobi Paris Singapore
Taipei Tokyo Toronto
and associated companies in
Berlin Ibadan

Oxford is a trade mark of Oxford University Press

Published in the United States
by Oxford University Press Inc., New York

British Library Cataloguing in Publication Data
Data available

Library of Congress Cataloging in Publication Data
McCall, Storrs.
A model of the universe:
space-time, probability, and decision/
Storrs McCall.
—(Clarendon library of logic and philosophy)
Includes bibliographical references and index.
1. Cosmology. 2. Metaphysics. 3. Space and time.
4. Probabilities. 5. Free will and determinism.
I. Title. II. Series.
BD511.M18 1994 113—dc20 93–5378
ISBN 0–19–824053–8
ISBN 0–19–823622–0 (Pbk)

Typeset by Pure Tech Corporation, Pondicherry, India
Printed in Great Britain
on acid-free paper by
Biddles Ltd. Guildford and King's Lynn

Preface

THALES said that the first principle of everything was water. Democritus maintained that there was nothing but atoms and the void. Each in his own way tried to account for the apparent diversity of things by putting forward a very general hypothesis about the nature of the world.

This book attempts to do the same. The problems addressed are less straightforward than those of Thales and Democritus: the direction and flow of time, the nature of scientific laws, the interpretation of quantum mechanics, the definition of probability, counterfactual semantics, transworld identity, essential properties, deliberation, decision, and free will. But the principle is the same. A very general hypothesis about the world is advanced, with reference to which the problems are explained and clarified. The work attempts to bring together a wide variety of philosophical issues.

I have been thinking about the book's subject-matter, beginning with the problem of temporal passage, since the early 1960s, and the students and colleagues to whom I am indebted for ideas and criticism are too numerous to mention. Nevertheless I will try to mention them. I began the actual job of writing in Australia during a six-month sabbatical in 1988, and without the impetus given by my delightful and generous Australian hosts the book would still be lying in limbo. I want to thank them for helping me put it on paper.

John Bigelow and Brian Ellis of La Trobe played a major role in getting me moving in the right direction, as did David Armstrong of Sydney, to whom I am indebted for many kindnesses and who is the only person in the world who has read the entire work. I had excellent discussions with Henry Krips, Graham Nerlich, Peter Forrest, Chris Mortensen, and Huw Price. Jack Smart kept me on the right path in time and space, except when walking through the Australian bush, and Allen Hazen saved me from philosophical error on more than one occasion.

Meanwhile, in North America, Nuel Belnap has been my intellectual sheet-anchor for many years. About a third of the ideas in the book originated in conversations in his office or living-room, even though we frequently disagree about things. Also in Pittsburgh, it would be fair to say that the idea of a dynamic, objective, branch-attritional, space-time model would never have occurred to me if I had not attended Adolf Grünbaum's lectures on the mind-dependence theory of temporal becoming in 1963–4. Despite the fact that my own views on time flow seem to have emerged in accordance with some Hegelian law of opposites, Adolf and I continue to be good friends. Others in Pittsburgh to whom I am indebted are Nick Rescher, Kurt and Annette Baier, Richard Gale, Jerry Massey, John Earman, and Christopher Hitchcock.

Probably the deepest and most lasting influence on my thinking about time and branching has been that of Arthur Prior. Reading *Time and Modality* when it appeared in 1957, and discussing tense logic with him on the banks of Manchester canals, set in motion ideas and images which eventually culminated in the branched model. I wish, in company with many others, that I could look forward to receiving one of his familiar blue air mail forms containing advice, criticism, corrections, and suggestions for further avenues to explore.

At McGill I want to mention especially the advice and support of my friend John Macnamara. The two of us have given inter-disciplinary courses together on abstruse topics with extremely good students: perhaps, as John remarks, the only courses at McGill where the non-attendance of the professors would raise the knowledge quotient of the class. Special thanks go also to Michael Makkai for comments on Appendices 1 and 2, and to Jim Lambek, Michael Hallett, Mario Bunge, Jim McGilvray, David Davies, Paul Pietrowski, Stephen Menn, and Bernie Margolis.

Others whose help and comments I would like to acknowledge include Twareque Ali, Roger Angel, Bill Anglin, Frank Arntzenius, Leslie Ballentine, Jonathan Bennett, Andy Botterel, Jeff Bub, Nancy Cartwright, Lori Clifford, Rob Clifton, Steve Davis, Roy Douglas, Gordon Fleming, Graeme Forbes, Bas van Fraassen, Michael Friedman, Michelle Friend, Brian Garrett, Giancarlo Ghirardi, Ian Hacking, Alan Hajek, Bill Harper, Geoffrey Hellman, Hans Herzberger, Andrew Holster, Paul Horwich, Colin Howson, David Lewis, Michael Lockwood, John Lucas, Brian MacPherson, Al Mele, Warren Neill, Calvin Normore, Jack Ornstein, Philip Pearle, Peter Riggs, Steve Savitt, Abner Shimony, Brian Skyrms, Quentin Smith, Howard Sobel, Richard Sorabji, Henry Stapp, Geza Szamosi, Kaoru Takeuchi, Alasdair Urquhart, Peter Vallentyne, Peter van Inwagen, and Ian Walker.

Last but not least, this book is dedicated with love and great affection to Ann, Mengo, Kai, and Sophie, who have stuck by

me through thick and thin and patiently listened to years of talk
on the details of time and branching.

S.McC.

McGill University

Contents

1

The Model

I SHALL ask the reader to conceive of the universe—all the objects and all the events to be found in the past, present, and future, together with the spatio-temporal manifold that holds them—in a way that is precise and specific, but which may be dismissed by some as outlandish. In the same way, perhaps, the ancient inhabitants of Miletus dismissed as outlandish Thales' speculations that everything was made of water, or the Thracians Democritus' view that there was nothing but atoms and the void. Speculation on a large scale like this is unpopular, probably even more so among philosophers today than in the time of Thales and Democritus. It may be justified, none the less, on two conditions: (i) that the outlandish model is not vague, but is specified precisely and in detail; (ii) that it may be used to throw a degree of light on otherwise refractory problems.

I ask, therefore, that the reader voluntarily suspend his or her disbelief in the picture of the universe about to be presented, at least for a few pages. In those pages I hope to show that the model can illuminate, perhaps even resolve, some difficult philosophical problems. I have in mind the problems of the direction and flow of time; what causation consists of; the nature of scientific laws; the interpretation of quantum mechanics; objective probability; counterfactuals and related conditionals; the identity of individuals across possible worlds; essential properties; and lastly the nature of practical reason and decision, and the problem of free will. Quite a mouthful, you will say. I agree. If the model I am about to describe stands a chance of throwing some light on even one or two of these problems, it will be worth a few pages' suspension of disbelief.

Here is the model.[1] Think of the universe, to begin with, as a four-dimensional space-time continuum, with every object and event occupying a position within it. Given a coordinate system, each event is located by three spatial and one temporal coordinate. Every enduring object occupies a volume of space-time that is

[1] The model is presented in my paper 'Objective Time Flow' (McCall 1976), and plays a role in McCall (1966; 1968; 1969; 1970; 1979; 1983; 1984a; 1984b; 1984c; 1985; 1987; and 1990).

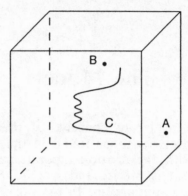

FIG. 1.1

relatively small in spatial extension and relatively large in temporal extension. (Enduring objects like thunderstorms, on the other hand, may be large in space and short in time.) This picture of the world may be embodied in a Minkowski space-time diagram (Fig. 1.1), which we shall refer to as the 'Minkowski world'. The Minkowski world is sometimes known as the 'block universe'. In the small sub-volume of the Minkowski world depicted, A might be the fall of Constantinople, B the accession of Elizabeth I, and the convoluted line C Columbus' voyage to America in 1492.

In the special theory of relativity, Minkowski diagrams also contain forward-pointing and backward-pointing light cones, and give a visual picture of how to transfer from one spatio-temporal coordinate system to another. The difference between a Minkowski world and the universe model being explored in this book is that a Minkowski world consists of a single space-time manifold, whereas the model to be presented consists of a branching set of many space-time manifolds.[2]

Think of all past events, and all past objects, as being represented in the conventional way on a Minkowski diagram. Then think of each physically possible future, relative to the state of the world at a given moment, as being represented on a separate Minkowski diagram which branches off that common past. In general there will be many such futures. If for example a draw for a lottery takes place on 31 December 1999, and a million

[2] It is not intended here that space-time, either in what is called a 'Minkowski world' or in the model to be presented, should necessarily be flat. It could be, and in general will be, curved in ways permitted by the general theory of relativity. In curved, branched space-time, the nodes at which branches split will be Cauchy surfaces rather than spacelike hyperplanes.

different people have purchased tickets for a prize of a million dollars, then, assuming that the procedure of drawing the winning ticket is a truly random one, there will be a million different physically possible outcomes, in each of which a different person wins. Every one of these futures branches off from a single space-time manifold, the date of the node or branch point being 31 December 1999. The universe, then, has in this model the shape of a tree, with a single four-dimensional trunk for the past and a densely branching set of four-dimensional manifolds for the future. Each of these manifolds in turn branches, so that the branching pattern is very complex and the number of branches very large. Any point on any branch above a given node is accessible from that node, but two points at the same level on different branches are inaccessible to one another.

Of all the possible futures represented by space-time manifolds which branch off from the first branch point on the model, one and only one becomes 'actual', i.e. becomes part of the past. The other branches vanish. The universe model is a tree that 'grows' or ages by losing branches.

Suppose that the tree at 12 noon on 15 March 1997 has the following shape as in Fig. 1.2. Then at 12.01 p.m. it may look like Fig. 1.3. Later trees are sub-trees of earlier trees, and since the branching continues into the future either for ever, if the universe has no end, or until Armageddon if it does, there is no fear of running out of branches. It should be emphasized that Figs. 1.2 and 1.3 are highly simplified pictures of the universe model, the branching of which is so dense that the number of branches a short temporal distance above every node is non-denumerably infinite. The need for so many branches is explained in Chapter 4, Section i.

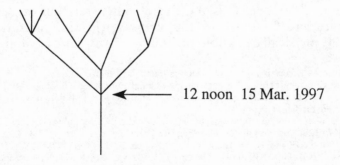

12 noon 15 Mar. 1997

FIG. 1.2

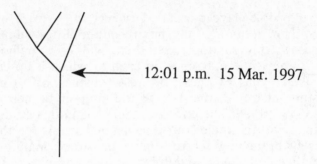

12:01 p.m. 15 Mar. 1997

FIG. 1.3

At each stage, as the present (the first branch-point) moves up the tree, it is a purely random matter which branch survives to become part of the trunk. There is no 'preferred' branch, no branch which is singled out ahead of time as the one which will become actual.[3] Instead, all branches are on a par. All are equally real and, together with the trunk, constitute the highly complex ramified entity I shall call the 'universe'. Since it is never exactly the same at any two times the universe is a dynamic not a static thing. Nevertheless it is the same universe throughout, just as a child can look very different at different times and yet remain the same person the whole of its life.

The species of universe model being proposed can be clarified by contrasting it with the different models which Fig. 1.4 depicts.[4] Each of the models A to D differs in important ways from the tree model. Model A is the Minkowski world. It includes all and only those objects and events which are actual and eschews possibility. In model B possible futures are represented but it is laid down in advance which of them is the 'real' future, the future that will come to pass. Models C and E, on the other hand, play no favourites. Model C is the Everett interpretation of quantum mechanics, also known as the many-worlds interpretation, in which all physically possible outcomes of any set of initial condi-

---

[3] In particular, there is no branch which is 'fated' to become actual, even in a harmless sense to be discussed below. The question of whether human agency or intentionality can single out a favoured branch or type of branch, and nudge it into actuality, is a difficult one, reserved for Ch. 9 below.

[4] Similar pictures of the future as branched and the past as unbranched are found in Prior (1967: 127); Thomason (1970; 1984); Rescher and Urquhart (1971); Lucas (1973: 210–71; 1989: chs. 8 and 9); McArthur (1974; 1976: ch. 3); Burgess (1978; 1980); Thomason and Gupta (1980); Lewis (1980: 94); Goldblatt (1980); van Benthem (1982); Earman (1986: 225); Horwich (1987: 26); Belnap (1991a; 1991b; 1992; 1993). Dummett (1973: 391–2), although containing no diagram, presents the same picture.

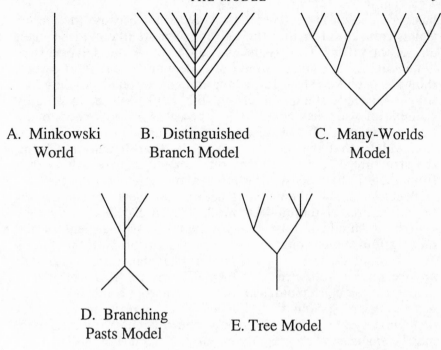

A. Minkowski   B. Distinguished   C. Many-Worlds
   World   Branch Model   Model

D. Branching   E. Tree Model
   Pasts Model

FIG. 1.4

tions are realized.[5] This model differs from the tree model in that none of its branches ever drops off, so that it has the shape of a bush rather than a tree. Model E can be seen as a variation of model C that is being continuously pruned from the bottom up, only one branch being permitted to survive. Finally, model D allows branching towards the past. In it, relative to any instantaneous state of the universe, there are a number of alternative past states from which that state might have come. Model D differs from the other models in that it abandons the principle that the past is unique. Instead it allows for the possibility that two incompatible historical accounts of the past might both be equally correct.[6]

Models A to E are very different. Yet each is, in its own way, a perfectly good model. The reason why, in this book, model E

[5] See DeWitt and Graham (1973). The differences between the many-worlds interpretation and the branched model are discussed below in Ch. 4, Sect. iv.

[6] In his story 'The Other Death' Jorge Luis Borges imagines a situation in which God rewards a soldier (named after Peter Damian, 1007–72) who lost his nerve in battle and then, in shame, lived a humble but moral life, by making it *to have happened* that the soldier behaved in a courageous way after all. (The text in which Damian asserts God's power to alter the past may be found in Wippel and Wolter (1969: 148).) If anything were to trouble the sleep of a conscientious historian, it would be model D.

has been chosen in preference to A–D, is not because the author claims from the start that the universe *is* more like model E than any of the others. The argument is different. *If* the universe were like model E, its being so would provide an explanation of certain things that philosophers for a long time have found puzzling and sought to understand. In effect, the book will be a lengthy example of what has come to be known as an 'inference to the best explanation'. If the inference is eventually made, its conclusion will be that the universe has, in its four-dimensional form, the structure of a tree. But at the moment we are a long way from that. I shall say a little more about inferences to the best explanation in this chapter, but before that there are some additional features of the model I would like to clarify.

First, it should not be thought that the arrangement of the model's four-dimensional branches, or the attributes of the objects and events that these branches contain, have anything to do with anyone's power to conceive of them. The model is an ontological model, not an epistemological one. Its structure is independent of our powers of imagination. Furthermore, there is no vagueness or ambiguity about what the trunk and branches contain. There may be vagueness or ambiguity in our minds, but not in the world. Every event on every branch (with certain possible exceptions in the quantum domain) has 'attribute-specificity'; it lacks indeterminacy or fuzziness.

Again, what objects and events occur where and on which branches is an *empirical* matter, a matter of fact. The events contained in each branch which passes through a given branch point are those which are physically possible relative to the set of conditions obtaining at the branch point in question. And what is physically possible is what is possible as a matter of scientific or empirical fact, not as an epistemic matter. What is physically possible is not what may happen 'for all anyone knows'. A little story will illustrate this distinction.

A chemical engineer has designed a modern plant which requires large quantities of hydrogen and oxygen, and these are stored in two tanks. On the night before the plant is due to begin operations the engineer has a vivid dream in which the gases mix prematurely and an explosion results. Awakening in a sweat, he telephones the superintendent and begs him to check the tanks. 'For all I know', he says, 'the gases may be mixing through that new pipe.' 'Don't worry,' the superintendent says, 'I checked the pipe and the valve is shut tight. The gases can't mix. Go back to sleep.' Note that the superintendent doesn't say that for all *he* knows, everything is OK. He says something much stronger, that because the valve is closed the gases not only *will* not mix, they *cannot*. It is no doubt epistemically possible that the gases should mix, but it is not physically possible. What is physically possible or impossible is not based upon what people think or believe, but upon conditions obtaining in the world.

At each branch point, then, the branches of the universe tree comprise the set of all physically possible courses of events, relative to the conditions obtaining at the branch point. If at a given time it is physically possible for John to cross the street, and at the same time it is possible for an atom of radium to decay, then there are at least four possible futures: one in which he crosses and the atom decays, one in which he crosses and the atom doesn't, one in which he doesn't and the atom does, and one in which neither does. The event-types instantiated on these branches, and the arrangement of the branching structure above each branch point, is an empirical matter, entirely independent of human powers of imagination and conception.

The universe tree can be regarded as a huge cosmic entity, depending neither for its existence nor for its nature upon being cognized by a conscious intelligence. Although on many of its branches, and at its first branch point, conscious intelligences presently do exist, there is no guarantee that this will always be the case. No doubt on many of the 'unused' branches which branched off early in the universe's history and were never followed, conscious beings never appeared, the physical environment being unsuitable for their emergence. Although the universe does in fact have conscious beings at its first branch point, 15 billion years ago there was no necessity that this would be the case.[7]

It should be emphasized that the universe tree contains no branches which are logically possible without being physically possible. For example, it is logically but not physically possible that I who am in Montreal should be in Melbourne one hour from now. No doubt a rocket or an extremely fast aeroplane could get me there in an hour, but no such rocket is presently available. Failing this, I can't get there. There is no branch on the universe tree in which I am in Melbourne one hour from now, though there are some in which I am in Melbourne twenty-four hours from now. If on the other hand the universe tree included branches for all the logical possibilities in addition to all the physical ones, there would be no difficulty about this. There are plenty of

---

[7] The 'anthropic principle' is discussed below, in Chs. 2 and 3.

logically possible branches in which I'm in Melbourne one hour from now, or one second from now, as well as branches in which Melbourne vanishes or expands overnight into a city of 100 million people. On a branch which is merely logically possible, anything can happen which does not involve a contradiction. The notion of physical possibility is much stricter, however, and embraces only a tiny fraction of the logically possible branches.

Expressed in formal logical terms, physical possibility and physical necessity are *relative* modalities, while logical possibility and necessity are not. It makes no sense to speak of an event or a state of affairs being physically possible in itself. On the contrary, it is physically possible only relative to another event or state of affairs at another time. Is it physically possible for someone to swim across the Mediterranean? Yes, given calm weather, warm water, an exceptional physique, and enough support services. Is it possible to jump 15 feet high? No on earth, yes on the moon. Et cetera. In the case of the logical modalities, on the other hand, it is possible for a child of 2 to swim across the Mediterranean in any weather, or to snuff out the sun like a candle. What is logically possible, unlike what is physically possible, depends in no way upon what other events or states of affairs obtain, with the exception of those that logically imply it or logically exclude it. But there is no logical implication or logical exclusion across times, meaning that no state of affairs obtaining at one time either implies or excludes any state of affairs obtaining at any other time.[8] Hence what is logically possible at any time is independent of what states of affairs obtain at other times, while what is physically possible is not. The physical modalities, unlike the logical modalities, are relative. What is physically necessary relative to the initial conditions obtaining at a node is what is on all branches above that node: what is physically possible is on at least one.[9]

[8] In *Tractatus* 2.061–2 Wittgenstein says that one atomic state of affairs is logically independent of all other atomic states of affairs. Although this can be disputed in the case of two *Sachverhalten* with the same time coordinate, since 'x is hot at t' contradicts 'x is cold at t', it cannot be disputed if the times are different. Hence at most an empirical connection, not a logical one, can link events at different times. Care is called for, however, since a 'temporally impure' description of an event can establish logical connections with events at other times. Thus the death of the last Mohican in 1850 (under that description) logically excludes the birth of a Mohican baby in 1950. If events at different times are to be logically 'loose and separate', temporally impure descriptions, which overtly or covertly appeal to times other than the time of the events being described, must be excluded. See Gale (1968: 155–64).

[9] It might be suggested, as an alternative, that a physically possible branch or state of affairs is definable as whatever is permitted by the laws of nature. This alternative is not available, however, since in Ch. 3 below the laws of nature emerge as themselves supervenient upon the structure of (physically possible) branches. As 'physical possibility' and 'law of nature' are interpreted here, both supervene upon the branched structure.

An important feature of the branched model is its *dynamic* quality. Because of branch attrition, its shape is never exactly the same at two different times. But in this there is a lurking difficulty. A three-dimensional object can change through time. But the universe model is a four-dimensional object, and it may be asked, how is it possible for *it* to change through time? The difficulty is well put by J. J. C. Smart, who confesses himself to be a 'Parmenidean' as opposed to a 'Heraclitean' in matters of time and change.[10] Here is how Smart understands the assertion that the four-dimensional universe tree changes.

Think (Smart says) of there being not just one universe tree, which changes through time, but of there being a vast multiplicity of trees-at-an-instant, each of which differs slightly from its neighbours. The image suggested is that of a film strip, or series of instantaneous images, which taken together collectively constitute a 'super-universe'. This 'super-universe', in Smart's words, is 'like a pack of continuum-many cards, one above the other, cards higher in the pack portraying a longer unbranched "trunk" compared with those lower in the pack'.[11] Smart's image of the changing universe tree as a deck of cards, or as a series of snapshots taken at different times, is indeed how a Parmenidean would choose to regard something that grows and changes. The procedure of picturing change in this way is in itself a perfectly good one, used by biologists to depict the process of cell division and development, by town planners to map the growth of cities, and by parents who keep a photograph album of their children. The only error to which it might lead would be the error of confusing the object pictured with the pictures of it—of supposing that the pile of snapshots *was* the object depicted. But this error can easily be avoided.

Consider a movie film, which is a sequence of instantaneous pictures of a changing scene. The film strip, when fed into a projector, creates the *illusion* of motion and change, whereas the objects being filmed *really are* moving and changing. There are exceptions of course, as in the case of animated cartoons where the camera snaps a series of static images each of which differs slightly from its neighbours. But here the exception proves the rule, for cartoons are 100 per cent illusion and the moving figures they seem to depict do not exist.

In the same way, Smart's deck of continuum-many cards can be used to represent change in the universe tree. In this case

[10] Smart (1980: 7).   [11] Ibid.

however there is no illusion: the branched structure really does change. We are speaking, therefore, of two quite distinct things. First there is the multiplicity of pictures or maps of the universe tree, one for each time. Secondly there is the universe tree itself, the thing of which the multiplicity of pictures or maps are instantaneous state-descriptions. This object is a branched four-dimensional object which changes, its mode of change being progressive loss of branches. Despite the loss of branches the tree retains its identity—is the same tree—from one moment to the next.[12] Each change which the model undergoes takes place *at* a time, but does not take place *in* time if this is thought of as requiring a second time dimension to change in. The image of the universe that all this conveys is certainly not a Parmenidean one, but it is not exactly a Heraclitean one either, since the model, unlike Heraclitus' river, remains one and the same throughout the continuous change it undergoes.

The picture given above of model E as a tree is oversimplified in one important respect. It does not take into account the special theory of relativity. The trunk and each of the branches of the tree are four-dimensional space-time manifolds, and the point of division which separates the trunk and the branches at the first branch point is an instantaneous three-dimensional cross-section of the trunk (a spacelike hyperplane). But the special theory tells us that, for any given time, there is no such unique hyperplane. There exists no absolute, global way of dividing events into past, present, and future, as there would be if space and time were as Newton conceived them. If two observers are in motion relative to each other, each is entitled to draw his own 'now' line, which is the line linking all events which have the same time coordinate in his frame of reference (see Fig. 1.5). Since the 'now' lines of the two observers O_1 and O_2 differ, the classes of events they regard as 'future' differ. In O_1's frame of reference, a future event is any event that lies above the line *AB*, while in O_2's frame of reference it is any event that lies above *CD*. Since in the tree model the future is branched while the past is single, the way in which the universe tree branches will have to be relative not only to a time but also to a frame of reference.

The tree model, then, is quite complex. Not only will the universe be different at different times, it will be different at different *frame-times*. At any given time, the shape of the universe tree depends upon the frame of reference or coordinate system

---

[12] Identity through time is discussed below in Ch. 7, Sect. iii.

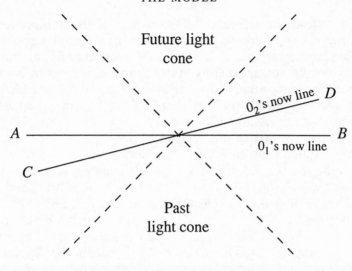

Future light
cone

0_2's now line *D*

A ———————————————————×——————————————— *B*

0_1's now line

C

Past
light cone

FIG. 1.5

used to describe it. This is not an idea to alarm geometricians, for the interesting geometrical properties of bodies that they study are precisely those that remain invariant under changes of co-ordinate system. Felix Klein, in his Erlanger Program, defined the non-perspectival properties of a geometrical object as those which remain constant under all rotations and translations of coordinate axes. The way in which the mode of branching of the universe tree changes from one frame of reference to the next at the same time will therefore be a 'perspectival' property of the tree, anal-ogous to the different aspects presented by a three-dimensional object when viewed from different directions. These differences can be 'transformed away' by shifting to a new coordinate system.

Of course, not all differences in universe trees can be trans-formed away. There is no transformation of coordinate axes that can take us in every case from a tree at one time to a tree at another. Past universe trees, or more correctly past snapshots of the universe tree, can be retrieved, and they are made use of in the discussion of semantics for counterfactual conditionals below in Chapter 6. But in order to find out what tomorrow's tree looks like there is no alternative to waiting and seeing.

It may be asked, if the universe is a tree, what is its topological structure? Each complete path through the tree, i.e. the trunk plus one of the branches, is a four-dimensional space-time continuum or manifold. Every manifold overlaps or coincides with every other manifold for a greater or lesser part of its temporal length, up to the point where they split. Given two such manifolds, the branch point where they divide is a three-dimensional spacelike hyperplane (an instantaneous state of the universe) in some co-ordinate frame or other. What is interesting about branch points is that above them on the tree there will be regions of space-time which are *inaccessible* to each other.

The idea of mutually inaccessible regions belonging to a single connected space is an unfamiliar one in topology, and is explored in the appendices. In Appendix 1, which attempts to give a precise mathematical characterization of the notion of a branched topological space, it is shown that there exist two quite different kinds of such spaces: those named 'lower cut' which are Hausdorff but not locally Euclidean spaces, and others named 'upper cut' which are locally Euclidean but not Hausdorff. In Appendix 2, a frame-invariant description of branching space-time is given, in which it is shown that underlying the set of all frame-dependent models there is an invariant topological structure $[\mathbf{W}, O, \leq]$ where $\mathbf{W}$ is the set of all point-events, O is the set of open subsets of $\mathbf{W}$, and $\leq$ is the causal relation of special relativity. The existence of this underlying structure allows us to speak of all the different frame-dependent models as models of a single object, a single branched world.

One essential, even glaringly obvious feature of the tree model has not yet been discussed, and that is the fact that it is a model of an indeterministic universe. Questions arise here. The model is indeterministic, yes, but under what definition of 'determinism'? Secondly, although the model is indeterministic, is it not at the same time consistent with the species of fatalism known as logical determinism? What exactly is the difference between fatalism and determinism? Finally, what happens to the model if determinism turns out to be true? These questions must be answered.

John Earman, in *A Primer on Determinism*, gives an excellent survey of attempts that various philosophers have made to define determinism. What determinism maintains has been eloquently stated by William James: 'What does determinism profess? It professes that those parts of the universe already laid down absolutely appoint and decree what the other parts shall be. The future has no ambiguous possibilities hidden in its womb: the part we call the present is compatible with only one totality.'[13]

---

[13] James (1897: 150).

Under James's characterization, determinism is clearly inconsistent with a proliferation of physically possible future courses of events, all branching out from a single set of initial conditions. Exactly this profusion is ruled out by Laplace's characterization of the world as knowable in its entirety by an infinite intelligence with access to a single one of its instantaneous states:

An intelligence knowing all the forces acting in nature at a given instant, as well as the momentary position of all things in the universe, would be able to comprehend in one single formula the motions of the largest bodies as well as the lightest atoms in the world . . . to it nothing would be uncertain, the future as well as the past would be present to its eyes.[14]

Earman rightly objects that any definition of determinism that makes reference to the powers of some predictor or other is flawed. If the predictor is endowed with the powers of a universal Turing machine we get the result that future states of the universe are almost certainly not Turing-computable from present ones. If on the other hand the predictor is endowed with divine powers may he not be foreseeing the future rather than predicting it? In this case the future might be 'predicted' by a divine intelligence whether the universe was deterministic in James's sense or not.[15]

A satisfactory definition of determinism should avoid mentioning predictability. Perhaps the clearest is Montague's, which is based on the notion of a possible world and is endorsed by Earman.[16] Let B be the set of all physically possible worlds (in the tree model, the set of all complete paths through the tree). Then the world $w \in B$ is deterministic if and only if for all $w' \in B$, if w and w' agree at any time, they agree at all times. This condition is plainly violated by the tree model, since every branch 'agrees' with every other branch at some time in the sense of coinciding with it, but any two branches agree up to a certain time but diverge thereafter.

Whether the universe is deterministic or indeterministic in Montague's (and James's) sense is, however, quite a different question from whether the theory of 'logical determinism' holds. Logical determinism, a thesis discussed by Aristotle and formulated precisely for the first time by Jan Łukasiewicz in the 1920s (Łukasiewicz 1922; 1930), is based on the law of bivalence which holds that every proposition is either true or false. This includes statements about the future, e.g. that X is in the town square of Warsaw at noon next Friday. Łukasiewicz, who wished to avoid what he took to be the undesirable consequences of the law of bivalence in

---

[14] Laplace (1820), translated in Nagel (1961: 281).   [15] Earman (1986: 7).
[16] Montague (1962: 321).

limiting human freedom, constructed a three-valued logic which permitted future contingent propositions (those that were not causally determined) to be neither true nor false. However, it is possible to avoid the undesirable consequences of bivalence without going to the lengths of abandoning two-valued logic. A 'supervenience' conception of truth, as will be seen, allows the indeterminism of the branched model to be compatible with the thesis that all propositions are either true or false.

Consider any proposition at all about the future, e.g. that X is in Warsaw town square at noon next Friday, or that a given atom of uranium decays and emits radiation in the coming year. These propositions, according to the principle of bivalence, have been true, or alternatively have been false, since the beginning of time, and will remain true (or false) for all future times. But the truth or falsehood of a future-tense proposition need impose no unwelcome limitations on what people can or can't do, or on what may or may not occur as a result of the chance selection of a single 'actual' branch on the branched model. We shall say that the truth of an empirical proposition *supervenes upon* events in the sense of being wholly dependent upon them, while at the same time events in no way supervene upon truth. Thus the truth of the proposition that X is in Warsaw town square at noon next Friday depends upon what happens next Friday, which in turn presumably depends on what X decides to do, and in this way the sting of 'logical determinism' is drawn. If X's actions next Friday somehow supervened upon, or depended upon, the truth of the proposition 'X is in the town square at noon on Friday, 30 March' then logical determinism *would* have a sting, but this is not so. According to what we have called the 'supervenience' theory truth supervenes upon events, not vice versa.[17] What is true today depends upon what happens tomorrow, not the other way round. The set of true propositions in no way determines what the future is like. Instead, what the future is like determines the set of true propositions.

The supervenience theory of truth, and the affirmation of bivalence, may appear to contradict what was said earlier about the difference between the tree model and the 'distinguished branch' model B of Fig. 1.4. The principle of bivalence, together with the principle that truth-values of propositions do not change with time, entails that there exists a unique set of true propositions about the future, very large and very detailed. The existence of

---

[17] The word 'supervenes' as used here emphasizes the asymmetric *dependency* of truth on events, and the *non-dependency* of events on truth. See Kim (1984) for supervenience as a relationship of dependency.

this set, whether or not any person or any thing cognizes it, might be thought to pick out or bestow a privileged status upon one future branch of the model (the 'actual' future) to the exclusion of others. This would be contrary to the 'democracy' principle which puts all branches on a par.

Concerning this objection, two remarks may be made. First, the notions of 'truth' and of 'true proposition' do not refer to things which have a spatio-temporal existence. Nor is there any feature of the branched space-time model which models them, unlike the notions of 'time flow', 'scientific law', and 'probability', which do correspond to features of the model. For this reason, nothing *in the model* picks out a preferred branch corresponding to the set of all true propositions about the future; the model as such exhibits branch democracy. Of other models, such as the distinguished branch model, this is not so. Secondly, because the model may contain two or more branches that are quantitatively identical but numerically distinct, not even the set of all true propositions about the future necessarily picks out a unique branch. As will be seen in Chapter 4, multiple sets of qualitatively identical but numerically distinct branches cannot be ruled out, since they may be needed for the definition of probability in terms of branch proportionality.

To return to the discussion of truth, in the branched model the truth of many (though not all) future-tense propositions depends upon which branch happens to survive branch attrition at each instant.[18] Truth supervenes upon events: upon past events in the case of past-tense propositions, and upon future events in the case of future-tense propositions. But nothing supervenes upon truth, or in any case no events do. Truth 'bakes no bread'. It simply floats above the world, influencing nothing.

Admittedly, some philosophers have thought otherwise. In the 1960s there was a lively debate about whether, given that some proposition about the future was true, it was possible to act in such a way as to cause it to be false.[19] If it were not possible, it was argued, people could do only what they were 'fated' to do, the determining factor in this case being the body of true propositions about the future. But, of course, it *is* possible to make a true proposition about tomorrow false. Since what makes it true

---

[18] Exceptions include propositions which are true because the events which make them true are on all future branches. This point, together with the failure of bivalence for counterfactual conditionals, is discussed in Ch. 6.

[19] See Taylor (1962); Saunders (1965); Cahn (1967); van Inwagen (1983: ch. 2, and the references contained therein).

in the first place is what we do tomorrow, and since there are
many different things that we can do tomorrow, we *can* make it
false. We won't in fact, as it turns out, but we can.[20] Hence the
type of fatalism forced on us by recognizing that there exists a
body of true propositions about the future is entirely harmless,
and in no way limits or restricts human freedom.

The alleged incompatibility of divine omniscience with human
freedom can be dealt with in the same way. It's not that our
future actions are determined by God's knowing what we are
going to do. In fact, it's just the other way round: our future
actions determine what is true, and therefore determine what God
knows. Human freedom, consequently, is in no way limited or
constrained by divine foreknowledge. The fact that God may
know ahead of time which branch is going to become part of the
model's trunk, and which branches are going to drop off, does
not imply that one of the branches is 'preferred' or 'distinguished',
any more than the fact that one of the babies born in the 1960s
will one day become Prime Minister of Canada implies that on
the day of its birth one baby is 'preferred' or 'distinguished'. Like
babies, and like future lottery-winners, all branches on the model
have exactly the same status before they drop off.

So much for fatalism, and 'logical determinism', but real determinism is another matter, *toute une autre histoire* as we in Montreal would say. If determinism were true, then relative to every
momentary state of the world there would be only one physically
possible future, not many. The universe tree would be, from the
start, deprived of all its branches but one, and the branching
model, whatever its explanatory power, would be a false picture
of reality.

Faced with this possibility, what can I say? Yes, determinism
may turn out to be true. Yes, if it is true, the tree model will join
a lot of other unsuccessful models in the land where interesting
but false theories dwell. And no, whether determinism is true or
false is not a question to be settled by philosophers, but an
empirical question to be settled by physical science. Philosophers
can *define* determinism, can specify under what conditions a
scientist or a philosopher would be justified in concluding that we

[20] Taylor argues that no one can do anything, a necessary condition for which is lacking.
If the necessary condition that is lacking is a presently obtaining empirical state of affairs,
then Taylor's principle is correct. No one can saw a plank in half without a saw. But if
what is lacking is the *truth of a future-tense proposition*, then Taylor's principle is not correct.
Truth and falsehood are not empirical states of affairs. What it is possible or not possible
for us to do is independent of what we *will* do, and hence of the truth or falsehood of
the corresponding propositions. We *can*, therefore, do things that we will not in fact do.

lived in a deterministic world. But whether in fact the world
we live in is deterministic is an empirical question the answer to
which is not yet known.

This having been said, there doesn't appear to be much more
that can be added. At present, indeterminism seems to be more
firmly entrenched in physics than at any time since Lucretius in
the first century BC. Modern quantum mechanics is, and has been
since the 1930s, the most all-encompassing and best-confirmed
physical theory ever devised. Yet very few of its predicted observations are of the form 'such and such *will* occur', as opposed to
'such and such *has a probability x* of occurring'. Each new set of
experimental results seems only to confirm the inherently probabilistic character of these predictions more solidly.[21] But of
course, all this could change. A hundred years from now, physics
may have turned its back on indeterminism and embraced a new
and subtle 'hidden variables' theory which replaced probabilistic
predictions with non-probabilistic ones. At the moment, all that
can be said is that such an eventuality seems unlikely. Nevertheless
it could happen, and if it did the tree model of the universe would
have to be discarded.

In this book, therefore, we shall trim our sails to the prevailing
winds in physics and see what can be done with indeterminism.
My own opinion is that it will turn out to be a steady trade wind
rather than a capricious breeze, but I may be wrong. Obviously,
this whole discussion is independent of the *explanatory* powers of
the tree model. They will be discussed in Chapters 2 to 9.
However, before we begin there is one final point concerning
determinism and the Minkowski world that must be made.

Despite the fact that both the Minkowski world and the deterministic universe have, as unbranched four-dimensional space-time
continua, the topological structure of a straight line (see model
A in Fig. 1.4), they differ fundamentally from each other. The
Minkowski world can contain chance events, the occurrence of
which is unconnected to earlier events by any but probabilistic
laws. Assuming for example that a truly random process exists
for selecting the winning numbers in a lottery, the Minkowski
world can contain the events which lead up to, and result from,
such a lottery. This a deterministic universe cannot do. In a
deterministic world the identity of a lottery winner before the

[21] In a recent summary of the present position in quantum physics, Abner Shimony
(1988: 40) asserts that 'the strange properties of the quantum world—objective indefinite-
ness, objective chance, objective probability and nonlocality—would appear to be permanently entrenched in physical theory'.

draw is made is at best epistemically uncertain, a function of our ignorance of the deterministic laws the world obeys. Although the Minkowski model and the deterministic model have the same unbranched topological structure, they should not be confused with each other.

This concludes Chapter 1's preliminary characterization of the universe model which the book discusses, and how it differs from other models. In the remaining chapters the model will be examined from the point of view of the light it can shed on a variety of philosophical problems, and in the process of this examination the precise nature of the model itself will come into better focus. As was said earlier, Chapters 2 to 9 constitute a long build-up to an 'inference to the best explanation'. In them it will be argued that the tree model does provide the best explanation of a number of features of the world, and of concepts like the notions of scientific law and decision, that philosophers have found to be important and puzzling. Since the concept of an 'inference to the best explanation' is itself a philosophical idea that needs clarification, a brief characterization of it will be given before Chapter 2 begins.

Noting its likeness to what has sometimes been called 'hypothetic inference', or 'the method of hypothesis', Gilbert Harman describes the notion of an inference to the best explanation in a short paper published in 1965.[22] From the fact that a certain hypothesis explains something that we accept, an inference to the best explanation allows us to infer that the hypothesis in question is true. Furthermore, since it may be that rival hypotheses would also explain the same thing, we must be able to reject the rivals, or at least be able to show that the favoured hypothesis is 'better' in some sense. If there are no rivals, inference is made to the only explanation available.

Harman offers four examples of such inferences. (i) A detective puts all the evidence together and decides it must have been the butler. No other explanation is plausible enough or simple enough to be accepted. (ii) A scientist explains certain data by postulating the presence of an elementary particle, the existence of which he then infers. (iii) When we infer that a witness is telling the truth, our inference goes as follows: (*a*) we infer that he says what he does because he believes it; (*b*) we infer that he believes what he does because he actually did witness the situation he describes. Our confidence in his testimony is based on the most plausible

[22] Harman (1965); see also Harman (1968). The 'method of hypothesis' is discussed in Laudan (1981a).

explanation for that testimony, and would be shaken if a rival explanation were to be discovered, e.g. that the witness had been suborned. (iv) An inference is made from a person's behaviour to some fact about his mental experience.

We may be pardoned if the inference to the best explanation seems familiar to us. It is familiar. Plato, for example, uses it in the *Phaedo* to argue the truth of the hypothesis that Forms exist:

Well, this is what I mean: it's nothing new, but what I've spoken of incessantly in our earlier discussion as well as at other times. I'm going to set about displaying to you the kind of reason [explanation] I've been dealing with; and I'll go back to those much harped-on entities, and start from them, hypothesizing that a beautiful, itself by itself, is something, and so are a good and a large and all the rest.[23]

Plato goes on to argue that the best explanation—the strongest *logos*—for a thing's being beautiful is that it partakes of the beautiful, hence the beautiful exists. Whatever our own views may be concerning a better explanation of a thing's being beautiful, plainly Plato is inferring to what he considers the best explanation. Other examples abound, from Lavoisier's oxygen to Copernicus's heliocentrism.[24]

The method I shall be following in subsequent chapters consists in showing that the branched model gives the best explanation, indeed in some cases the only explanation, of things such as the flow of time, the probabilities of chance events, etc. At the end of the last chapter, readers will be able to judge for themselves (i) whether the various explanations offered have been successful, and (ii) whether they wish to make the inference to the conclusion that the model gives a true picture of the world. The book makes no physical or metaphysical assumptions concerning the overall structure of space-time. Rather it suggests a certain type of spatio-temporal structure as an explanation.

---

[23] *Phaedo* 100b, tr. David Gallop (Oxford, 1975). David Bostock, in *Plato's Phaedo* (Oxford, 1986), notes on p. 135 that the Greek word *aitia* in the middle of the passage can be translated as 'cause', 'reason', or 'explanation', and I have added the last in square brackets.

[24] Inference to the best explanation is strongly criticized in van Fraassen (1989: 131–69). However, van Fraassen's arguments all concern attempts to arrive at inductive generalizations or hypotheses which best explain the evidence in cases such as rolling a biased die, and have little or no relevance to attempts to explain things like choice and decision, or non-locality in quantum mechanics, which will be the concern of this book.

2

The Direction and Flow of Time

ONE of our deepest and most closely held convictions is that time carries us successively through all the events of our lives, from our birth in the past to our death in the future. No Eastern seer, wrapped in meditation, no lover can free himself from this process:

> Thus, though we cannot make our Sun
> Stand still, yet we will make him run.

However, the idea of time moving from the past to the future is not one that is generally shared by scientifically minded philosophers. In the Minkowski world there is no division into 'past', 'present', and 'future'. Nor could there be, since the special theory of relativity tells us that there is no unique global simultaneity class of events which constitutes 'now'. Instead there are as many 'nows' that can be drawn on the Minkowski diagram as there are times and frames of reference. There is a 'now' for Cleopatra. There is a 'now' for us. There is a 'now' for Mary, travelling in her rocket ship near Alpha Centauri. But where is *the* present? In the Minkowski world, there is no such thing.

Instead of finding past, present, and future, and the flow of time, in the world itself, modern science finds these qualities in the mind of the observer. Thus Weyl:

The objective world simply *is*, it does not *happen*. Only to the gaze of my consciousness, crawling upward along the life line of my body, does a section of this world come to life as a fleeting image in space which continuously changes in time.[1]

And Eddington:

Events do not happen; they are just there, and we come across them. 'The formality of taking place' is merely the indication that the observer has on his voyage of exploration passed into the absolute future of the event in question.[2]

Paul Davies gives a nice statement of contemporary thinking about time flow in the scientific community:

[1] Weyl (1949: 116).　　[2] Eddington (1920: 51).

Relativity theory has shifted the moving present out from the super-structure of the universe, into the minds of human beings, where it belongs. . . . Present day physics makes no provision whatever for a flowing time, or for a moving present moment. Those who might wish to retain these concepts are obliged to propose that the mind itself participates in a novel way in some form of physical activity that is not manifest in the laboratory, a suggestion that meets with a great deal of reserve among the scientific community. Eddington has written that the acquisition of information about time occurs at two levels: through our sense organs in a fashion consistent with laboratory physics, and in addition through the 'back door' of our own minds. It is from the latter source that we derive the customary notion that time 'moves'.[3]

Of those who have written on the direction and flow of time in recent years, unquestionably the most thorough is Adolf Grün-baum.[4] Let us begin with temporal direction. Many of the import-ant philosophical issues surrounding this notion may be brought out by examining a recent debate between Grünbaum and John Earman.

(i) *The Direction of Time*

Eddington's famous image of time's arrow conveys the idea of one of the two directions of time, namely the direction from past to future, being picked out or selected as structurally different from the other. What does the picking out is a physical process, or type of physical process, that takes place in the past-to-future direction but not in reverse. Typical examples (remembered with affection by Grünbaum's colleagues at Pittsburgh) include such things as sitting in a bath and experiencing the mixing of hot and cold water into lukewarm. We *almost never* (Grünbaumian em-phasis) have the experience of sitting in a bath and finding that our head boils while our feet freeze into a solid block of ice. Such entropy reversals can occur, and in a permanently closed stable system reversals do occur with the same frequency as increases. But Reichenbach's 'branch systems', which branch off either natur-ally or by human intervention (as when air is separated into warm and cool volumes by the action of a refrigerator) exhibit, once they are formed, entropy behaviour which is statistically

---

[3] Davies (1974: 3). The reference to Eddington is to ch. 5 of *The Nature of the Physical World* (1928), where Eddington suggests that entropy increase might provide a direction (but not a flow) to physical time.
[4] See Grünbaum (1974; 1967a, 1967b: ch. 1; 1969; and 1973).

irreversible.[5] Another temporally asymmetric example is a steadily expanding spherical shell of photons formed by lighting a match at the top of a mountain on a dark moonless night. Although the reverse of this, a steadily *contracting* spherical wave-front, is theoretically possible given the right boundary conditions, it is to all intents and purposes never seen. It would be rare indeed to stand on the top of the Cathedral of Learning in Pittsburgh and have a matchstick ignited by a contracting spherical shell of this kind.

So far so good, but although the existence of irreversible processes serves to distinguish one temporal direction from the other, it does not indicate which is *the* direction of time. Asymmetric physical processes yield temporal anisotropy, but not a preferred direction. This was a crucial distinction which Grünbaum was the first to make explicitly, though Eddington came very close to it.[6] Consider the Minkowski world once more. The effect of temporally asymmetric processes is to create a 'grain' along the time axis, so that the universe has a different internal structure when looked at in one direction than when looked at in the other. The universe is, in respect of an important class of its physical processes, temporally anisotropic. But this is not equivalent to the existence of *the* direction of time. Temporally asymmetric processes may have given time an arrow, but as Grünbaum remarks an arrow has both a head and a tail, and to speak of 'the' direction is to focus on the head to the exclusion of the tail. The Minkowski universe does not come labelled 'This end up'.

At this point in the discussion, enter John Earman.[7] Earman makes two important points. The first concerns the temporal orientability of the universe, and the second questions whether anisotropy in Grünbaum's sense has anything to do with the direction of time.

With regard to orientability, take any point p in space-time, and consider some small four-dimensional region around p. The infinitesimal 'tangent vectors'[8] at p which reflect the space-time curvature of this region in the general theory of relativity fall into three disjoint categories: timelike, spacelike, and null. The set of null vectors defines the surface of the familiar double light cone at p. Timelike vectors are those which fall inside one of the other of the two lobes of this cone, while spacelike vectors fall outside

[5] Grünbaum (1967a: 161).   [6] Eddington (1928: 87).
[7] Earman (1972: esp. 636–7; 1974).   [8] See Friedman (1983: 35 ff.).

it. Timelike vectors link points between which there is at least the theoretical possibility of some causal connection: spacelike vectors join points between which any causal influence would have to travel faster than light.

Of the two lobes of the light cone, one will be 'future' and the other 'past'. The problem of finding *the* direction of time is to determine which is which. But as Earman points out, any attempt to do this would be defeated if space-time turned out not to be temporally 'orientable', in the following sense.

In any affine space (which we assume physical space to be) it is possible to 'transport' vectors in parallel fashion, i.e. move them in a continuous way along some curve so that they preserve their direction from one infinitesimal region to the next. Let a tangent vector in the 'future' lobe of the light cone of some point be called 'future directed', and a vector in the 'past' lobe 'past directed'. Space-time would turn out not to be temporally orientable if parallel transport of a future directed vector v at some point p, along some continuous path, eventually resulted in v sticking backwards out of p's 'past' light cone. That is, v would be transformed in parallel fashion from a future directed to a past directed vector (or vice versa). Provided that for no point, no vector, and no path in space-time is this possible, we say that the universe is 'orientable' in time.[9]

If the universe is not temporally orientable, no consistent division of timelike vectors into future directed and past directed is possible, and the problem of finding a direction of time is insoluble. On this point both Grünbaum and Earman agree. Where they disagree is over Earman's 'Principle of Precedence':

Assuming that space-time is temporally orientable, continuous timelike transport takes precedence over any physical method of fixing time direction: that is, if the time senses fixed by the entropy method (or the like) in two regions of space-time disagree when compared by means of transport which is continuous and which keeps timelike vectors timelike, then if one sense is right, the other is wrong.[10]

Since the direction of entropy increase in branch systems can vary, even though in the vast majority of those with which we are acquainted it takes place in one direction only, Earman is right to object that the temporal orientability of the universe implies

---

[9] The definition I have given is essentially Grünbaum's (1974: 795–6). Earman's definition is slightly different in that he stipulates that any type of continuous vector transport is allowable provided timelike vectors are kept timelike.

[10] Earman (1972: 637).

that in a certain percentage of cases the entropy method must give us the wrong direction of time. Grünbaum's reply, that a unique direction results if we take observed branch systems, such as ice cubes melting in glasses of ginger ale, to be typical of branch systems in the universe as a whole, rests on the assumption that they *are* typical.[11] This assumption can be questioned. However, Earman is wrong to suggest that continuous timelike transport of vectors provides a better method of establishing temporal direction. As Grünbaum points out, it provides no method at all. At most, temporal orientability is a *necessary condition* for global temporal anisotropy. If we confine ourselves to a single manifold, empty space-time is not anisotropic. Only physical processes, like entropy increase, can endow the two different directions of time with a structural difference. And even then, who is to say which is 'the' direction of time? Taking the argument up to this point, Grünbaum would appear to be right in holding that the picking out of one of the two structurally anisotropic directions of time as 'the' direction is a matter of convention, of *nomos* rather than *physis*.

This is where the debate rests at present. But if we look at the problem of temporal direction in the light of the universe model introduced in Chapter 1, the situation alters radically. In the tree model there is a clear difference between the two directions of time, and this difference is part of the topology of the model. The past is single, and the future is branched. This is not a difference constituted by asymmetries in the physical processes which give the world its content, but is a structural difference which characterizes the world in a formal way. The branched universe itself is anisotropic along its time axis, and this anisotropy derives not from physical processes, but from the topology of space-time.

The universe model under consideration, therefore, provides an answer to the problem of the direction of time. The direction is *from* an unbranched past *to* a branched future. Furthermore, when combined with the notion of flow to be discussed shortly, it is the direction in which time is moving or progressing. In a perfectly clear sense, the universe of the branched model does come labelled 'This end up'.

---

[11] Grünbaum (1973: 791–3). A good short assessment of how consideration of thermo-dynamical branch systems, the formation of which is due to boundary conditions not laws of nature, can lead to statistical time asymmetries, and hence to anisotropy in Grünbaum's sense, is given in van Fraassen (1970: 86–95).

(ii) *Time Flow: The Mind-Dependence Theory*

Let us turn to the problem of the flow of time. Is there such a thing as temporal passage? If so, what does it consist in? If time flows, how fast does it flow? If it carries everything with it, including ourselves, how does time flow differ from a state of rest? These questions have alternately fascinated and irritated philosophers since the days of Heraclitus. Up to now, no agreement on how to deal with them has been reached.[12]

A particularly poetic, and for many particularly irritating, picture of temporal passage is the following, given by Bradley:

Or we seem to think that we sit in a boat, and are carried down the stream of time, and that on the banks there is a row of houses with numbers on the doors. And we get out of the boat, and knock at the door of number 19, and, re-entering the boat, then suddenly find ourselves opposite 20, and, having there done the same, we go on to 21. And, all this while, the firm fixed row of the past and future stretches in a block behind us and before us.

Bradley himself was not satisfied with this picture, for he continues:

If it is really necessary to have some image, perhaps the following may save us from worse. Let us fancy ourselves in total darkness hung over

---

[12] Despite, or perhaps because of, the long-standing intractability of these problems, many recent studies of space and time fail even to mention time flow. Of thirty-eight books published since 1970 that the author has seen, twenty-two discuss time flow and sixteen do not. Those that do include: Davies, *The Physics of Time Asymmetry*, 1974; Davies, *Space and Time in the Modern Universe*, 1977; Denbigh, *An Inventive Universe*, 1975; Denbigh, *Three Concepts of Time*, 1981; Fraser, *Time, the Familiar Stranger*, 1987; Grünbaum, *Philosophical Problems of Space and Time*, 2nd ed., 1974; Harris, *The Reality of Time*, 1988; Hinckfuss, *The Existence of Space and Time*, 1975; Horwich, *Asymmetries in Time*, 1987; Jaques, *The Form of Time*, 1982; Kroes, *Time: Its Structure and Role in Physical Theories*, 1985; Loizou, *The Reality of Time*, 1986; Lucas, *The Future*, 1989; Mellor, *Real Time*, 1981; Morris *Time's Arrows*, 1985; Oaklander, *Temporal Relations and Temporal Becoming*, 1984; Park, *The Image of Eternity*, 1980; Schlesinger, *Aspects of Time*, 1980; Seddon, *Time: A Philosophical Treatment*, 1987; Sorabji, *Time, Creation and the Continuum*, 1983; Whitrow, *The Natural Philosophy of Time*, 2nd ed., 1980; Zwart, *About Time*, 1976. Those that do not: Akhundov, *Conceptions of Space and Time*, 1986; Angel, *Relativity: The Theory and its Philosophy*, 1980; Chapman, *Time: A Philosophical Analysis*, 1982; Costa de Beauregard, *Time: The Physical Magnitude*, 1987; Earman, *World Enough and Space-Time*, 1989; Friedman, *Foundations of Space-Time Theories*, 1983; Hawking, *A Brief History of Time*, 1988; Lucas, *A Treatise on Time and Space*, 1973; Newton-Smith, *The Structure of Time*, 1980; Salmon, *Space, Time and Motion*, 1975; Sklar, *Space, Time, and Spacetime*, 1974; Swinburne, *Space and Time*, 2nd ed., 1981; Szamosi, *The Twin Dimensions*, 1986; Torretti, *Relativity and Geometry*, 1983; van Fraassen, *An Introduction to the Philosophy of Time and Space*, 1970; Whitrow, *What is Time?*, 1972.

Chapman (1982: 62) reflects the general uncertainty: 'As things stand at present the concept of temporal "passage" seems to me too obscure a concept for anyone to be sure whether it has any content or not.'

a stream and looking down on it. The stream has no banks, and its current is covered and filled continuously with floating things. Right under our faces is a bright illuminated spot on the water, which ceaselessly widens and narrows its area, and shows us what passes away on the current. And this spot that is light is our now, our present.[13]

Though expressed in a more literal and less metaphorical way, Bradley's conception of time flow is accepted by C. D. Broad:

It seems to me that there is an irreducibly characteristic feature of time, which I have called 'Absolute Becoming'. It must be sharply distinguished from qualitative change, though there is no d(ibt a connexion between the two. In the experience of a conscious being Absolute Becoming manifests itself as the continual *supersession* of what was the latest phase by a new phase, which will in turn be superseded by another new one. This seems to me to be the rock-bottom peculiarity of time, distinguishing *temporal sequence* from all other instances of one-dimensional order, such as that of points on a line, numbers in order of magnitude, and so on.[14]

Broad's view would seem to accord with the conception of absolute time in classical mechanics which, in Newton's words, 'flows equably, without relation to anything external'. But the general view today of scientifically minded philosophers concerning temporal passage is that it is a subjective illusion. All of us, as conscious beings, are aware of the transient quality of sense experience, in which the sensation of sitting down to breakfast succeeds upon listening to the 7.30 news, and is in turn replaced by the smell of toast and coffee. We are led by this transient quality to attribute temporal passage to the world. But this, according to these philosophers, is a mistake. In the world itself there is no temporal passage. There is no 'moving present'. The very notions of 'past', 'present', and 'future' themselves apply, not to the real world, but to the world of conscious experience. If there were no conscious beings, events would continue to be related to one another by the relations of 'earlier' and 'later', but the categories of 'past', 'present', and 'future' would find no application:

It is of the utmost importance not to confuse time-relations of subject and object with time-relations of object and object; in fact, many of the worst difficulties in the psychology and metaphysics of time have arisen

---

[13] Bradley (1883: 54).

[14] Broad (1959: 766). A witty attack on the notion of time flow is Williams (1951); Smart (1949; 1967) and Prior (1962) are serious and balanced analyses. The first philosopher to have explicitly put forward the idea of time as flowing seems to have been Iamblichus in the 4th c., although Aristotle in the *Physics* 4. 11 may also be suggesting such an idea. On this see Sorabji (1983: 33–51).

from this confusion. It will be seen that past, present, and future arise from time-relations of subject and object, while earlier and later arise from time-relations of object and object. In a world in which there was no experience there would be no past, present, or future, but there might well be earlier and later.[15]

This view, that the notion of time flow, and the associated categories of past, present, and future, owe their sense and content to the presence in the world of conscious beings in a way that the categories of earlier and later do not, has been named the 'mind-dependence' theory of temporal becoming. The mind-dependence theory, according to which the passage of time and its division into past, present, and future occupy roughly the same status as the secondary qualities of colour, sound, and taste did for John Locke, has been laid out persuasively by Grünbaum. I shall show that within the context of the Minkowskian picture of the universe that he accepts, Grünbaum's arguments are incontrovertible. Any attempt to defeat them can be turned aside. But if the Minkowskian picture is replaced by something different, a mind-independent theory of temporal becoming can be consistently introduced.

In his discussion of the direction of time Grünbaum showed, in my opinion conclusively, that neither space-time geometry nor the existence of irreversible physical processes (if there are any) singles out one of the two temporal directions as preferred. At most, such processes confer anisotropy, not a distinguished direction. In establishing therefore which of the directions shall be earlier-to-later and which later-to-earlier, a (non-arbitrary) human choice is needed. Physical anisotropy differentiates the two directions of time *in all but name*, and the role of the conscious observer is to bestow the name 'earlier-to-later' upon one of them.[16] In this sense, therefore, there is a human or 'mind-dependent' factor which enters even into the concepts of 'earlier' and 'later', despite Russell's assertion that these arise not from time relations of subject and object but from time relations of object and object. It is a consequence of Grünbaum's position that not even two successive celestial events a billion years ago could be related by the relation 'later than', without some conscious being existing at some time to choose which direction 'later than' lies in.

[15] Russell (1915: 212).
[16] Grünbaum (1974: 790). Without the anisotropy deriving from irreversible or quasi-irreversible processes, Grünbaum maintains that the best human beings could do would be to single out one of the directions *in name only*. In this case the choice would be a purely arbitrary one.

In discussing time flow and the categories of past, present, and future, Grünbaum finds the same element of choice and 'singling out' entering into their definition. Just as a necessary condition of A's being later than B is that someone, somewhere, has laid down which direction 'later than' denotes, so a necessary condition of a given moment being 'the present' is that a conscious being should so denote it. No conscious beings, no present. No present, no flow of time, and no division into past and future.

What Grünbaum is saying is that there is no objective way of singling out 'the present' from amongst the infinity of different instants on the time axis. There can be only a subjective way. In fact the situation is worse than in the case of time direction, for if there is physical anisotropy then the two directions differ from each other objectively, in a non-arbitrary way. But if we look at the Minkowski world, the choice of a moment in time as 'present' cannot be anything but arbitrary. Nothing in the Minkowski world comes labelled as 'the present'. There *is* no present. So when we speak of 'the present', we can only mean 'the moment we are now experiencing', or 'the time simultaneous (in some frame of reference) with some physical event currently entering into our awareness', or the like. To repeat, no moment in the Minkowski world is preferred as 'now', hence 'now' must be arbitrarily picked out by some observer without whose conscious awareness 'now' would not exist.

Critics of Grünbaum[17] have invariably focused on the irreducibility of 'tensed' to 'untensed' language, how 'now' and other temporal indexicals cannot be defined by appealing only to times and dates, and how 'past', 'present', and 'future' cannot be analysed into 'earlier' and 'later'. But all these criticisms are irrelevant to the point Grünbaum is making, which has nothing to do with language. Grünbaum's point is that in the world, not in language, there is nothing that corresponds to 'the present'. This point, it is true, does concern language to the extent that if someone thinks she is denoting something objective by the word 'now', she is mistaken. But that is the extent of it. Grünbaum's thesis is not primarily about language at all, but about the world, or rather about something lacking in the world. For him, past, present, and future, and 'now', and time flow are lacking in the world. Hence objections based on linguistic considerations, which fail to address the ontological issue, invariably miss the point.

[17] See e.g. Gale (1968).

One recent criticism of Grünbaum is sufficiently ingenious to warrant comment.[18] This is the argument, based upon the anthropic principle, that there is something special about the present era in the history of the universe. What is special is that the Hubble age of the universe T (the length of time since the big bang) could not have been much less than it now is, nor much more. Constraints upon T are that it could not be less than the time at which heavy elements necessary for life were 'cooked' within stars, nor more than the time after which stars which warm planetary systems should have consumed all their fuel and ceased to exist.[19] Hence there is indeed something non-arbitrary about the choice of our cosmic epoch as 'present'. No other epoch *could* be present, since no conscious beings could exist in it. Without conscious beings, it could not be designated as 'present'.

This argument, though initially appearing to confer some objective status upon 'the present', in fact confers none. The universe remains non-anthropocentric. The argument shows, to be sure, that the Hubble age in AD 1994 cannot be less than, say, 10 billion years nor more than 20 billion. But Dicke's argument would show just as well that in AD 25 billion, the Hubble age could not be less than 35 billion nor more than 45 billion. The only thing special about the era in the neighbourhood of the year 2000 is that intelligent beings exist who can formulate Dicke's argument. The argument itself, however, holds for any time. Nothing, in any way, ties 'present' to the year 2000, nor 'present era' to the period from 2 billion BC to AD 2 billion, except the existence of conscious beings. But this is exactly what Grünbaum himself maintains. Presentness is mind-dependent. No argument based upon the anthropic principle, therefore, can refute him.

(iii) *Objective Time Flow*

If the Minkowski model of the universe is adopted, no viable alternative to the mind-dependence theory suggests itself. But replacing the Minkowski model by the branched model of Chapter 1 permits time flow, and the divisions into past, present, and future, to be objective features of the world. The 'present' is the first branch point, the time of which is measured along the time axis of some suitably selected frame of reference. The 'past' is

---

[18] Quentin Smith (1985), 'The Mind-Independence of Temporal Becoming'.
[19] See Dicke (1961). The anthropic principle is discussed further in Ch. 3 below.

the universe's trunk, below the first branch point, and the 'future' is the set of branches above it. Time flow, or temporal becoming, is constituted by the dropping of branches. All these features of the world are perfectly objective, and none depends upon the existence of conscious observers. The dynamic character of the model, constituted by branch attrition, was a feature of it billions of years in the past, and will presumably continue to be a feature of it billions of years into the future. The universe has the dynamic character it has (according to the model) quite independently of the powers of any rational being to conceive it. Time flow would not be a 'phenomenon', in Kant's sense.

It may be asked, if the model is a dynamic one, does it not change? And if it changes, being itself a four-dimensional thing (strictly speaking a set of intersecting four-dimensional branches), does it not need a fifth dimension to change in? Does acceptance of the model not imply acceptance of a second time dimension? This is not an easy question, and in the process of answering it the dynamic character of the model will emerge more clearly.

Briefly, the answer is in the negative. No second time dimension is needed, nor would there be any place for it in the model if we tried to insert it. Each branch, including the trunk, has its own internal time dimension built into it, in virtue of being a four-dimensional manifold. Imposing a metric upon the branched structure provides a way of assigning three spatial coordinates and one temporal coordinate to each point on each branch. No extra time dimension, permitting the assignment of a fifth coordinate to points in space-time, can be accommodated.

Nevertheless, it may be said, if the tree model changes, then it must change *in time*. An apple tree, which blossoms in the spring and bears fruit in the autumn, changes in time. (It may even lose branches.) If we represent this change by depicting the apple tree as a four-dimensional object, then an instantaneous three-dimensional slice of that four-dimensional object will have one appearance in the spring, and a different appearance in autumn. But to represent the four-dimensional universe as changing in this way would require depicting it as a five-dimensional object, a four-dimensional slice of which has one appearance in 1994, and a different appearance in 1995. Hence a second time dimension, according to this line of thinking, would be needed.

This is a persuasive argument, but not a conclusive one. An apple tree, it is true, changes *in* time. But the universe tree, though it changes, does not change *in* time. Rather, its change constitutes the flow of time. Branch attrition, in the model, is what time flow *is*. Therefore branch attrition cannot take place in time, any more than time flow can take place in time. To suppose that it can would be to allow that the question, how fast does time flow, makes sense.[20] No, there is a great difference between the kind of change undergone by an apple tree and the kind of change undergone by the universe tree. Change in the apple tree takes place *in* time. Change in the universe tree *constitutes*, not time itself, but what Broad calls its 'rock-bottom peculiarity', i.e. time flow. Time is what is measured along the time axes of the different branches in the model. Time flow is progressive branch attrition.

In Chapter 1, the dynamic quality of the branched model was referred to, and a distinction was made between (i) the model itself, change in which constitutes the flow of time, and (ii) successive pictures or 'snapshots' of the model, which depict its shape at different times in the past. The difference between (i) and (ii) is important for the theory of time put forward in this book, and is explained more clearly in McCall (1984*b*) than in McCall (1976). Peter Kroes, in his comments on the branched model in his paper 'Objective versus mind-dependent theories of time flow' (1984), points out that the notion of *flow* or temporal *passage* is precisely what is left out if the branched model is regarded in Smart's Parmenidean way as nothing but a collection of instantaneous universe-pictures or state-descriptions. Insofar as it encouraged this latter way of looking at the model my 1976 paper was deficient, and the deficiencies are, I hope, made up for here and in the 1984 *Analysis* paper. The branched model is not a collection of instantaneous pictures, like a movie film, it is the changing dynamic thing that the pictures depict. Its dynamic character is the feature of the universe tree that corresponds to (i.e. 'models') the passage of time. We shall return to the difference between the model and the successive pictures of it in Chapter 6, p. 174.

The species of time flow or temporal becoming that we are attempting to explicate is one described as 'objective', or 'independent of the observer'. But before it can be justifiably described

---

[20] To the question 'How fast does time flow?' Arthur Prior (1958) answers 'One second per second'. The same answer is given in Zwart (1972: 140). But to suppose that branch attrition is a process which takes place in a second time dimension, independent of the first, is to allow for the possibility that time might flow at a different rate: e.g. an interval during which the model lost X branches might extend two seconds when measured by the model's 'internal' time axis, while the same interval might extend one second when measured on a putative 'second' time axis, leading to the absurd conclusion that time was flowing at the rate of two seconds per second. If on the other hand measurements on the two different time axes could never disagree, they would not be independent axes. Insuperable difficulties attach to the idea of having two distinct time axes.

as such, care must be taken to remove any suspicion that the notion of 'objective time flow' conflicts with the special theory of relativity, which denies the existence of a privileged simultaneity class of events, and might be thought to make impossible any coherent notion of non-observer-dependent becoming. Special relativity was discussed in Chapter 1, in connection with the 'perspectival' shape of the branched model. Here it will be discussed in connection with becoming.

What is it for an event to 'take place'? For Eddington and Weyl, it is for that event to swim into the absolute past of some observer. What is it then for an event to take place when there are no observers? Suppose we were to try and capture the idea of an objective non-observer-dependent type of becoming or taking place as follows, using as our matrix the Minkowski world.

Add to the Minkowski world the following dynamic feature: a tiny light that switches on whenever an event 'becomes', or 'takes place'. Each space-time point has its own light, and each light goes on once and only once. Whether a light goes on or not has nothing to do with the presence of an observer. Now, could such a set-up provide a model of objective time flow?

The quick answer is, no. Plainly no random or incoherent pattern of illumination of events in the Minkowski manifold could represent time flow. The only acceptable representation would be an orderly progression of 'happenings' up the manifold: a regimented march of becoming. But in order for a march of this kind to be regimented, events must be placed in simultaneity classes, and the theory of relativity tells us there is no privileged way of doing this. Instead there are many ways, one for each frame of reference, where we take as 'simultaneous' all those events in a single spacelike hyperplane orthogonal to the time axis. In short, no frame-invariant process of becoming could possibly represent time flow, and hence there exists no single consistent way of representing time flow by the switching on of point-like lights in the Minkowski world.[21]

This problem is solved in the branched model by making each instantaneous state or picture of the universe relative to a frame-time, i.e. relative both to a time and to a coordinate frame. A

---

[21] Cf. Gödel (1949: 558): 'The existence of an objective lapse of time . . . means (or, at least, is equivalent to the fact) that reality consists of an infinity of layers of "now" which come into existence successively. But, if simultaneity is something relative in the sense just explained, reality cannot be split up into such layers in an objectively determined way. Each observer has his own set of "nows", and none of these various systems of layers can claim the prerogative of representing the objective lapse of time.'

frame and a time together determine a spacelike hyperplane orthogonal to the time axis, which in the model we have called the 'first branch point' of the structure. Becoming, then, is represented in the model by branch attrition within every frame-dependent or hyperplane-dependent branched structure. This does not make becoming observer-dependent, but it does make it frame-dependent. Time flow is not absolute, but is relative to a reference frame.

Having avoided the danger of absolute pre-relativistic becoming, we must take care not to succumb to a 'becoming' that is trivialized in such a way that it applies to future events as well as present ones. An argument of Hilary Putnam's, dating back to 1967, claims to show that such a trivialization is implied by special relativity.

Suppose, Putnam says, that we are wondering about whether there will be a space-fight tomorrow.[22] The region of space-time in which it might take place is in our absolute future, and hence whether or not there will be a space-fight is, for us, a contingent matter. But consider an observer O some distance away, who has spacelike separation from us (is in neither our absolute past nor our absolute future), and who falls within the 'present', as determined by our coordinate system. Let us suppose that O is moving relative to us, and that in his coordinate system, which is different from ours, the space-fight is taking place 'now' (see Fig. 2.1).

If, for us, the existence of O is determinate or 'real' (because O is in our 'present', not in our 'future'); and if, for O, the existence of the space-fight is determinate or real (because the space-fight is in O's present); then whether there will be a space-fight tomorrow can have only the appearance of contingency. It cannot be genuinely contingent, since it would be impossible for

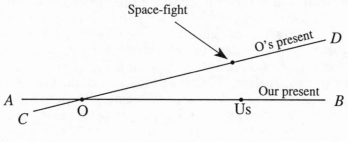

FIG. 2.1

[22] See Putnam (1967). Similar arguments are found in Rietdijk (1966) and Fitzgerald (1969). For discussion see Sklar (1974: 272–5); and Penrose (1989: 303–4).

the space-fight to be taking place today for O and not take place for us tomorrow. Therefore whether or not the space-fight will take place is already 'fixed' or 'determined' by its taking place or failing to take place in O's coordinate frame. Since, as is pointed out by Howard Stein,[23] for any two points a and b in space-time there is a point c in spacelike relation to both, it can always be argued that events at b are 'determined' relative to a in virtue of a's being simultaneous with c in some coordinate system, and of b's being simultaneous with c in some other coordinate system. So, in a trivial sense, all events are decreed by special relativity to be 'determined' or 'real' relative to all other events.

Do these arguments show that the branched model, according to which it may be a matter of objective chance whether or not there is a space-fight tomorrow, is inconsistent with the theory of special relativity? No. Relative to the spacelike hyperplane AB the occurrence of the space-fight tomorrow is indefinite or indeterminate, as is reflected by the fact that the space-fight is found on some branches of the hyperplane-dependent model whose first branch point is AB. But relative to the hyperplane CD, which is O's present, the occurrence of the space-fight is not indefinite but 'fixed'. In the branched picture of the world there is no contradiction here, since each branched model is hyperplane-dependent and an event which forms part of the trunk in one model may lie in the branches of another. If we reserve the adjective 'actual' to describe objects and events which lie in the present or past of the branched model, then what is 'actual' depends upon the choice of hyperplane.

Unlike what obtains in Minkowski space-time, where a description of events in one coordinate system can be transformed at will into a description of the same events in another coordinate system, such transformations cannot be performed in the branched model. A description of the world relative to one hyperplane cannot be transformed into a description relative to another hyperplane. In particular, the only 'transformation' that can be made from a universe-picture at a hyperplane H to a later picture at a parallel hyperplane H', both orthogonal to the same time-axis, is the transformation of waiting and seeing. Contingency is inherent in the branched model, and is progressively removed (though

---

[23] Stein (1968: 12). Stein gives an excellent critique of Putnam's and Rietdijk's arguments, which aim to show that special relativity is incompatible with the notion of temporal passage as a becoming real, or becoming definite, of what is not yet real or definite. But Stein's own view of becoming is that, for any observer O, those events which have 'become' are just those within O's absolute past (p. 14), and this makes becoming observer-dependent.

never exhausted) by the process of relativistic branch attrition, i.e. time flow.

Nicholas Maxwell, in a recent paper, maintains that what he calls 'probabilism'—the view that the future is open with many ontologically real alternative possibilities—is incompatible with special relativity.[24] Maxwell bases his argument upon the fact that special relativity asserts the equivalence of all inertial reference frames, denying the existence of a preferred absolute 'present' which (Maxwell claims) is needed in order to make sense of probabilism's open future. In his words: 'Ontological probabilism . . . asserts that future events have physically, ontologically real alternative possibilities associated with them which are progressively annihilated as the future becomes the present and the past.[25] If indeed the idea of an ontological difference between a single fixed past and a multiple open future required a Newtonian non-relativistic present, then Maxwell would be right in saying that probabilism and special relativity were incompatible. But one of the principle theses of this book is that the distinction between the (single) past and the (branched) future is a frame-dependent or better hyperplane-dependent one. In particular, as the imaginary example of the tiny lights switching on in the Minkowski manifold showed, no attempt to represent the 'progressive annihilation' of future possibilities in a non-frame-dependent way could conceivably meet with success. Special relativity in fact, far from being incompatible with probabilism and the idea of alternative future possibilities, is consistent with them. In the branched model, every complete path through the tree is a Minkowski space-time, and the division between past and future hinges upon the choice of a hyperplane which intersects *here-now*. A description of the branched structure which underlies *all* hyperplane-dependent models—what may properly be called the 'Universe'—is given in Appendix 2.

(iv) *McTaggart's Alleged Proof of the Inconsistency of the A-Series*

In preceding sections it was seen that on the branched model well-defined regions correspond to past, present, and future, and that the borderline between these regions is constantly altering.

[24] Maxwell (1985). See also Dieks (1988); Maxwell (1988); and an excellent critical discussion in Stein (1991).
[25] Maxwell (1985: 27).

The present—the first branch point—moves steadily up the tree. If, however, what McTaggart tried to demonstrate in the early years of this century were correct, no such consistent model of past, present, and future could exist. McTaggart claimed to have shown that the A-Series, the characterization of times and events as 'past', 'present', and 'future' which for McTaggart constituted the essence of time, engendered a contradiction. The arguments by which McTaggart produced this contradiction, though looked on with suspicion by many, have been endorsed by some.[26] If therefore there exists an objective model of past, present, and future which permits these categories to form part of a consistent theory, there must be something wrong with McTaggart's proof.

McTaggart's proof runs as follows. The determinations 'past', 'present', and 'future' apply to events, and they are incompatible. Every event must have one or the other, but no event can have them all. Nevertheless, every event *does* have them all. If e is past, it has been present and future; if future, it will be past, etc. Letting P stand for past, N for present, and F for future, we have Pe and Ne and Fe which are incompatible.

The instinctive reply is that e is not *simultaneously* past, present, and future, but (for example) *is* present, *was* future, and *will be* past. We thus obtain NNe, PFe and FPe, so that the apparent incompatibility is removed. But, according to Dummett and Mellor, the move to compound tenses will not help, for there are nine of these, viz. PP, PN, PF, NP, NN, NF, FP, FN, FF, and no event can have all of them. In particular, the predicates PP and FF are incompatible. If we attempt to counter by saying that PPe and FFe are not simultaneously true, although $FPPe$ and $PFFe$ are, then this move to triple compound tenses will not help either, for $PPPe$ and $FFFe$ are likewise incompatible. Every attempt to get rid of the incompatibility by moving to a more complex compound tense is apparently met by a further incompatibility at the higher level.[27]

This argument is a sophism. At *no* level are all the tensed attributions simultaneously true of any event e, so that although $PPPe$ and $FFFe$ are incompatible, no one would ever wish to assert more than one of them at any one time. 'X is hot' and 'X

[26] Notably by Michael Dummett, D. H. Mellor, and Paul Horwich. See Dummett (1960); Mellor (1981*a* and 1981*b*); and Horwich (1987). Those who reject McTaggart's argument include Broad (1938: ii. 313–17); Smart (1967: 128); and Prior (1967: 4–7). For a variant of McTaggart's argument see Schlesinger (1980: ch. 3). A lengthy discussion is found in Gale (1968).

[27] McTaggart (1927: 329–33); Dummett (1960: 497–8); Mellor (1981*a*: 92–4).

is cold' are incompatible, but no contradiction results if they are not asserted about the same thing at the same time. McTaggart's argument may appear to produce the contradiction he desires, but this appearance is illusory. The existence of the branched model provides a clear proof of the consistency of the concepts of 'past', 'present', and 'future', and of 'temporal passage', in the same way as model-theoretic semantics provides a clear proof of the consistency of first-order logic.

It would not be worth making more than passing mention of McTaggart's proof were it not for the far-reaching consequences that its correctness would entail. Mellor, for example, uses McTaggart's conclusions as the basis for an extensive and well worked out theory of time that provides truth-conditions for both tensed and untensed utterances. And Horwich bases his criticisms of temporal passage on McTaggart's argument. Since the theory of time implied by the branched model in this book differs fundamentally from Mellor's, and from Horwich's, it is worth while making the differences explicit.

Mellor distinguishes two separate schools of thought on the subject of time, which he calls the 'tensed' and the 'tenseless' views. This difference should not be interpreted as involving merely language, for Mellor maintains that both tensed and untensed linguistic forms are legitimate, indeed indispensable in their proper context.[28] Instead, the tensed and the tenseless conceptions are different metaphysical theories, different views about what time and temporal relations are. The 'tensed' school of thought maintains that the distinctions of past, present, and future reflect real non-relational differences between past, present, and future things and events. This is denied by the tenseless camp. What makes 'The train will arrive in ten minutes' true is merely the fact that its utterance takes place ten minutes before the train arrives. This relation, 'ten minutes before', is a tenseless relation. The tenseless school denies that there is anything at all that makes an event 'future', apart from its being later than the time at which someone refers to it as future. Similarly for 'now'. What makes 'It is now two o'clock' true is just the simultaneity of the utterance of that sentence with what Mellor calls a 'date', viz. a particular chronological (tenseless) time. When exposed to tenseless analysis, the distinctions of past, present, and future disappear.

And disappear they should, according to Mellor, for McTaggart's proof has shown them to engender contradictions. But this is the

[28] Mellor (1981a: 73 ff.).

point where the argument leaves the rails. We may grant that tenseless truth-conditions can be provided for tensed utterances. However, the further step of saying that there is no such thing as past, present, and future, that these characterizations are ultimately incoherent because contradictory, is false. It is demonstrated to be false by the fact that the branched model furnishes a consistent picture of a universe with a clearly delineated past, present, and future, a world with changing but real non-relational differences between past, present, and future events. Since the model is clear and *widerspruchsfrei*, so is what Mellor calls the tensed theory of time. No logical contradiction ensues from picturing time as flowing, with a constantly changing present separating the future from the past.

Paul Horwich, despite his endorsement of McTaggart's proof, recognizes that the tree model, if adopted as an objective or metaphysical representation of the world, is capable of depicting what he calls the 'moving *now*'. But at the same time Horwich maintains that the tree model is on no account to be interpreted objectively. Instead it is to be taken epistemologically, as a representation not of the world but of what human beings can know about the world. Specifically, the tree model according to Horwich reflects a verificationist theory of meaning and truth, the roots of which, he claims, are to be found in Aristotle's *De interpretatione* and the modern flowering of which are seen in contemporary antirealism. This theory, in one of its variants, holds that we can understand a sentence only to the extent that we know how to recognize if it is true.[29] In the specific case of future-tense propositions, we can understand a prediction that some event will occur tomorrow only if we are *now able to tell* whether or not the prediction in question is true. But typically, in the case of predictions, this is just what we are *not* able to do. Instead, we have to wait and see what happens. These considerations lead to Aristotle's views about future contingents, to three-valued logic, and (according to Horwich) to the tree model.

The tree model, in Horwich's eyes, furnishes us with an excellent picture of how the verificationist sees the world. Since there may be nothing in the state of the world today which serves as a present determinant of the truth or falsehood of a statement about tomorrow, the tree model treats the future not as unique but as multiple, and puts in all possible future courses of events without

[29] Horwich (1987: 32); cf. Dummett (1978: pp. xxii, xl, 146). One instance of the present book's divergence from contemporary antirealism emerges if we ask, what makes statements about the past true or false? See e.g. Dummett (1969), McDowell (1978).

any indication of which is *the* future. Propositions about the future are neither true nor false, nor are they even intelligible, in the absence of any method of determining whether they are true or false today. But, says Horwich, the verificationist sets the standards for truth and for intelligibility extraordinarily high. Why should the truth of a future contingent, as the verificationist maintains, depend upon conditions or events that exist *now*? Why shouldn't it depend upon what happens *tomorrow*? If the verificationist were to relax his strict requirements, a single future, according to Horwich, would suffice, the need for multiple possible futures would disappear, and the tree model could be dispensed with. The main reason for adopting antirealism about the future, that is to say a multiplicity of possible futures rather than just one, may be removed by attacking antirealism's verificationist roots.

The conception of truth which Horwich proposes as an alternative to verificationism is similar to, if not identical with, the conception described in Chapter 1 as the 'supervenience' theory. This theory holds that truth supervenes upon events, that the truth today of the proposition that I shall tie my shoelace at noon tomorrow depends upon what I do tomorrow, that what I do tomorrow does *not* depend upon what propositions about the future are true today, and that worries over fatalism or 'logical determinism' are unfounded.[30] But adopting the supervenience theory of truth, and upholding the principle of bivalence which

[30] See Horwich (1987: 30). It is significant that even though Aristotle's discussion of future contingents has led some philosophers to propose abandoning the principle of bivalence, on the grounds that the truth of a true future-tense proposition *p* today makes it impossible for anyone to act in such a way as to make *p* false tomorrow, Aristotle's own conception of truth in the *Categories* and *De interpretatione* supports the 'supervenience' theory. In ch. 5 of the *Categories*, on substance, he asserts that only substances can, in his words, 'receive contraries' by changing, and he then goes on to consider a hypothetical objection to the effect that statements and beliefs might also 'receive contraries' in the same way: e.g. the statement that someone is sitting might be true at one time and false at another. But Aristotle rejects this possibility. 'Statements and beliefs . . . remain completely unchangeable in every way; it is because the *actual thing* changes that the contrary comes to belong to them . . . For it is because the actual thing exists or does not exist that the statement is said to be true or false.' (4a34–7, 4b8–9, translation in Ackrill (1963)) Further on, he reiterates the insight that statements and beliefs are passive and unchanging, depending for their truth or falsehood on the active world of substances: 'whereas the true statement is in no way the cause of the actual thing's existence, the actual thing does seem in some way the cause of the statement's being true; it is because the actual thing exists or does not that the statement is called true or false' (14b18–22). And again later on, in the *De interpretatione*, 'it is not because of the affirming or denying that it will be or not be the case' (18b38–9), to which we might add, completing Aristotle's thought, 'but because it will be or not be the case that we can truly affirm or deny today'. Put in contemporary terms, what Aristotle appears to be saying is that truth supervenes upon objects and events in the sense of depending upon them, but that objects and events do not supervene upon truth.

asserts that every proposition is either true or false, does not entail rejecting the tree model. As will be seen in the next section, the tree model as an objective structure can be combined with a great variety of different semantical truth-conditions, including bivalent ones. Branching structures play a central role as 'frames' or 'model structures' in semantics, without being tied to any particular set of meaning postulates. One can therefore reject a verificationist interpretation of future-tense propositions while still retaining the tree model as an objective picture of reality, the acceptability of the model resting ultimately upon its explanatory capacities.

For Horwich, there is no metaphysical or ontological difference between past and future; no inherent asymmetry to time.[31] Despite this, there exist many varieties of temporally asymmetric phenomena. Horwich distinguishes at least ten of these (1987: 4), and to account for their asymmetry he postulates at the origin of the universe the existence of the following set of conditions:

1. an uneven, i.e. asymmetric, distribution of energy constituting a high degree of *macroscopic order*; as well as
2. the highest possible degree of *microscopic disorder* compatible with the aforementioned macroscopic order (1987: 72, 201).

The overall thrust of Horwich's book consists in arguing that these conditions provide for the asymmetry of all other temporally asymmetric phenomena, including entropy increase, the fact that we know more about the past than the future, etc. But how much more perspicuous to argue, if it were possible, that these asymmetries derived from an asymmetry in the topological structure of space-time itself. Such an asymmetry, if Horwich could accept it, would explain a great deal. The asymmetric tree model is our central theme, and it will be the burden of this and of succeeding chapters to argue its explanatory virtues.

(v) *Truth-Conditions for Temporal Discourse*

Although this book is primarily about the universe and how to model it, about ontology and metaphysics rather than logic, something should be said about those logics of temporal discourse that seem to fit most naturally with the branched model.

The word 'model' can be used, as we have been using it, to denote a representation of something, as when an economist puts

---

[31] Horwich (1987: 1, 26, 37–8, 42, 57).

forward a model of ideal competition. It is also used, in formal semantics, to refer to a means of providing a set of sentences or well-formed formulae with truth-values. We say, in formal semantics, that a sentence S is 'true in a model M'. Let us see if the branched model can also serve as a semantic model in this sense.

A semantic model consists of two parts. The first is a *non-linguistic component* which can be, as in quantification theory, a domain of individuals, or, as in modal logic, a set of 'worlds' or 'points' ordered by a relation R. The second component is an *assignment function* which links items of language to elements or sets of elements in the non-linguistic domain or model structure.

The branched universe modal, complete with all the individuals and events located within its four-dimensional branches, can serve admirably as the non-linguistic component of a semantic model. Given a language, an assignment function can be found which links items in this language with items in the universe model. Perhaps the simplest such language would be one that contains only predicates describing the spatio-temporal curvature, fields, forces, and distribution of matter at points in space-time. In that case the assignment function would assign to each predicate an appropriate set or sets of space-time points in the model. If the language were a language such as English, the assignment function would associate names like 'Socrates' and predicates like 'snub-nosed' with features of the model, in a natural way.

In addition to a model M, a semantics needs a set of rules for assigning truth-values to sentences. These rules are sometimes collectively referred to as a 'valuation function over a model M', and constitute a third quite independent component of the semantics, distinct from the non-linguistic component and the assignment function. The choice of what method to use in giving truth-values to sentences is not dictated by the model. This means that there is a large variety of different ways in which propositions can be evaluated, each of which is compatible with a number of different models, and each of which leads to a different semantics for temporal discourse. Here are some of the alternatives.

First, we have a choice of how many truth-values to employ. Our semantics can be bivalent (every sentence either true or false), or trivalent (true, false, or 'indefinite'), or multivalent, or it can be such that the valuation function is a partial function, not everywhere defined. In the last case there exist 'truth-value gaps'. The most popular contenders for the third truth-value, or for truth-value gaps, are of course propositions about the future. We shall return to such 'future contingents' shortly.

Secondly, we have a choice of where or when on the branched model to evaluate sentences, or more precisely with respect to which parts or components of the model to evaluate them. For example, we might choose always to evaluate sentences at a point on a branch, or at a time (a spacelike cross-section) on a branch, or at a time but on no particular branch. Or again, we could choose to evaluate sentences with respect to 'histories', i.e. complete paths through the tree.[32] Each of these different alternatives yields a different semantics.

Finally, there are the truth-conditions themselves; the conditions under which a temporal proposition is true or false. These may differ widely. For example, if the proposition 'The dam will burst tomorrow' is evaluated at a branch point over a branched model structure, should it be counted as 'true' if the dam's bursting figures on at least one future branch, or should it be true only if the bursting figures on all branches? Obviously there are many alternatives.

We should not be surprised, therefore, if a given semantic model can serve as the basis for many different methods of valuation in correspondingly many different semantics. This is so in the case of the branched universe model. As an illustration, consider the different truth-values obtained for (i) tenseless sentences, and (ii) tensed sentences, using either (a) bivalent valuations, or (b) non-bivalent valuations, all on the tree model and all adopting the option of evaluating sentences at the present, i.e. at the model's first branch point.

Beginning with (ia), what are the truth-conditions for the tenseless sentence 'Miroslava Griffin is elected prime minister of Canada in 2031', evaluated in 1995? In Chapter 1, it was seen that it is quite consistent with the indeterminism of the branched model to suppose that there exists a complete set of true propositions about the future, so long as it is realized that the composition of this set depends entirely upon what happens in the future, i.e. upon branch attrition between 1995 and 2031. What is true now depends upon future events, not future events upon present truth, and in this way fatalism about the future is avoided, or at least the species of fatalism known as 'logical determinism'. It follows

---

[32] This last alternative is adopted by Richmond Thomason (1970), who uses van Fraassen's method of supervaluations to distinguish between *what will be*, and *what will inevitably be*. His semantics are non-bivalent. A different approach is taken in the author's (1979), in which bivalent truth-conditions are given for *Fp* ('*p* will be') and *Sp* ('*p* will inevitably be'). Although they are both based on the same branching model structure, the resulting tense-logics are very different.

that bivalent truth-conditions for tenseless sentences are perfectly feasible over the branched model, and lead to no philosophically unpalatable consequences.[33]

The same holds for (ii*a*), bivalent truth-conditions for tensed discourse. For past-tense statements there is no problem. The truth-value of 'The Israelites crossed the Red Sea dry-shod' hinges on events to be found in the branched model's trunk. (Note that though there is no question of inspecting the model's trunk by travelling to it, its objective character makes it at least theoretically knowable. There is a 'fact of the matter' about whether specific events are located in it or not.) Concerning future-tense statements, 'It will rain tomorrow' is given the value true, or false, by future events in exactly the same way as in the case of tenseless propositions.

One of the advantages of the branched model is that it provides a changing semantic reference-point for tensed discourse. 'Now', i.e. the present, is always located at the first branch point, and the dynamic character of the model ensures that the denotation of 'now', and hence the truth-value of 'It will rain tomorrow', varies with time. Of few if any other semantic model structures is this so.

Turning now to non-bivalent truth-conditions, in category (i*b*), the tenseless proposition 'Rwanda and Burundi reunite in 2000' may be assigned a value other than true or false, or may be assigned no truth-value at all, if on some branches the two countries unite, and on some they do not. Exactly the same holds in category (ii*b*) for 'It will rain tomorrow'. In all these cases suitable truth-conditions, based on the branched model, can be constructed.

Perhaps enough has now been said to show that the branched model, when taken as a model in the semantic sense, is compatible with a wide variety of valuation functions. This is not the place to list the large range of chronological logics that can be based upon it. The most distinctive feature that it possesses, its dynamic quality, which sets it apart from all other semantic models hitherto studied, has not yet been explored.[34] An obvious area of application

---

[33] The author confesses to having changed his mind on this matter. Up to 1976, when I wrote 'Objective Time Flow', I was a follower of Łukasiewicz and regarded future contingents as neither true nor false. But one day I realized that in the field of human action the notion of free choice amongst physically possible alternatives could coexist peacefully with sempiternal truth, and since then have happily embraced bivalence, except in the case of counterfactuals and some future-tense conditionals (see Ch. 6 below).

[34] All the models discussed in vol. ii of Gabbay and Guenther's *Handbook of Philosophical Logic* (1983–9) appear to be static rather than dynamic. See esp. the sections by John Burgess on tense logic, and by David Harel on dynamic logic.

for dynamic models would be in the semantics for formal systems which describe the behaviour of Turing machines, and for cognitive and biological systems, where the question of which future states of a given device or organism are realizable depends on which states have already been attained. But these are questions for the future. The purpose of this section has been only to demonstrate the extreme liberality of the constraints placed on a semantics by the adoption of a dynamic tree structure as a model. No reasonable method of assigning truth-values to either tensed or tenseless discourse on the basis of such a model seems excluded.

(vi) *Cambridge Change and Real Change*

The kind of change that things and events undergo when they move from the future to the present to the past, and which they continue to undergo as they recede ever farther into the past, is sometimes deprecatingly referred to as 'Cambridge change'. The expression comes from Peter Geach, who in a book on the philosophy of McTaggart presents his readers with a beautiful metaphysical problem about change that deserves to be better known.[35] Far from being unimportant, the Cambridge change of things and events in time is a necessary condition of all other change, and (in the branched model) derives from changes that are more than just Cambridge changes.

In the early 1900s philosophers at Cambridge, including Russell and McTaggart, shared an account of change which applies quite generally to things whether in time or out of it. This is what Geach calls 'Cambridge change'. An object O undergoes Cambridge change if and only if there are two propositions about O, differing only in that one mentions an earlier and the other a later time, and one is true, the other false.[36] A poker which is cold at t_1 and hot at t_2 undergoes Cambridge change, but so does the Great Pyramid of Egypt when a sandcastle falls on a beach in England, and so does even the number two, as Geach points out, for '2 is the number of X's children at t_1' may be true, and '2 is the number of X's children at t_2' may be false. Although Aristotle is long dead, he undergoes a Cambridge change each time a new student comes to study him; indeed, he changes Cambridge-fashion whether anyone thinks of him or not, for each day the date of his birth recedes farther and farther into the past.

[35] Geach (1979: 90 ff.).   [36] Cf. Russell (1903: 469); also Geach (1968: 13).

Now we may not think that Cambridge change, which affects everything all the time, is worth very much. In particular it may not seem to be worth much in comparison with real change, which affects relatively few things and for limited periods of time. The poker suffers real change; Aristotle and the number two do not, though Aristotle did when he was alive. The death of Socrates effected a Cambridge change in Xanthippe when it made her a widow. As Cambridge change is defined, instances of real change form a proper subclass of instances of Cambridge change. But though real change may appear genuine, while Cambridge change has a phoney look to it, in actual fact Cambridge change, according to Geach, is the only clear, sharp conception of change we have.

The definition of Cambridge change is precise and unambiguous. But how is real change to be defined? Geach confesses that he does not know, and I don't think anyone else knows either.

Consider the problem. Since I can effect a Cambridge change in the Pyramids by moving my little finger, or simply by thinking of them, it might seem that Cambridge changes were purely relational. But what does this mean? Suppose that at t_1 Theaetetus is not taller than Socrates, while at t_2 he is. Has the change in truth-value of the relational statement induced a 'purely relational' Cambridge change in Socrates? Yes if it is Theaetetus who has grown, no if it is Socrates who has become stooped. (Since a thing that suffers real change also thereby suffers Cambridge change, but not necessarily vice versa, we shall speak of something that undergoes the latter but not the former as suffering *mere* Cambridge change.)

Next, if Jack becomes fat he undergoes real change. But what if he becomes famous? Or 40?[37] Plainly someone can become famous without undergoing real change. Just as the relational predicate 'taller than' may become or cease to be true of the pair (Theaetetus, Socrates), so various monadic predicates may become or cease to be true of Jack. In neither case can any inferences be drawn from these facts alone about whether the changes induced are real or merely Cambridge.

Mellor has attempted to solve the problem by saying that real changes must have causes and effects that are contiguous in space and time, whereas mere Cambridge change can be brought about by causes operating far away, or by the passage of time alone. This can't be entirely right, as Mellor himself realizes, since the decay of a radioactive atom is a real change, yet uncaused, and

[37] Mellor (1981a: 107–9).

the real change of Socrates' death can have remote but none the less important effects on the marital status of Xanthippe. To the best of my knowledge, the elusive difference between Cambridge change and real change can be made to emerge only by comparing the four-dimensional representation of objects in the branched model when they lie in the past, with their representation when they lie in the present. When this comparison is made in Chapter 7 (p. 223 below) Geach's problem will, I hope, have been resolved.[38]

To return to the principal subject of this chapter, both Geach and Dummett fix upon an important point, one which has troubled everyone who has thought at length about time. There is no difficulty in imagining someone, a god perhaps, who could contemplate the whole of space without being located in space. But could a god contemplate the whole of time without being located in time? Could there be a complete description of physical reality, a picture of the universe, which was not a picture *at* or *from* a particular time?

Aquinas, as reported by Geach, appears to have thought that there could be. What on the level of us travellers in time and space looks like a succession of events, from the 'high citadel of eternity' looks like a series spread out and simultaneously viewable.[39] It looks, that is, like a Minkowski world. But, Geach says, this will not do. It's not possible, as he puts it, 'to get a view of the Universe which is complete and which does not specify *when* it is, what bit of the Universe is *now*'. However, even if we agree with Geach, saying what he says raises a problem, which we have encountered before. According to the special theory of relativity, there are no privileged observers and no unique global 'now'. For Geach, and for Dummett, this constitutes a dilemma. We cannot accept Aquinas's view of the universe *sub specie aeternitatis* without implying that time as human beings perceive it is unreal. On the other hand, to embrace the transient qualities of time, to focus on time as human beings perceive it, seems to involve rejecting special relativity.

This is Geach's and Dummett's dilemma, and will be the dilemma of anyone who attempts to reconcile the desire for a complete description of the universe with the recognition of the reality of transience, and with the truth of special relativity. However, as

[38] The question of the status of mathematical objects is interesting in this connection. A classical mathematician would say that they could participate only in Cambridge change, whereas an intuitionist might hold that mathematics underwent not just Cambridge but real change every time a new proof was constructed.

[39] Geach (1979: 101). Cf. Dummett (1960: 502–3).

we saw earlier, there is a way of resolving the dilemma. We can conceive of the universe as characterized objectively by the divisions of past, present, and future, and by relativistic time flow, with the help of branched frame-dependent models as described in section (iii) above. Such a universe permits of a complete description, as laid out in Appendix 2. To picture the world in this way is to see time and its distinctive character of transience not as incompatible with our desire for a complete description of reality, but as answering to it.

Finally, it seems evident that the branched world undergoes real change. This change, constituted by branch attrition, forms the basis of the Cambridge changes suffered by things and events as they pass through time. To picture the universe as changing in this way is diametrically opposed to the manner in which it is pictured by the 'tenseless' school, or by those who conceive of the world *sub specie aeternitatis*. For them time flow and the accompanying Cambridge changes in things, are regarded either at best (i) as a harmless triviality, arising out of the use of tensed discourse but unrelated to the real world, or (ii) as a subjective illusion, or at worst (iii) as masking a contradiction. But there is an alternative, one that makes time flow and the divisions of past, present, and future neither trivial, nor subjective illusions, nor self-contradictory. This alternative is the objective status conferred upon them by the branched model, of which they are integral components. The model 'explains' our ideas of past, present, and future, and of the passage of time, not (as Kant would say) because our minds are so constituted that we are unable to conceive of the world without these ideas, but because they really do characterize the world, whether we conceive of it or not.

3

Causation and Laws of Nature

DAVID HUME argued persuasively two hundred years ago that
we look in vain in the physical world for any necessary connection
between cause and effect. Given his doctrine of ideas, and the
requirement that to every idea there must correspond an impression, Hume's request that an impression of sensation be produced,
from which the idea of power or necessary connection is derived,
is both reasonable and unanswerable: 'The small success, which
has been met with in all the attempts to fix this power, has at
last oblig'd philosophers to conclude, that the ultimate force and
efficacy of nature is perfectly unknown to us, and that 'tis in vain
we search for it in all the known qualities of matter.'[1]

Instead of finding the original of the idea of efficacy, power,
or necessary connection in the world, Hume found it not in an
impression of sensation but in an impression of reflection. The idea
of necessary connection derives from the easy transition that the
mind makes from the idea of the cause to that of the effect, after
it has experienced the constant conjunction of the corresponding
impressions. (In the case of burning one's finger in a candle flame,
one such conjunction suffices.)

Thus as the necessity, which makes two times two equal to four, or
three angles of a triangle equal to two right ones, lies only in the act
of the understanding, by which we consider and compare these ideas;
in like manner the necessity or power, which unites causes and effects,
lies in the determination of the mind to pass from the one to the other.
The efficacy or energy of causes is neither plac'd in the causes themselves, nor in the deity, nor in the concurrence of these two principles;
but belongs entirely to the soul, which considers the union of two or
more objects in all past instances. 'Tis here that the real power of causes
is plac'd, along with their connexion and necessity.[2]

(i) *The Link between Cause and Effect*

If the world is looked upon as a single four-dimensional course
of events, or for that matter as a three-dimensional state of affairs

---

[1] *A Treatise on Human Nature*, I. iii. 14, ed. Selby-Bigge (1888: 159).

[2] Ibid. 166. Hume is sensible that, of all the paradoxes he might be taken to have
advanced, 'the present one is the most violent'.

with only one successor-state, then Hume's conclusion seems justified. In the case of two events believed to be causally connected we observe only that they are contiguous in space and time, that one is prior to the other, and that they are constantly conjoined. Not only is this all we observe; this is all there is to observe. Furthermore, given the single-manifold Minkowskian picture, it is all the world contains.[3] In these circumstances, it is not unreasonable that Hume should have concluded that the source of any 'necessary connection' between cause and effect should be the mind's own activity, plus the laws of association that govern its workings. These laws are very powerful; witness a person's inability (in Hume's charming example) to keep from trembling when hung out from a high tower in an iron cage, though he knows very well that the solidity of the iron will keep him from falling.[4]

But though in Minkowski space-time contiguity, priority, and constant conjunction may be the sum total of the world's contribution to the relation of cause and effect, the case is very different with the branched model. If some set of initial conditions *A* is found at a branch point, and if all the branches above *A* contain *B*, then *B* is *causally necessitated* by *A* (see Fig. 3.1). That is, there is no physically possible future, relative to this occurrence of *A*, that does not contain *B*. Hume in several passages in the *Enquiry* makes much of the observation that although *in fact A* may always be followed by *B*, it could for all we know be followed

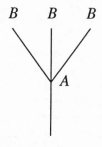

FIG. 3.1

[3] Cf. A. J. Ayer (1972: 10–11): 'I think we can make sense of [the issue of whether there are synthetic necessary connections] if we understand Hume to be denying that apart from relations of comparison there could be anything more than spatio-temporal relations between events. . . . It amounts in fact to the claim, which is scientifically most respectable, that everything that happens in the world can be represented in terms of variations of scenery in a four-dimensional spatio-temporal continuum.'
[4] *Treatise*, I. iii. 13, Selby-Bigge (1888: 148).

by something very different.[5] Bread could conceivably fail to nourish us, and the face of our best friend might mask our bitterest enemy. But if all branches above A contain B, then, in this case if not always, such eventualities are ruled out. And if *all* instances of A have nothing but B-branches above them, then such eventualities are ruled out forever. The branched model therefore contains a topological or structural feature which is itself empirical, but which has the effect of transferring or re-locating the concept of necessary connection from the mind back into the physical world.

Much more, of course, remains to be said about this topological feature. In its simplest form, it consists of all the branches above an A-node being B-branches. To make sense of this, we need to have the idea of event-types, or universals, or 'repeatables', being instantiated at different space-time points on different branches. Without the idea of something being *repeated*, no meaning can be attached to causation, or law, or statistical correlation. In a world composed entirely of particulars, without recurrent qualities or features, spatio-temporal relations among events would be definable but causal relations would not. For causation, we need the notion of 'the same again'. Even in the case of truth-conditions for singular causal statements, event-types are required, since the same event-type B needs to be instantiated on all branches above A.

As an integral part of the branched model, therefore, we shall suppose that properties or universals are instantiated at many places and times throughout the universe tree. The topological or structural features of the model which constitute causal connections can then be viewed as the way branches instantiating certain properties are arranged, or alternatively as the way property-universals are distributed on branches.

Although we have stressed universals and repeatables in introducing causation, particulars or individuals are also indispensable. To ask what caused Harry's heart attack is not the same as to ask what causes heart attacks in general, and there are important questions concerning causation that hang not just upon the overall pattern of instantiation of universals, but upon the way universals are instantiated at different times by the same or different individuals. An extended discussion of individuals in the branched model, which is difficult and full of pitfalls, is found in Chapter 7.

To return to the notion of causal necessitation and the link between cause and effect, we shall say that if a node or branch point x of the model is such that it features A, and if all branches

<hr />

[5] *An Enquiry Concerning Human Understanding*, iv. Selby-Bigge (1894 29, 34, 35).

above x feature B, then B is 'causally necessitated by A at x'. This is an instance of an event A causally necessitating another event B at a single branch point. On the branched model, it is possible for this pattern of event-type instantiation to occur only once, without a similar pattern recurring at any other point. Alternatively, the pattern may repeat itself throughout the tree. If all branch points which contain A are such that all branches above them contain B, we shall say that B is 'causally necessitated by A' without restriction. Note the double quantification in the unrestricted case. The difference between the two cases corresponds to Anscombe's separation of singular causal statements from universal causal laws, and more recently to Elliott Sober's distinction between 'token causation' and 'property causation'.[6] The branched model provides so-called *truth-makers* for singular causal statements of the type 'The assassination of the Archduke Ferdinand caused the outbreak of the Great War', as well as for universal causal propositions such as 'Malaria is caused by the bite of the *Anopheles* mosquito'.

The irreducible nub of all questions concerning causation is the question of necessitation. Does a cause necessitate its effect? *Must* the effect follow? Hume says not. For him, 'All events seem entirely loose and separate. One event follows another; but we never can observe any tie between them.'[7] By contrast, the branched model furnishes a clear criterion of causal necessitation. If event-types or property-universals are instantiated on the branches in the right way, in the right spatio-temporal locations, then a particular event A can necessitate an event of type B. Given uniformity of this pattern throughout the tree, any event of type A will necessitate an event of type B. Whether or not universals are in fact instantiated in the right patterns for such necessitation to obtain is unknown but irrelevant: *if* they are so instantiated, necessary corrections exist in nature. The branched model furnishes an *empirical criterion* for the existence of necessary connections between events.

In order for an event A to be properly described as causing B, Hume required not only that a necessary connection (in his understanding of the term) should extend between them, but also that A and B should be contiguous in space and time, and that A should precede B. Although most discussion of Hume's doctrine has focused on his analysis of necessary connection, the second and third requirements of contiguity and precedence have also

[6] Anscombe (1971); Sober (1985).   [7] *Enquiry*, vii. 2, Selby-Bigge (1894: 74).

received their share of criticism. There are good reasons to doubt the requirement of contiguity, but the branched model provides support for those who argue that Hume's criterion of temporal precedence should be retained.

If A and B are separated in space and time, A may still be the cause of B if there exists a causal chain of contiguous events linking cause and effect. For example, the sinking of the *Titanic* caused the *Carpathia* to alter course because of a telegraphic message sent from one to the other. But in some cases A may cause B in the absence of a chain of intermediate events, e.g. if A is a 'negative' event.[8] A person's death in the desert may be caused by no one's remembering to check the water canisters. To maintain that the absence of something, or the failure of an event to occur, cannot be the cause of another event without that absence or failure being 'contiguous' to the effect in question is implausible. Another possible example of non-contiguous causation is the EPR experiment (named after Einstein, Podolsky, and Rosen), where the measurement of one of a pair of electrons as 'spin-up' at point a may cause the other electron to be measured 'spin-down' at point b without any chain of intermediate events.[9] Such examples of apparent 'action at a distance' lead us to question Hume's criterion of contiguity.

The case is otherwise, however, with the criterion of temporal precedence. Beginning with Kant, a series of arguments have been produced which attempt to show that causes may fail to precede their effects, by dint of either (i) following, or (ii) being simultaneous with them. I shall not discuss the possibility that an effect might actually precede its cause, the objections to which probably strike the reader as overwhelming,[10] but instead concentrate on the less controversial idea that some causes may be simultaneous with their effects. Whatever initial plausibility this idea may have had, I shall argue that the weight of the philosophical evidence is against it.

Kant criticized Hume's criterion of temporal priority by giving the example of a lead ball which depresses a cushion simultaneously with its lying on top of it.[11] There are, however, two processes involved here, first the movement downwards of the

---

[8] See Brand (1979: 254).   [9] See Ch. 4 below, sect. ii.

[10] See e.g. Dummett (1954); Flew (1954); Black (1955); Pears (1956); Dummett (1964); Gorovitz (1964); Gale (1965). For a different and more up-to-date view, see Horwich (1987: chs. 6 and 7).

[11] *Critique of Pure Reason*, A203, ed. Kemp Smith (1933: 228). Simultaneous causation is discussed in Brand (1979 and 1980).

lead ball which is the cause, and second the movement downwards of the cushion's surface which is the effect. Although the two processes overlap in time and are consequently simultaneous with one another for a good part of their temporal length, the first process must begin before the second and in that sense 'precedes' it. To prove this, note that if a downward force is applied at some instant t to a stationary part of the cushion's surface, then that part will be in motion at each instant later than t, but not at t itself.[12] But the lead ball, in order to exert the downward force, must itself be in motion at t; hence the motion of the ball which is the cause must precede the motion of the cushion which is the effect.

On the branched model, a necessary condition of A's causing B is that A be located at a node all branches above which contain B. Alternatively, if the branching is upper cut (see Appendix 1), A may hold throughout an open set of space-time points, while B holds on each of the branches which contain one of that set's limit points. In either case A precedes B. Hence the branched model supports Hume's criterion of temporal precedence and, in addition, the thesis that causation is asymmetric: if A causes B, B cannot cause A.

So far it has been assumed that if an event of type A causes an event of type B, then A is *always* followed by B. (On the branched model, A is followed by B on *every* branch.) But the word 'cause', in the way in which it is normally used, permits two different kinds of case in which this is not true. The first of these is causal *interference*, and the second is *probabilistic* causality.

In instances where something interferes with a cause A, preventing it from bringing about the effect B which usually follows it, we do not necessarily jump to the conclusion that we were mistaken in ever thinking that events of type A cause events of type B. Instead we may conclude that events of type A do cause events of type B, but that some third factor H was interfering with the relationship. The bite of the gaboon viper causes death,

---

[12] To insist that the cushion's surface might move at the very instant the force is first applied to it is to be faced with the impossible question, how fast? In order to avoid discontinuity, it seems necessary to require that every body B which starts to move is in motion throughout an open set of instants, in contrast to the body which moves B and which is in motion at the earlier limit point of this open set. Tooley (1987: 208) also offers the following nice argument: If the surfaces of two objects are in contact, and if either surface has any thickness, then relativity theory forbids that any movement of one surface should be instantaneously transmitted to the other surface. If on the other hand the two surfaces have no thickness, they must be represented by closed sets of points and hence cannot be in contact.

for example, but only if no antidote is administered. Consequently, it may be that even though we wish for various reasons to say that A causes B, it is only A in the absence of H, I, J, K, etc. that is followed by B on all branches of the branched model. In that case we can say either that A in the absence of H, I, J, K, etc. causes B, or (more normally) that A causes B, but when interfered with may not.

Secondly, A may cause B in cases where the relationship between the two is probabilistic, and where A is followed by B on only relatively few branches of the branched model. Such cases are described in Anscombe (1971), and in Dretske and Snyder (1972). Suppose that a bomb is attached to a Geiger counter, so that it will go off when the Geiger counter registers a certain reading. If the counter is placed near some radioactive material, it may or may not register and the bomb may or may not explode. Nevertheless, placing it near the radioactive material causes it to go off, *when* it does go off. Cases like this show that A may cause B even though there is only a small probability that A will be followed by B, and the most straightforward way to describe them is to say that the probability of A's causing B is p. Other cases require more sophisticated treatment. As will be seen later in this chapter, and in Chapters 4 and 5, the branched model is particularly well adapted to providing objective 'truth-makers' for statements of the form $p(B \mid A)$, i.e. 'the conditional probability of B, given A, is p'. Use of the branched model provides a firm ontological starting-point for the analysis of probabilistic causality.[13]

Returning to the central theme of causal necessity, we have seen how the branched model supplies an objective necessary connection between cause and effect, this connection residing in the model's branching structure. At the same time, the relationship between cause and effect is *contingent* in the sense that no logical necessity attaches to the structure's being the way it is. Its particular shape, and the way universals are instantiated at spatio-temporal points and throughout regions, are matters of empirical

---

[13] More sophisticated examples involve Reichenbach's and Salmon's notion of 'screening off'. The probability of a storm on a day when my barometer executes a sudden drop is greater than on a day when it does not drop, yet we would not say that the fall in the barometer reading causes the storm. The factor which is the common probabilistic cause of both barometer drop B and storm S is widespread decline in atmospheric pressure D: we say that D screens off B from S and consequently it, not B, is a probabilistic cause of S. The formal definition is as follows. D screens off B from S if and only if: (i) B is statistically irrelevant to S in the presence of D, i.e. $p(S \mid B \& D) = p(S \mid D)$, and (ii) D is statistically relevant to S in the presence of B i.e. $p(S \mid B \& D) \neq p(S \mid B)$. See Salmon (1971: 55).

fact. Hence the branched model's philosophy of causation combines elements of necessity and contingency. Hume defied anyone
to produce a genuine instance of a necessary connection in nature.[14] This the branched model is able to provide, at least in
theory, although we do not have sufficiently detailed knowledge
about the branching structure to be sure that necessary connections, so defined, exist. But if they do exist, they are part and
parcel of physical reality.[15]

(ii) *Laws of Nature*

In his paper 'On the Notion of Cause' (1913), Bertrand Russell
recommended the 'complete extrusion' of the notion of cause from
the philosophical vocabulary. Noting that in the motions of
bodies under gravity 'there is nothing that can be called a cause,
and nothing that can be called an effect', Russell replaced causation by the notion of functional dependency. He remarks, as
against the authority of Mill, that the 'uniformity of nature' is
not a law of causation, but merely asserts the principle of the
permanence of such functional dependencies or laws. Since the
appearance of Russell's paper, the notion of cause in the philosophy of physics has been in eclipse compared with the notion of
law, and it is only recently that a trend in the opposite direction
has started.[16]

In Chapter 1, it was said that the structure of the branched
model is determined by what is physically possible, relative to a
given set of initial conditions. Each branch represents a distinct
set of mutually compossible outcomes of the conditions obtaining
at a branch point. It might be asked, what does 'physically
possible' mean here? What determines whether or not a given
branch is physically possible relative to a set of initial conditions?
And a natural answer would be that this depends upon the laws
of nature. For example, if two electrons collide, according to the
laws of quantum mechanics there is a certain probability that they
will be deflected at this or that angle, but no probability that
one of them will turn into a proton, or into a water molecule. In

---

[14] 'This defiance we are obliged frequently to make use of, as being almost the only
means of proving a negative in philosophy' (*Treatise*, I. iii. 14, Selby-Bigge (1888: 159)).
[15] The title of Putnam (1984) asks 'Is the causal structure of the physical itself something
physical?', and the answer of the branched model is 'yes'. The causal structure of the world
consists in a particular pattern of instantiation of event-types on branches.
[16] See e.g. Cartwright (1983: 21); Salmon (1984: 19).

Mackie's words, 'the universe needs to know where to go next', and what Mackie calls 'laws of working' cater to this need.[17]

This view, that physical laws dictate what is physically possible, and hence determine the structure of the branched model, is one that is natural and recommends itself to good sense. Nevertheless, it is not the view that will be taken in this book. Instead, a different view will be proposed which is formally more elegant, and which is made available by the richness of the information data base contained within the branched structure. According to this way of looking at things, instead of the laws determining the branched structure, the branched structure determines the laws.

This proposal, to make the structure of physically possible courses of events the primary thing, and to make the notion of law depend upon it, reverses the usual order. The usual order is to base the notion of what is physically possible upon the notion of law, rather than vice versa.[18] But here we are standing the usual order on its head. Instead of laws determining physical possibility, the branched structure of the universe comes first. It is our ontological primitive. Laws supervene upon it. To know in detail its topological structure, and how event-types and other universals are instantiated on its branches, is to know what is and what is not a law of nature.

Imagine the world at the moment of creation—at the big bang. Did it come into existence *together with* its laws? Or did the laws antedate it? Is the universe governed by the same laws now as it was then? The more one thinks about these questions, the less likelihood there is of answering them. If laws exist apart from the universe, serious questions arise as to their status. For example, what language are they written in? The language of mathematics? If so, which mathematics? Was God a classical mathematician? An intuitionist? What did he think about the axiom of choice? Axiomatization in general? If natural laws aren't written in *any* language, in what sense can they be said to be laws? These questions are impossible to answer. What I propose is an entirely different approach.

This time, imagine the world at the moment of creation, but without any laws. What does it look like? According to current theories of the big bang, the world is a point, a singularity in space-time. According to the branched model, the world at the moment of the big bang is an enormous profusely branching set of spatio-temporal manifolds in the shape of a tree but lacking any trunk. At its lower end, it narrows down to a single point,

[17] Mackie (1974*a*: 225).   [18] Cf. Pargetter (1984).

or to an open set with a single limit point.[19] All possible future histories are in this structure, and it is the fact of their being in it that makes them possible, not the other way round. Physical possibility, and physical law, are determined by the branching structure. All the information that is contained in all the laws, and a great deal else besides, is concealed within it. The principles which guide the universe's development are not imposed upon it from without, but are built into its structure.

What are the laws then? There are two different answers to this question. Objectively, or ontologically, laws are patterns of branching in the universe model. These patterns reflect the mode of instantiation of event-types on the branches. They are complex items of information, but they are not written in any language. Alternatively, laws of nature can be looked on as human beings' attempts to make sense out of the tree; to formulate, in mathematical language or whatever language is available, various characteristics of the branching that they find significant. Human attempts to formulate the laws can be relatively crude, or relatively refined. Given the complexity of the task, and given the tiny area of the tree that is open to inspection, it is unlikely that any full, adequate statement of the laws will ever be achieved.

At the moment of the big bang, to ask Mackie's question, how did the universe know where to go? This is where arbitrariness enters in, or (perhaps) symmetry-breaking. Branching out above the initial singularity were a number of alternative paths, any one of which could have been followed. The selection of whichever one was in fact followed was a random matter. No doubt this choice has profoundly influenced the shape of the world as we see it today. And in fact the process still continues: choices made today determine which areas of the branched structure still ahead of us will contain the first branch point of the universe in years to come. These choices are random (though see section (iv) below).

Consider a concrete example of a law of nature, say the law that copper conducts electricity. What does this law consist in? According to the 'ontological' account provided by the branched model, the law consists in the fact that, relative to any branch point containing a set of initial conditions in which an electric

[19] In Ch. 7, p. 217, there occurs a speculative hypothesis to the effect that the universe can be equivalently described *either* as a 3-dimensional spatial cosmos which evolves and endures through time, *or* as a branched 4-dimensional spatio-temporal entity. If this hypothesis is correct, the universe immediately after the initial singularity is at its smallest if viewed 3 dimensionally, but at its largest if viewed 4 dimensionally. The two correspond in the sense that the 4-dimensional universe is the setting-out in space and time of the 3-dimensional universe's capacities or potentialities.

potential is applied between two different parts of a piece of copper, every branch at a slightly later moment exhibits a current flowing in the copper. This obtains whether the branch point features the copper at the Equator or the North Pole, hot or cold, in a vacuum or under pressure. The object of scientific experimentation is to discover what types of events lie on all branches which share a common set of initial conditions, and returning many times to those initial conditions while varying other parameters eventually yields knowledge of the common features of the branches. To discover a law of nature is to discover what kind of branching above a given set of initial conditions is permitted by the branched model.[20]

The example considered above, that copper conducts electricity, is an example of an exceptionless law. If one case were ever found of a piece of copper through which no current flowed upon applying an electric potential to its ends, the law would be falsified. But not all laws are of the exceptionless variety. In Fig. 3.2, if a branch point contains A, and if 60 per cent of the branches above A contain B, there can still be a statistical or probabilistic law linking A and B. If the 60 per cent pattern repeats itself throughout the tree, there will be a law to the effect that any event-type A has a probability of .6 of being followed by a B, i.e. $p(B \mid A) = .6$. One of the virtues of the branched model is that it provides a physical interpretation of probability values

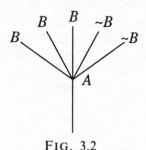

Fig. 3.2

in terms of proportions of sets of branches, in this case the proportion of B-branches to the set of all branches immediately above the node containing A. In the branched model:

$$probability = proportionality,$$

---

[20] Vallentyne (1988) contains a similar account of laws, the main difference being that Vallentyne proposes that lawhood be explicated in terms of a whole forest of possible branched structures, not just one.

and the probability values of probabilistic or statistical laws are built into the structure of the objective world. There are, however, certain problems that must be resolved before the identification of probability values with branch proportionality can be officially made. Although the matter is not dealt with fully until Chapter 4, some preliminary scouting of the ground can be done.

If the conditional probability of a B-event following an A-event, $p(B\,|\,A)$, takes a rational value like .6, then this value can be represented on the branched model by proportionality between *finite* sets of branches. Any rational number has the form n/m, where n and m are finite whole numbers. But not all probability values in modern science are rational numbers. As will be seen in the next chapter, the probability in quantum mechanics that a vertically polarized photon should pass through a piece of polaroid, the axis of polarization of which is tilted at an angle ϕ to the vertical, is not in general a rational number. Consequently, in order to represent this number, proportions between infinite sets of branches are needed. But Georg Cantor showed, in the nineteenth century, that there exist no fixed proportions between infinite sets. Fortunately a way to resolve this dilemma (originally put to the author by Bas van Fraassen) can be found, and in Chapter 4 precise definitions of probability values in terms of proportions among sets of branches are given with the help of a new concept, that of a 'decenary tree'.

Let us return for a moment to the question asked several pages earlier; whether the laws of nature exist *in addition to* the universe, or whether they are part and parcel of it. The answer that coheres most elegantly with the branched model is that they are part of it. Being no more than patterns of the branching, they came into existence at the moment the world came into existence, and will cease to exist when it ceases to exist. Consonant with this conception of laws is the proposition that, at the moment of creation, the four-dimensional universe was much bigger than a point of singularity. It was a large trunkless treelike structure, at the root of which was a point. But the laws of nature, which govern all physical processes then and now, were part of the universe's structure from the beginning.

At the stage in history that we have reached, about 15 billion years after the big bang, the universe is considerably less profusely ramified than it was. How do the laws of the present slimmed-down structure compare to those of 15 billion years ago, or for that matter to those of 1 billion years ago? If 'law' is interpreted to mean (i) pattern of branching that obtains *throughout the whole of the tree*, including the far-distant future, then the laws of today will be at least as strict, in the sense that there will be at least as many of them, as they were in the past. But if 'law' means (ii) pattern of branching that holds *in our era*, that is in the part of the tree that lies in our medium-term future, though not necessarily in the distant future, then the laws of today may be the same as they used to be, or they may be stricter, or they may be less strict. Where the probability of *A* being followed by *B* used to be X per cent, today it may be Y per cent, and a billion years from now it may be Z per cent. Of course, if 'law' is taken in sense (i), then if at any time it were a law that the probability of *A* being followed by *B* was X per cent then it would remain X per cent forever, though earlier than that time there might have been no law at all, no constant value for $p(B \mid A)$. In sense (ii), therefore, laws vary; in sense (i), they stay the same, but the number of laws increases monotonically. In either case the laws of nature are contingent; they might have been different from what they are, and doubtless *would* have been different if in the early history of the universe the first branch point had moved to a different area of the tree.

(iii) *Cosmology*

Just as the laws might have been different from what they are, so the fundamental constants of nature, such as the gravitational constant, might have differed from their actual values. Such possibilities give rise to intriguing problems in cosmology.

If the average energy density ρ of matter in the early universe had differed by more than a small amount from a specific critical value ρ_{crit}, the universe either would have expanded unchecked ($\rho < \rho_{crit}$), in such a way as to inhibit the formation of galaxies, or would already have collapsed back on itself ($\rho > \rho_{crit}$) through the force of gravitational attraction.[21] Cosmologists estimate that in order to bring about the type of evolution that the universe has in fact undergone, the actual value of ρ shortly after the big bang, when the general form of cosmic evolution was established, must have differed from ρ_{crit} by no more than one part in 10^{60}. Since there is, it seems, no a priori constraint on the choice of a value for ρ, the question arises of why a value was selected which

---

[21] Davies (1982: 88).

led to the present configuration of the universe rather than a quite different configuration.

An answer to this question is provided by the anthropic principle, which in its most general form states that the universe which an observer observes will inevitably be one which permits that observer to exist.[22] Care is needed in dealing with arguments based on the anthropic principle, since its general form can be given either a weak or a strong interpretation. In its weak form, the anthropic principle may be applied to the question of the difference between ρ and ρ_{crit} in a natural and uncontroversial way. If a universe in which ρ differed from ρ_{crit} by more than one part in 10^{60} would not be such as to permit the subsequent emergence of living creatures, then plainly such a universe, if it existed, would not be *our* universe. In this way the weak anthropic principle permits inferences to be made from our present existence to properties of the world 15 billion years ago. Since the principle can be seen as selecting one type of universe from an ensemble of competing types, Barrow and Tipler call it a 'self-selection principle'. The weak anthropic principle 'expresses only the fact that those properties of the universe we are able to discern are self-selected by the fact that they must be consistent with our own evolution and present existence'.[23]

In contrast to the weak principle, the so-called strong anthropic principle is much more controversial. As stated by Brandon Carter, it holds that 'the universe (and hence the fundamental parameters on which it depends) must be such as to admit the creation of observers within it at some stage'.[24]

What is meant here is not that the universe must *in fact* have been such as to admit the emergence of observers, but that it was so *of necessity*, presumably in virtue of some law. Ordinary language is misleading on this point. If I look at the cupboard and say, Ann must have forgotten to take her umbrella, I mean only that she *in fact* forgot it, not because of some inexorable law of forgetfulness. In the same way, the weak anthropic principle states only that the universe *in fact* is life-sustaining. But the strong principle uses the word 'must' not epistemically, as I do in saying that Ann must have forgotten to take her umbrella, but

---

[22] See Carter (1974); Barrow and Tipler (1986).

[23] Barrow and Tipler (1986: 16). Davies (1982) notes that the principle is profoundly un-Copernican: whereas Copernicus succeeded in displacing the earth from the centre of the universe, the anthropic principle restores human beings, or intelligent creatures of some kind, to a special place within cosmology.

[24] Carter (1974: 294).

ontologically or nomically, as an astronomer does when he says that the heavy elements must have been formed out of hydrogen and helium in the interior of stars. The astronomer means that by necessity, because of the laws of physics, the heavy elements could not have been formed in any other way. Similarly the strong anthropic principle holds that as a matter of necessity the universe could not fail to be life-sustaining.

Philosophers, sensitized by modern modal logic, will require powerful arguments to be convinced that the universe is necessarily life-sustaining. But the strong anthropic principle may not be needed in cosmology. If the idea of an ensemble of universes is introduced, its work may be done by the weak principle alone.

The question we started with was this. If the probability of the universe being life-sustaining was only one in 10^{60}, because of the fine-tuning demanded of ρ and other parameters, is there any explanation of why it in fact turned out to be so? An explanation would be provided by postulating an ensemble of universes, that is to say a set 'characterized by all conceivable combinations of initial conditions and fundamental constraints' (Carter 1974: 295). If instead of a single universe a multitude of universes is assumed, then the fortunate circumstance of our universe's turning out to be one in 10^{60} which supports life is explained, for (using the self-selection principle) had it not been life-supporting it would not be ours. If there is only one universe the question of improbable coincidence will always arise—if there are many universes the improbability disappears.

Universe ensembles are of two kinds, successive and coexistent.[25] At the end of the nineteenth century Ludwig Boltzmann hypothesized that the universe had arrived at its present low-entropy state by a rare chance fluctuation from its normal thermodynamical equilibrium. Given enough time (at least $(10^{10})^{80}$ years) whole galaxies, stars, and planets containing life could be expected to form spontaneously, for a brief period, before the universe once more lapsed into chaos.[26] The life-sustaining character of the world would be 'explained' in the sense that it was highly likely that such a world would one day arise by chance, and when it did it would be self-selected. In exactly the same way, an ensemble of coexistent universes of all possible types 'explains' the nature of the physical world by the fact that only life-supporting worlds get self-selected. In neither the Boltzmann nor the coexistent

[25] Hacking (1987: 132).   [26] Davies (1983: 168).

universes model should we be surprised by the improbable coin-cidence of our world's being life-supporting.

The situation of the branched model *vis-à-vis* the improbability of the world's being life-supporting is somewhat different. Unlike the Boltzmann model the branched model goes through only one 'cycle', and if during that cycle events in the early universe had been different, i.e. if branch attrition had taken the first branch point into a different area of the tree, then doubtless the universe would not have turned out to be life-supporting after all. Unlike Boltzmann's universe, the branched universe might *never* have supported life. In its early stages, the probability of the universe's supporting life might in fact have been quite low. We have therefore something to be thankful for—chance led the first node into a life-sustaining part of the branching structure.

With respect to the coexistent universes ensemble, the branched model again differs (see Fig. 3.3). Whereas in the ensemble there are many actual worlds, in the branched model the actual world is unique. Only the set of possible futures is multiple.

If a probability measure could be defined on the many universes ensemble, or on the set of branches in the branched model which existed at the time of the universe's creation, then it would be possible to quantify and obtain a value for the degree of improb-ability of a life-sustaining universe. In the case of the many worlds ensemble no natural measure presents itself, and any number of arbitrary probability measures can be imagined. But in the case of the branched model, sufficient structure exists for there to be a natural measure of the set of all future nodes at the time of the big bang which admit life, as compared to the set of nodes which

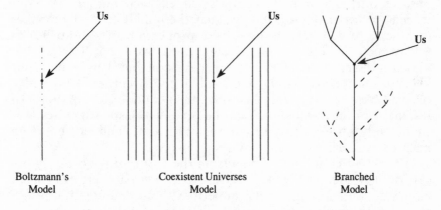

Boltzmann's
Model

Coexistent Universes
Model

Branched
Model

FIG. 3.3

exclude life. This measure is described in Chapter 4. An advantage which the branched model consequently enjoys over the other models is that it yields, at the time t_0, an exact probability value for the subsequent emergence of intelligent beings.

(iv) *Scientific Knowledge*

Consider any simple low-level chemical law, say the law that electrolysis of water with a little salt in it yields hydrogen and oxygen. On the basis of the branched model, how could such a law come to be known?

Let's assume the law was not originally deduced from some more general law, but was subjected to experimental confirmation on its own. Such confirmation would take the form of repeatedly immersing electrodes in salty water, connecting them to a power supply, and collecting the gases given off. This experiment has been performed under a wide variety of external conditions, and always yields the same results. Consequently we are now reasonably confident that, whenever the universe returns to a branch point which manifests the initial conditions of electrolysis, every branch above it will exhibit the subsequent production of hydrogen and oxygen.

From this example two conclusions may be drawn. The first is that repeated experimentation can reveal the mode of distribution of a universal or repeatable B on a fan of branches above A, even though on every experimental occasion only one of the branches is actually inspected. Despite the limitation of being able to observe only *one* branch, centuries of experimentation has convinced us, for many As and Bs, that there exist fans of branches in which every A-branch, or a fixed proportion of A-branches, is a B-branch. Where the initial conditions involve rolling an unbiased die, centuries of experience have convinced us that a four appears on approximately one-sixth of the branches. Knowledge of the shape of the universe tree, then, and of the pattern of instantiation of event-types, can be obtained despite the fact that we actually observe only one branch at a time, and can never go back to an earlier branch point.

The second conclusion stems from the fact that we can *perform* experiments such as the electrolysis experiment, and are not forced merely to wait around in the hope that the initial conditions will re-create themselves at some future date. If that were the case, scientific experimentation would be a slow business. John Bigelow, in a discussion in Melbourne, remarked that one group of scientists who have to sit and wait for initial conditions to recreate themselves are astronomers. But the majority of us do not, and if we wish to make one more test as to whether there is a physically possible branch on the universe model where hydrolysis takes place and hydrogen and oxygen do not result, we have only to go out and procure a tank, an old car battery, some wires, and collecting bottles. The interesting question is this. In our future right now, above the model's first branch point, there are physically possible branches on which water tanks, car batteries, etc. are assembled in this room, or in a nearby laboratory, and there are branches on which they are not. How is it that we have the power to select which type of branch the universe follows? How can scientific experimenters *choose* to move upward on the tree to a given type of branch point?

This question is too difficult, and raises too many fundamental philosophical issues, to be answered here. A full discussion of it appears in Chapter 9. But the problem exists, and cannot be side-stepped. To say that scientists can choose which kind of experiment to perform, or for that matter to say that I can choose to stoop and tie my shoelace, appears flatly to contradict what has been said all along, namely that the selection of which branch at the first branch point of the model becomes actual is a purely random matter. If it's a purely random matter, how can I tie my shoelace when I want to? The branched model should be able to provide a basis for answering this question, and if it can't, it deserves to be relegated to the land of interesting but false theories. Much, therefore, hangs on Chapter 9.

In the mean time there is more to be said concerning our ability to know the shape of the branched model, or at least the part of it that is accessible to us. To obtain such knowledge is the aim of scientific understanding. In the branched world the framing of scientific hypotheses, and the testing of them, proceeds exactly as it does in the Minkowski world. Exactly the same Humean doubts obtain, as to whether a hypothesis which has so far been consistently confirmed by observation (like the hydrolysis result), will continue to be confirmed in the future. There is, however, a significant difference between the epistemic state of a scientist seeking to understand the Minkowski world, and the epistemic state of a scientist seeking to understand the branched world. This difference lies in the nature of the objects that the scientist seeks to know, namely the laws. Laws have a dual character. Ontologically, they give the world pattern and structure. Epistemologically, they are the vehicle of scientific understanding. If there is an overall difference between laws in the Minkowski world and laws in the branched world, the impact of this difference on our ability to acquire scientific knowledge will not be negligible.

I shall return to the general difference between 'branched' and 'unbranched' conceptions of laws in section (vi). In the next section I examine three views concerning laws, those of (i) David Armstrong, Fred Dretske, and Michael Tooley, (ii) David Lewis, and (iii) Bas van Fraassen, each of which differs considerably from the branched theory and from each other.

(v) Three Differing Views: Armstrong, Dretske, and Tooley; Lewis; and van Fraassen

A theory which has been the subject of much recent discussion, and which differs in important ways from the branched theory, is the theory that laws are relations between universals. This theory was put forward independently in 1977–8 by David Armstrong, Fred Dretske, and Michael Tooley,[27] and a clear and well-reasoned presentation of it appears in Armstrong's *What is a Law of Nature?* published in 1983.

The Armstrong–Dretske–Tooley (henceforth ADT) theory maintains that laws of nature are relations between universals. Like the 'branched' theory, it is a realist theory of laws, meaning that it assumes that laws of nature exist independently of the minds which attempt to grasp them. Consequently, Armstrong says, 'Laws of nature must . . . be sharply distinguished from law-*statements*. Law-statements may be true or (much more likely) false. If they are true, then what makes them true is a law.'[28] Armstrong's distinction between laws and law-statements corresponds to the difference noted above between patterns of instantiation of event-types in the universe tree, and human beings' attempts to write down, or otherwise reduce to words, instantiation-patterns that they find significant. A law of nature is a feature of the physical world, the truth-maker of a true law-statement. Law-statements, on the other hand, are human creations.

Yet another realist theory of laws, which ADT take pains to distance themselves from, is the regularity theory. This theory holds that laws are nothing but regularities in the behaviour of things. In its simplest form, the regularity theory is based upon what

[27] See Armstrong (1978); Dretske (1977); Tooley (1977).   [28] Armstrong (1983: 8).

Armstrong calls 'unrestricted' or 'cosmic' uniformities. A statement p states a cosmic uniformity if (i) p is universally quantified, (ii) p is true, (iii) p is contingent, and (iv) p contains only non-local empirical predicates, apart from logical connectives and quantifiers.[29] For example, 'Electrons are deflected by an inhomogeneous magnetic field' states a cosmic uniformity, while 'All the vegetables in Bill's garden are potatoes' does not. I shall not describe the criticisms directed by ADT against the regularity theory, except to say that they are pretty devastating. Whatever a law of nature is, it would appear not to be a cosmic uniformity.

To replace the regularity theory, ADT propose that laws of nature be regarded as relations between universals of the form $N(F,G)$, where 'N' stands for a relation of non-logical, contingent, 'nomic' necessitation. $N(F,G)$ implies $(x)(Fx \supset Gx)$ but not vice versa, and $N(F,G)$ implies 'If x had been an F, it would have been a G'. For ADT, therefore, to assert that it is a *law* that all Fs are Gs is to say something stronger than to assert that all Fs are *in fact* Gs (the regularity theory). If it's a law that Fs are Gs, it will be nomically impossible for an F not to be a G. In this respect, the theory that laws are relations between universals and the branched theory agree.

Another way of bringing out the differences between the regularity theory and ADT is to consider how the latter cope with inductive scepticism. The problem of inductive scepticism is Hume's legacy, the problem of how to proceed from the premiss that all observed As have been Bs to the conclusion that the next A will be a B. Without some assurance of uniformity—that the future will resemble the past—such an inference is unjustified. But experience gives no such assurance;[30] hence for Hume the inference cannot be justified.

Since Mill's *Logic*, the standard way of coping with inductive scepticism has been to expand Hume's one-step inference of the form

(1) observed instances $\Rightarrow$ unobserved instances

into a two-step inference:

(2) observed instances $\Rightarrow$ law $\Rightarrow$ unobserved instances.[31]

---

[29] (1983: 12).

[30] As Russell remarks, we may have observed past futures to resemble past pasts, but future futures are another matter.

[31] Armstrong (1983: 56). In *A System of Logic*, III. iv. 1, Mill says, 'every well-grounded inductive generalization is either a law of nature, or a result of laws of nature, capable, if these laws are known, of being predicted from them. And the problem of Inductive Logic may be summed up in two questions: how to ascertain the laws of nature; and how, after having ascertained them, to follow them into their results' (1973: 318).

Since, on the regularity theory, a law or unrestricted uniformity is simply the conjunction of the observed and unobserved instances, this inference becomes, for a regularist:

(3) observed instances $\Rightarrow$ observed instances + unobserved instances $\Rightarrow$ unobserved instances.

But this reduces to (1), hence for a regularist the appeal to laws leaves the problem of inductive scepticism exactly where it was. For Mill, and for ADT, and for the branched theory, the problem has shifted to the problem of discovering the laws of nature. Once a genuine law has been found, it is analytic that unobserved instances will conform to it. But sceptical doubts can (and will) arise as to the genuineness or universality of any suggested law.

In view of this, have ADT made any progress? In the sense of laying to rest Hume's doubts, no. If we could be sure that F-ness $\Rightarrow$ G-ness, i.e. $N(F,G)$, denoted a genuine law, that is to say a connection of necessitation between universals, then there would be no doubt, the next time an F, appeared, that it would be a G. But, when we are confronted with an F, the doubt will always persist, whether F-ness $\Rightarrow$ G-ness *is* a law. Analogously, in the branched model, the doubt will always persist whether Gs occur on all branches above Fs at all future branch points. Such things, by their nature, will never be certain. As long as there exists an intelligent and inquiring mind, there will always exist a degree of inductive scepticism in it.

So far, no great dissimilarities have emerged between the theory that laws are relations between universals, and the 'branched' theory that laws are constancies in the proportionality of B-branches above A-nodes. ('Law' is understood here in sense (i) of p. 60 above, not in sense (ii).) There are, however, some basic differences between the two theories, and these must be made clear.

In the first place, ADT assert that laws are relations between *universals*, whereas the branched theory holds that laws concern the *mode of instantiation of universals* or event-types in the universe tree. For the latter, the question of how universal F is related to universal G is determined by where and when—at which spatio-temporal locations and on which branches—F and G are instantiated on the tree. The branched model, in so far as it licenses any theory of universals, licenses an extensional, not an intensional theory. It is true that Armstrong takes an Aristotelian rather than a Platonic view of universals, holding that they live not in the world of Being but with us, in the world of Becoming. For him, as for Aristotle, universals are non-existent unless instantiated. The branched theory carries this thought a step further, holding that relations among law-of-nature universals are non-existent except as supervenient upon their mode of instantiation.

Secondly, in the branched theory the notion of *time* is associated in an intimate way with the notion of law. It is true that many laws, such as Boyle's law, contain no reference to time, and may be described as 'laws of coexistence'. Boyle's law simply says that one variable quantity, the volume of a gas, is inversely proportional to a second, its pressure, and directly proportional to a third, its temperature. But there is another class of laws, namely dynamical or causal laws, in which the time variable plays an essential role. For example, if x is a compass needle located near a wire through which a current is flowing, then x will be deflected according to the right-hand rule. However, when we switch on the current and observe the needle, an obvious fact is that it takes time for the needle to move. This time lag is itself lawlike, but it is not easily dealt with by the theory that laws are relations between universals. On the branched model, on the other hand, the time delay is represented in a perfectly natural fashion on the branches which fan out above the point containing the initial conditions.

Thirdly, many if not most laws of nature concern relations between universals that hold not necessarily but probabilistically. Armstrong's theory deals with these laws by introducing what amounts to a relation of contingent probabilification in addition to the earlier relation of contingent necessitation, the degree of probabilification being indicated by a number between zero and one. Representation of probabilities on the branched theory, on the other hand, is in terms of proportions of branches, and a probabilistic law is one where the proportionality remains constant. Here the number which denotes the probability value is built into the branched structure, whereas in the relations-between-universals theory the number is put in by hand, without any apparent ontological grounding. In the branched theory, the truth-makers of probabilistic law-statements are what in the next chapter will be called *prism stacks*, namely structured fans of branches. What makes probabilistic law-statements true, and what gives probabilities between zero and one their precise value, are the fixed proportionalities among sets of branches in a prism stack.

Very different from ADT's view of scientific laws is that of David Lewis. Lewis takes as his starting-point a view that F. P. Ramsey held in 1928: 'Even if we knew everything, we should still want to systematize our knowledge as a deductive system, and the general axioms in that system would be the fundamental laws of nature.'[32] Ramsey in turn, as is pointed out by John Earman, may have taken his cue in 1928 from Mill, who wrote that 'According to one mode of expression, the question, What are the laws of nature? may be stated thus:—What are the fewest and simplest assumptions, which being granted, the whole existing order of nature would result?'[33] Lewis's restatement of Ramsey's theory is this: 'A contingent generalization is a *law of nature* if and only if it appears as a theorem (or axiom) in each of the true deductive systems that achieves a best combination of simplicity and strength.'[34]

The important ideas in Lewis's definition are those of *true deductive system, simplicity*, and *strength*. To these must be added *balance*, which is a separate criterion, since only systems which achieve a satisfactory balance between simplicity and strength qualify as law-containing. Unfortunately, however, as Lewis recognizes, the existence of two or more competing standards allows an element of subjectivity to enter into the judgement of which systems qualify as law-containing and which don't; it makes lawhood, as he puts it, 'depend on us', a conclusion he does not welcome.[35] I shall show that this difficulty does not arise in the case of the 'branched' theory of laws, which provides an objective criterion of lawhood. I shall then go on to argue that the branched model resolves a nice problem propounded by Lewis concerning the supervenience of chances.

In the construction of deductive systems there is, as is noted by van Fraassen, generally a trade-off between simplicity and strength. A system can frequently be strengthened by the addition of a new axiom, but only at the expense of reducing its simplicity.[36] Who is to say what is the correct balance, the 'best' system? Two different people with different standards of simplicity, strength, and balance may, consistently with Lewis's position, arrive at quite different beliefs concerning the laws of nature.

The root of the difficulty seems to be this. Lewis's theory requires that laws be formulated propositionally, in some suitable syntax that enables them to be items in a deductive system.

[32] Ramsey (1990: 143). See also Lewis (1973: p. 73; 1986a: pp. xi, 121–4). Lewis tells us that Ramsey later went on to develop a different theory of laws.
[33] Mill, *System of Logic* III. iv. 1. (1973: 317). See Earman (1984: 196–7).
[34] Lewis (1973: 73). In fact, Lewis's position is not identical with either Ramsey's or Mill's, as he points out (Lewis 1986a: 122 n.).
[35] Lewis (1986a: 123).   [36] van Fraassen (1989: 41).

Harking back to Armstrong's distinction between *laws* and *law-statements*, Lewis's laws seem to be constrained by their role in deductive systems to be law-statements. Laws, on the other hand, are what in Armstrong's words make true law-statements true. They need not be expressed in any language, mathematical or otherwise. On the branched model, they consist in patterns of instantiations of universals or event-types on the branches, in such a way that if *A* is instantiated at any node, then *B* is instantiated on all or a fixed proportion of branches above *A*. Such a pattern is not a proposition. It cannot figure as an axiom or a theorem of a deductive system. But, in the branched model, it is a law.

Bas van Fraassen, in *The Scientific Image*, distinguishes between what he calls the 'semantic' as opposed to the 'syntactic' approach to theories. Traditionally, in the history of philosophy and mathematics, a theory is regarded as a body of propositions or sentences, expressed in a syntax, which may be axiomatized or alternatively which may hang together in deductive fashion more loosely (e.g. simply be closed under *modus ponens*). But there is another way of looking at a theory. A theory may be regarded, *more semanticae*, as a collection of models or structures which, given a suitable interpretation, makes a set of propositions true. In van Fraassen's words:

The syntactic picture of a theory identifies it with a body of theorems, stated in one particular language chosen for the expression of that theory. This should be contrasted with the alternative of presenting a theory in the first instance by identifying a class of structures as its models. In this second, semantic, approach the language used to express the theory is neither basic nor unique; the same class of structures could well be described in radically different ways, each with its own limitations. The models occupy centre stage.[37]

It is in this second, ontological or semantic sense that we should understand the theory of laws of nature presented here. The 'theory' (in the new sense of the term) is in effect the set of all frame-dependent branched models. According to this view an individual law, as distinct from a law-statement, is not a proposition. It is, instead, a feature of the models which, given an interpretation, makes the law-statement true. And the collection of all laws is the collection of all such models. For example the law-statement to the effect that the probability that an event of type *A* will be followed by an event of type *B* is p, i.e. $p(B \mid A) = p$, is true just in case every *A*-node on every future hyperplane-dependent

[37] van Fraassen (1980: 44).

model has B on a proportion p of the branches above it. That is to say, the *law* is the set of all such models: intuitively, the set which makes $p(B \mid A) = p$ true.

This view of laws differs from Lewis's in not requiring judgements of simplicity and strength—in fact not requiring the notion of a deductive system at all—in order to distinguish laws from non-laws. In the next few paragraphs the branched theory will be put to the test by applying it to the problem of accidental uniformities, the results being compared with Lewis's treatment of such uniformities. Accidental uniformities are important because they are laws' closest mimics: they are the things that most nearly resemble laws without being laws themselves. To get clear about one, we must get clear about the other, and the line between the two is not easy to draw.

A frequently encountered example of an accidental uniformity or accidental regularity is the following:

(1) Every solid gold sphere is less than one mile in diameter.

It might be accidentally true that every gold sphere is less than one mile in diameter, but we would be reluctant to think that (1) stated a law of nature, or held because it was a consequence of a law. In this respect it differs from

(2) Every solid enriched uranium sphere is less than one mile in diameter.

which may very well be true in virtue of being a consequence of laws of nature.[38] Lewis deals with accidental uniformities by putting them in an elastic category that may be expanded or contracted according to the demands of the standards of simplicity and strength. If, as he says, it would complicate the otherwise best system to include (1) as an axiom, or to include premises that would imply it, and if (1) would not add sufficient strength to pay its way, then it is best left as a merely accidental regularity.[39] For Lewis, the distinction between laws and accidental regularities is flexible, and hangs on judgements of simplicity and strength.

The treatment of accidental regularities that fits best with the branched model is quite different. Suppose that every gold sphere in the entire history of the universe, in every galaxy, from the big bang to doomsday, turned out to be less than one mile in diameter. Then indeed (1) would state a true, cosmic, accidental uniformity.[40] What makes (1), in the model, true by accident, not

---

[38] van Fraassen (1989: 27).   [39] Lewis (1986a: 122).
[40] Cosmic uniformities are those which contain only non-local predicates. See p. 000 above.

of necessity, is the fact that (1) fails on at least some of the model's branches, even though it may hold everywhere on the one branch that eventually constitutes the actual history of the world. Nothing in the branched model today, of course, distinguishes that one branch from all the rest. There is no preferred branch. But to say that (1) is an accidental uniformity, rather than a law, is to say that the truth of (1) depends on the eventual surviving branch being of a non-large-gold-sphere-containing variety rather than the opposite. That is to say, it depends on branch attrition.

In contrast to (1), the truth of a law, or of a non-accidental uniformity like (2) which is a consequence of laws, does not depend on branch attrition.[41] Instead, it depends on the way universals are instantiated in the branched model. For it to be a law that all A-events are followed by B-events, every branch above every A-node must be a B-branch. For it to be a true accidental uniformity, on the other hand, two conditions must be met: (i) some branches above some A-nodes are not B-branches; and (ii) on the one branch which eventually turns out to constitute the actual history of the world, every A-event is followed by a B-event.[42] Since it is the vanishing of branches on the model which progressively reveals this branch, accidental uniformities depend upon branch attrition in a way that laws do not. The line which separates the two is not flexible, but is based upon an objective dynamic characteristic of the branched model.

A few pages earlier, mention was made of a problem raised by Lewis concerning the supervenience of chances. The problem is

---

[41] Strictly speaking, in the 'ontological' approach to laws of nature adopted here, there is no distinction between laws and their consequences. Both consist of patterns of instantiations of universals in the branched model. But if all the statements and mathematical equations made true by these patterns could be collected together and axiomatized, then the 'laws' which science seeks could be identified with the axioms. They would be, in Mill's words, 'the fewest and simplest assumptions, which being granted, the whole existing order of nature would result'.

[42] This characterization of the difference between laws and uniformities on the branched model applies to *dynamical* or *causal laws*, rather than to *laws of coexistence* such as Boyle's law or the non-accidental uniformity (2). A law of coexistence to the effect that all As are Bs holds if all As on all branches are Bs, whereas an *accidental uniformity of coexistence* holds if (i) some As on some branches are not Bs, but (ii) every A on the branch comprising the actual history of the world is a B. An example of an *accidental dynamical uniformity* would be this: a radioactive substance X is discovered in which, by chance, every nucleus lasts for exactly its expected lifetime, neither more nor less (see Lewis 1986a: 125). Although such a uniformity would be both simple and strong, it could not be a law, since the chance of its holding would be infinitesimally small. On the branched model, its truth depends not on the fan of branches, all with different lifetimes, that confronts each X atom on its creation, but on branch attrition, which by pure chance gives the X atom in every case its expected lifetime.

this. Lewis wishes to construct an overall philosophical theory which respects the thesis he calls *Humean supervenience*.[43] This thesis maintains that the world consists of nothing but a vast mosaic of particular facts, an arrangement of qualities in space and time. All else—modal properties, laws, causal relations, counter-factuals, probabilities and chances—supervenes upon the arrangement of qualities in the following sense. If two worlds match perfectly in all matters of particular fact, they match perfectly in all other ways too. There can be no differences in modal properties, laws, etc. without concrete differences in the point-by-point arrangement of qualities in space-time. The latter are primary: all else is secondary.

Lewis is able to satisfy himself that the thesis of Humean supervenience is tenable for all categories save one, namely chance. The obstacle which stands in the way of a supervenience account of chance is the following. Suppose we make the not unreasonable assumption that the chances for alternative futures that obtain at a moment depend on whatever conditions prevail at the moment in question, together with whatever conditions prevailed at previous moments.[44] We have then a series of what Lewis calls history-to-chance conditionals: if history is so-and-so then the chances are such-and-such. If we knew exactly which history-to-chance conditionals were true and which false, then we would know, given the history of the universe up to now, precisely what the chances would be of every possible future continuation of that history.

Now, Lewis asks, what is the status of these history-to-chance conditionals? Do they supervene upon events, i.e. upon the spatio-temporal arrangement of qualities in the world? If they did, then chance itself would supervene upon events. But Lewis concludes that, regrettably, this cannot be:

The trouble is that whatever pattern it is in the arrangement of qualities that makes the conditionals true will itself be something that has some chance of coming about, and some chance of not coming about. What happens if there is some chance of getting a pattern that would undermine that very chance?[45]

For Lewis, trying to make chances supervene upon events involves unavoidable circularity. The very pattern of events upon which chances might be made to supervene, and which includes, for Lewis, future as well as past events (1986a: 129), is itself some-

---

[43] See Lewis (1986a: pp. ix–xvii, 111–13, 124–31).   [44] Lewis (1986a, p. xiv).
[45] Ibid.

thing which depends on chance. Chance therefore does much more than supervene upon the arrangement of qualities: it determines what the arrangement of qualities is, upon which it itself allegedly supervenes.

Lewis's conclusion is that a supervenience theory of chance is desirable, but out of reach. However, such a theory is not out of reach if the spatio-temporal arrangement of qualities which is taken as our basis is the branched model. Given the history of the world up to the present moment, which is represented in the model by the four-dimensional trunk of the tree structure, the branches above the first node define clearly and unequivocally which future courses of events do, and which do not, have a chance of becoming actual.

The precise value of the chance of a future event occurring, relative to the initial conditions at a node, is given by the proportion of branches above that node on which the event is located. This chance, relative to the node and hence to the whole history of the world up to that node, is the objective single-case probability of the event in question. Both the probability and its value supervene entirely upon the spatio-temporal distribution of qualities on the branches of the model, and consequently the theory of chances which fits most naturally with the model is indeed one of Humean supervenience.

Lewis's difficulty, if I understand it correctly, is this. Suppose we consider the chance or the probability that tomorrow we shall go on a picnic, and ask whether there is any way that this chance can supervene upon events, i.e. upon the arrangement of qualities in space-time. Well, if tomorrow we wake up and the sun is shining, then the chances of going on a picnic are excellent. But if we wake up and it is raining, the chances are poor. So the chances depend upon events tomorrow. But these events in turn— the arrangement of qualities tomorrow—are things that themselves have some chance of coming about, and some chance of not coming about. So there is no definitive set of events on which chances supervene, and hence no well-defined value to the chance that we shall go on a picnic tomorrow. Chances depend not only on events, but on other chances as well.

On the branched model, this tangle of chances-depending-on-events-depending-on-chances can be sorted out. As will be seen in Chapters 4 and 5, the tree model provides a straightforward answer to most if not all questions about the future probabilities of events, and about conditional probabilities. Granted that relative to today, because of the chance of rain tomorrow, the probability of going on a picnic tomorrow, although well-defined, is neither high nor low. But this probability changes as the branches drop off, and by the time we awake tomorrow and look at the sky, the probability will be either very high, or very low. The branching structure, together with branch attrition, provides a complete picture of how chances evolve with, and supervene upon, events.

Turning now to the views of Bas van Fraassen, we shall see that he is not friendly to the idea of objective chance, or to the idea that probabilistic laws can be based upon a branched courses-of-events structure. However, we shall also see that a structure of this kind can provide answers to a number of problems that van Fraassen raises concerning laws and probability, and consequently it should not be ruled out that even by van Fraassen's own lights an understanding of probabilistic laws in the deepest sense flows from precisely such a branching structure.

Van Fraassen's position concerning probabilistic laws and objective chance is contained in his book, *Laws and Symmetry* (1989). In the first part of that work, entitled 'Are There Laws of Nature?', van Fraassen brings forward two major problems that in his view bedevil all proffered philosophical accounts of laws, the *problem of inference* and the *problem of identification* (1989: 38). To resolve one of these, he claims, spells serious trouble from the other. The problem of inference is this: that its being a law that all *A* is *B* should imply that all *A* is *B*. The problem of identification is the problem of identifying the relevant sort of fact about the world that gives 'law' its sense. These problems do not have the appearance of being insoluble, from the perspective of what has already been said about the branched model, but they must be examined with care.

The problem of inference emerges in its starkest form in the case of ADT's theory of laws as relations between universals, though it also gives rise to what van Fraassen calls the 'horizontal–vertical' problem for branched structures. Beginning with the former, supporters of universals are faced with the following question. Why does the identification of 'It is a law that all *F*s are *G*s' with '$N(F, G)$', i.e. with a relationship of contingent nomic necessitation between the universals *F* and *G*, permit us to infer, from the fact that it is a *law* that all *F*s are *G*s, that all *F*s *are*, in fact, *G*s?[46] Why, for example, would the existence of a contingent, nomic necessitation relationship between the universals *copper* and *conductivity* entitle us to infer that all copper conducts

---

[46] van Fraassen (1989: 107).

electricity? If the relationship were a logical or conventional one, the case would be different: the existence of a logical or conventional relationship between the universals *red* and *coloured* enable us to infer that all red things are coloured. But the copper/conductivity case seems different. What would stand in the way of $N(F, G)$ being true, and $(x)(Fx \supset Gx)$ being false?[47] That is van Fraassen's problem of inference.

On the branched model, the problem of inference is easily resolved. If it is a law that all Fs are Gs, then (if the law in question is a law of coexistence) every F is a G on every branch, while (if it is a dynamical law) then every branch above every F-node is a G-branch. This holds throughout the model. Therefore from the fact that it is a law that all Fs are Gs it follows, purely extensionally, that all Fs are Gs.

It also follows, which is not the same thing, that every F we shall observe will be a G. By contrast, in the case of the dynamical probabilistic law $p(B \mid A) = p$, the ontological truth-maker of which on the branched model is that a proportion p of all branches above all A-nodes are B-branches, it does *not* follow that a proportion p of all observed As will be Bs. However, it does follow that the *single-case probability* of each A being a B is p, which is stronger than saying merely that the average or statistical probability of As being Bs is p. The name which van Fraassen gives to the problem of relating chance as a proportion among possible futures, to frequency of occurrence in our *actual* future, is the 'horizontal–vertical' problem.[48] This is the probabilistic version of the problem of inference.

The name arises in the following way. Suppose that a sample consisting of a single atom of a radioactive substance R, with a half-life of thirty minutes, is repeatedly prepared, and observed after one hour. On the branched model, every node at which such a sample is prepared will have a fan of branches above it, on exactly $\frac{1}{4}$ of which the atom is undecayed after one hour, and on $\frac{3}{4}$ of which it has decayed (see Fig. 3.4).

The proportion of branches above each node on which the atom decays is measured over the horizontal dotted lines, while the

---

[47] 'Don't try *defining* N in terms of there being a law and hence a regularity—we're trying to *explain* lawhood. And it's no good just giving the lawmaker a name that presupposes that somehow it does its stuff, as when Armstrong calls it "necessitation". If you find it hard to ask why there can't be F's that are not G's when F "necessitates" G, you should ask instead how any N can do what it must do to deserve that name' (Lewis 1986a: p. xii).

[48] van Fraassen (1989: 84).

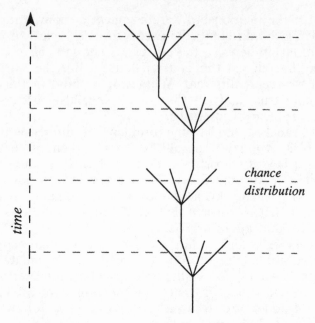

FIG. 3.4

relative frequency of decay *v.* non-decay observations in the actual history of the world is measured along the vertical line. Van Fraassen's problem is: What does one have to do with the other? Why should the chances of each individual atom decaying at each node help determine the actual frequency with which decayed states are observed?

On the branched model, the answer to this problem is simple. Since branch selection is random at each node, i.e. every branch has an equal chance of being selected, and since a proportion of exactly $\frac{3}{4}$ of the branches above each node are 'decay' branches, the probability of the actual or selected branch being a decay branch is $\frac{3}{4}$. If this is so, one would expect that the observed relative frequency of decay over a sufficiently long series of trials would be $\frac{3}{4}$. On the branched model, the horizontal–vertical problem is resolved.

Let us turn to the second of van Fraassen's problems concerning laws, the problem of identification. Suppose we are trying to define laws of nature in terms of the notions of physical possibility and physical necessity, and advance the following suggestion: *L* is a law of nature in a world *w* if *L* is true in all worlds *w'* which

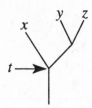

FIG. 3.5

are possible relative to w. The question then arises, which relation among worlds is the relation of relative possibility—the *access* relation?[49] If no such relation can be clearly picked out, then the identification problem for laws defined in this modal way will be unresolved.

Now a branching structure of worlds—where a 'world' is defined as a path through the branching structure, i.e. what I would call a complete possible history of the world—does contain a clear and unambiguous relationship of accessibility, as van Fraassen points out.[50] Call two worlds w and w' *t-equivalent* if and only if they have the same history through time t. In Fig. 3.5, x, y, and z are all t-equivalent. The time-indexed relation of t-equivalence can serve as the access relation in the definition of law:

> A general proposition L is a *law of nature* in x at t (or, in van Fraassen's terminology, *settled* in x at t) if L holds in every world which is t-equivalent to x.

With one important modification, this is the definition of law in the branched model that we have adopted. But for van Fraassen, the definition suffers from two fatal weaknesses. In the first place something can be settled without being a law (in some other, intuitive sense of the word), and in the second place something can be a law without being settled. These objections must be answered, if the 'branched' theory of laws is to be viable.

In the branched model, owing to the progressive slimming down of the structure through branch attrition, laws may be created by historical accident. At a certain point (Fig. 3.6) there may be a choice between two alternatives, one of which leads to a structure

---

[49] van Fraassen (1989: 72); see also Pargetter (1984).

[50] It should be emphasized that the definition of a 'world' as a single path through the branched model is given here solely for purposes of discussing van Fraassen's views on the identification problem. On the branched model there is only one 'world', namely the whole branched structure. The distinction is important for the definition of laws: see p. 81 below.

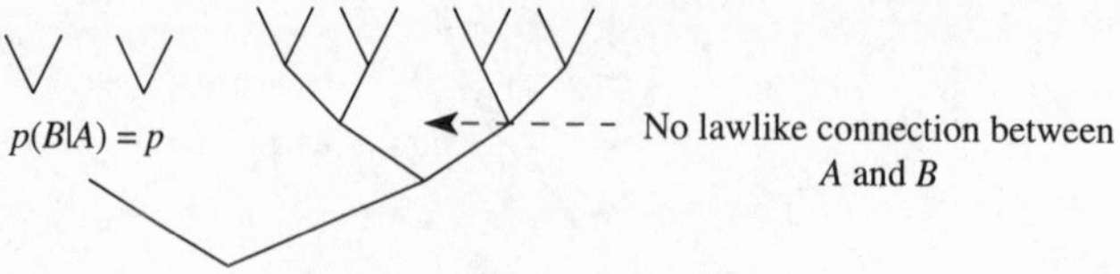

$p(B|A) = p$

No lawlike connection between
A and B

FIG. 3.6

in which $p(B \mid A) = p$ at all nodes, and the other to a structure
with no lawlike connection between A and B.

Since it is a matter of chance which branch is chosen, pure
historical accident will determine whether or not it is settled that
$p(B \mid A) = p$. To eliminate the element of historical accident, which
he apparently feels sits ill with the notion of law, van Fraassen
proposes the stronger concept *fully settled*. A proposition L is
fully settled in x at time t if it is settled in x at every time
$t' \leq t$.[51] If the laws of nature are those propositions which are
fully settled, then two worlds which shared the first two seconds
(or nanoseconds) of their history would have the same laws. But
this, van Fraassen says, is absurd. Historical fact cannot be so
powerful as to determine what the laws shall be after so short a
time. The intuitive concept of law requires that there are other
laws our world could have had, consistently with having had exactly
the same history for a short period after coming into being.

Van Fraassen's arguments against identifying what is lawful
with what is *fully settled* are justified, but they have no weight
against the identification of lawfulness with being *settled*. In fact,
if the laws are constituted by what is settled progressively at each
moment in time, then we have just the result he desires: events
which took place in the early stages of the universe's history
shaped the structure of the laws that hold in the world today,
and different events would have produced different laws. We may
call those early events *symmetry-breaking*; if indeed they consisted
of choices within fans of equipossible identical segments of branches in the branched model, each leading to qualitatively different
consequences, the expression is appropriately descriptive. Today,
on the other hand, when it may be assumed that the major choices
leading to different possible sets of laws have already been made,
the role of historical accident in bringing new laws into being is
minimal, through perhaps still not negligible. However, to eliminate

[51] van Fraassen (1989: 76).

that role entirely is to rule out the possibility that early contingent events in the history of the universe should have helped produce the laws we now live under, and I take it that van Fraassen would not wish to deny this possibility.

The second objection to our definition is that it is possible, according to van Fraassen, for something to be a law without being settled. This can come about as follows. Suppose that x and y are two worlds which have the first two minutes of history in common, but which have different laws of nature.[52] For example, L may be a law in x but not in y. But then L will not be settled in either x or y during those first two minutes: in x, L will be a law but not be settled.

This argument, however, cannot be applied to the branched model. On that model, laws of nature are objective features not of individual branches, but of the branched structure as a whole. Therefore it does not make sense to speak of L being a law in branch x but not in branch y. If L is a law which comes to hold in the universe as a whole during the first two minutes, then it becomes settled at the same time, and if it holds in the universe at any time then it holds on every branch of the universe, at that time and thereafter. On the branched model, it is misleading to think of the separate branches as 'worlds'. There is only one world, and it is branched. To say that L is a law of nature is not to say that L is true on a particular branch of the model, but that L is true, at a certain time, of the whole model, i.e. on every one of its branches.

This being the 'branched' conception of a law, it is not difficult to see how van Fraassen's identification problem is resolved. A law is an objective pattern of instantiation of universals or event-types on the branched structure, and this pattern is easily identified. Van Fraassen, who is an antirealist, examines branched structures with the purpose of attacking the idea that laws of nature can be based upon them. However, the straightforward, objective, realist account of laws that such structures provide will appeal to those who seek to understand how laws can necessitate, and at the same time be grounded in the physical world. In the case of probabilistic laws, this grounding extends to the provision of exact probability values, as will be seen in the next chapter. Few other accounts of laws yield comparable benefits, or satisfy the desire for objectivity to the same degree.

As mentioned earlier, van Fraassen doubts that analysis of the structure of the branched model will produce the concept of

[52] Ibid.

objective chance. Although a solution to this problem must await the next chapter, the challenge laid down by van Fraassen is clear. If objective chance is to be a measure of proportion among possible futures, it must be quite unambiguous what this measure consists in. If one particular measure is adopted, then the objective probability that a certain bridge will collapse within a year takes one value; if a different measure is adopted then it takes quite a different value. Who is to say which is *the* measure? Van Fraassen gives a diagram containing different probability measures for sets of possible futures, it being plainly impossible to say a priori which measure is better than the others.[53] Hence the importance of being able to isolate, in the branched model, a single, unique measure of proportionality among sets of branches, in terms of which the concept of objective chance may be defined.

(vi) *Branched Laws*

It was stated above, in sections (ii) and (iv), that there exist important differences between laws of the branched universe and laws of the Minkowski world, and the time has come to sum up these differences in a precise way. In what follows I shall use the words 'unbranched theory' to denote any theory in the ontology of which branching space-time continua do not occur.

A paramount fact about the branched model is that it is an indeterministic one. There can therefore be nothing, in Mackie's words, that 'tells the universe where to go'. On the other hand, the degree of freedom permitted may be quite small. What establishes what is and what is not possible, relative to any branch point, is the fan or prism of branches at that point, and the overall differences between the members of the fan may be relatively minor. Not all physical processes are indeterministic, but in the case of those that are, their indeterminism stems both from the fact that the outcomes open to them form a fan, and from the fact that the selection of which outcome becomes actual is stochastic. On the branched model, therefore, indeterminism enters into physical processes in two different ways.

The dynamical laws of the branched world are all of the following form: given initial conditions A, the proportion of branches in all fans above A which contain B is x. This law may be stated as $p(B \mid A) = x$, where $0 \leqslant x \leqslant 1$. Furthermore, as we saw

[53] van Fraassen (1989: 80).

above, the number of branched laws (in sense (i) of p. 60) increases monotonically with time, in consequence of the progressive elimination of more and more areas of the tree by branch attrition.

The preceding paragraph points to three distinctive characteristics of 'branched' laws. Consider $p(B \mid A) = x$. If $x = 0$ or $x = 1$, this formula states a 'deterministic' law; if $0 < x < 1$, a probabilistic law. In unbranched theories, problems arise as to what feature of the world makes probability statements true. What are the 'truth-makers' of such statements? The answer given by frequency theorists, following Venn and von Mises, is that what makes true the statement that the probability of an A being a B is x, is that the relative frequency m/n of Bs in a reference class of As approaches the limiting value x as $n \to \infty$. The difficulties of this answer have been discussed by many authors. Hacking and Salmon, for example, point out that *any* observed frequency of Bs in *any* finite sequence of trials, no matter how long, is compatible with *any* value for the limit.[54] There is no guarantee that the relative frequencies converge to any limit at all. It is therefore far from obvious what are the truth-makers for probabilistic laws in the Minkowski world. In the branched world the truth-makers, being topological features of the model, are clear enough.

Another difference between branched and unbranched models concerns single-case probabilities. If it is difficult to give any physical meaning to the idea of long-run or limiting frequencies in unbranched situations, it is even more difficult to give a meaning to the idea of single-case probabilities. What is the probability that a particular atom of radium will decay in the next half-hour? What is the probability that the coin I hold in my hand will come up heads the next time I toss it? How are such single-case probabilities to be defined? Unbranched theories yield no clear answers. The branched model, on the other hand, gives single events like the decay of a particular radium atom a definite probability value, the value being determined by the proportionality of a certain type of outcome on a single fan or prism.

A third respect in which branched and unbranched laws differ is that the number of the former increases with time, whereas on every non-epochal interpretation of 'law' in unbranched theories, the set of laws remains constant. (In the case of the Armstrong–Dretske–Tooley theory, of course, laws are fixed relations between universals and cannot change.) Let us see how progressive branch attrition in the branched model creates a sense in which laws 'evolve' with time.

[54] Hacking (1965: 5–6); Salmon (1967: 85).

At time zero, the moment of creation, there were perhaps very few laws. This is not to say that the constraints upon the universe's development at that time were very lax, but that because of the vast areas of branching it contained, it may have been that very few conditional probability values $x = p(B \mid A)$ were constant over the whole branched structure. Today, however, in our slimmed-down model, the product of a long period of branch attrition, there may be many more conditional probability values that remain constant for the rest of the universe's life. That is, there may be more laws than there used to be. In this sense, new laws could be said to 'come into being' on the branched model.

Summing up the differences between 'branched' and 'unbranched' theories of laws, three areas of dissimilarity have been identified:

1. different truth-makers for both exceptionless and probabilistic laws,
2. objective single-case probability values,
3. evolution of laws with time.

These differences will be further discussed in subsequent chapters.

4

An Interpretation of Quantum Mechanics

IT is not easy, in quantum mechanics, to construct a picture in space and time of what is going on. An advantage of the branched model is that it provides both a picture and an explanation of some of the features of quantum mechanics that have up to now resisted attempts to understand them. The picture is a spatio-temporal one, and the interpretation that the branched model affords consists in laying out quantum phenomena in space and time, in such a way that their inherently probabilistic character becomes manifest. The branched model can explain why the quantum world behaves probabilistically. It can explain how it is that, relative to given initial conditions, quantum events come to have precise probability values, and it provides an ontological basis for these probability values by locating them in the spatio-temporal topology of the model. The key concept, in the branched interpretation of quantum mechanics, is the concept of objective quantum probabilities.

The difficulties of interpreting quantum probabilities are compounded by the distant correlations of quantum events that are observed in the EPR experiment. In this case too the branched model provides an explanation. Two events which are widely separated in space may both occur on the same branch, and the observed correlations which their joint occurrence obeys are explained by the relative proportionality of the sets of branches on which they jointly occur. The apparent way in which one distant event can instantaneously affect the occurrence of another distant event, without any causal connection between them, turns out in the branched model to be explicable by branch attrition.

Thirdly, the model has something to say about measurement. In the branched model there exists a clear and objective difference, based on what kinds of outcomes are permitted on the branches above given sets of initial conditions, between unitary or Schröd-inger evolution and what we may call 'measurement'. In the branched interpretation, there exists an unambiguous definition of this term. This does not fully resolve the measurement problem inasmuch as it may still be unclear, in particular cases, whether a given quantum process is an instance of Schrödinger evolution or an instance of 'measurement'. But although there may be processes regarding which we are at present unable to answer this question, for the branched model there is always one and only one correct response. What constitutes a 'measurement', on the branched model, is a purely objective feature of the world, having nothing to do with human observation or consciousness.

Finally, and most importantly, the interpretation given here is a realist one. In the words of one recent author, quantum theory challenges 'the basic belief, implicit in all science and indeed in almost the whole of human thinking, that there exists an objective reality, a reality that does not depend for its existence on being observed' (Squires 1986: 2–3). The branched model, in distinction to this point of view, *is* objective. It does not depend for its existence on being observed. And it is in terms of this model that it is proposed that quantum mechanics should be interpreted. In it, the 'role of the observer' is of secondary importance.

The branched interpretation is an interpretation, not a new quantum theory, and does not replace the elegant theoretical apparatus of state vectors and operators in Hilbert space that we use to calculate and predict quantum probabilities. If it does anything at all, the branched interpretation yields understanding of how quantum events come to have determinate probability values. The interpretation is via the concept of probability, this concept having both a theoretical and an ontological aspect. It attempts to give one possible answer (not the only answer, perhaps, but one answer) to the metaphysical question, 'What would the world have to be like, in order for quantum mechanics to be true?'[1]

(i) *Quantum Probabilities*

Quantum probabilities are represented on the model in the following way. It will be simplest if we illustrate the mode of representation by an example. Let an HV polarizer be a two-channel polarization analyser which separates incident light into two

[1] Different versions of this question, and also different responses, are found in two recent books by philosophers on quantum mechanics: Healey (1989: 6) and Hughes (1989: 82, 175, and 296). Van Fraassen (1985: 128) puts it this way: 'To understand quantum mechanics means to understand how the world could possibly be the way quantum mechanics says it is.'

beams, one containing photons which are polarized vertically relative to the floor of the laboratory, and the other containing photons which are polarized horizontally. Suppose that a photon has emerged in the 'vertical' channel, and that it is moving towards a second analyser which is tilted at an angle ϕ to the vertical. This analyser has two exit channels, ϕ^+ and ϕ^-. Quantum theory tells us that the photon is in a superposition of states, with an amplitude to be measured ϕ^+, and an amplitude to be measured ϕ^-. The probabilities of these two events, obtained by squaring the absolute value of their amplitudes, are $\cos^2\phi$ and $\sin^2\phi$ respectively.

The way in which the two probabilities $\cos^2\phi$ and $\sin^2\phi$ are represented on the branched model is shown in Fig. 4.1. Relative to the time at which the photon enters the polarizer, there are two different kinds or sets of branches. Let A be the event of the photon entering the polarizer. On the model, A occurs at a point where four-dimensional histories branch. Relative to that point, there are two different kinds of outcome: branches on which the photon exits in the ϕ^+ channel, and branches on which the photon exits in the ϕ^- channel. The two different sets of branches are depicted in Fig. 4.1, which is simplified in that it represents not merely two branches, but two *sets* of branches. One is the set of ϕ^+ branches, and the other the set of ϕ^- branches.

The set of ϕ^+ branches, as will be argued shortly, constitutes a fixed and definite proportion of the branches above the point A. The value of this proportion is the number $\cos^2\phi$. Similarly, the set of ϕ^- branches constitutes a proportion $\sin^2\phi$ of the branches. Note that $\cos^2\phi + \sin^2\phi = 1$. Since outcomes of these two different kinds occur on branches belonging to sets with the indicated proportions, and since it is a random matter, on the model, which branch survives as the actual branch, the chance that it is a ϕ^+ branch is $\cos^2\phi$, and the chance that it is a ϕ^- branch is $\sin^2\phi$.

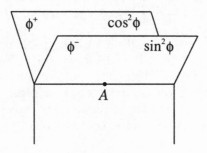

FIG. 4.1

These, then, are the probabilities that the photon will emerge in one or the other channel. On the model, the *probability* of a quantum event is given by the *proportion* of branches on which that event occurs. In the branched interpretation of quantum mechanics,

probability = proportionality.

Defined in this way, quantum probabilities have a physical meaning, their values being built into the branched spatio-temporal structure of the universe. But more needs to be said about this. It is not yet clear how the branched structure allows a fixed and definite proportion of the set of all branches above A to be ϕ^+ branches, and therefore it is not yet established that the probability values for the two different outcomes ϕ^+ and ϕ^- are well-defined. This point, which is crucial to the proposed interpretation, must be dealt with before we can proceed further.

If every quantum probability were a rational number of the form n/m, where n and m are whole numbers, then the problem would be simple: the value of the probability would be captured by a numerical ratio between finite sets of branches. But in general the values of quantum probabilities, e.g. $\cos^2\phi$, will not be confined to rational numbers, and therefore no ratio of finite sets of branches can represent them. Infinite sets of branches would be required. But there is a difficulty here, because Cantor showed long ago that there exist no well-defined proportionalities among infinite sets. For example, it might be thought that in the set of all natural numbers the proportionality of odd and even numbers was the same, namely $\frac{1}{2}$. But if the integers are ordered as follows:

$$1\ 2\ 4\ 3\ 6\ 8\ 5\ 10\ 12\ 7\ldots$$

then the proportion of even numbers appears to increase to $\frac{2}{3}$. In Cantorian set theory, fixed proportions among infinite sets do not exist. How, against the authority of Cantor, can it be argued that branch proportionality yields exact values for quantum probabilities?[2]

Fortunately, there is a solution to this problem, which lies in the concept of what I shall call a *decenary tree*. (For *Homo sapiens*, who counts by tens, a decenary tree is the simplest solution, although in nature a binary tree, or indeed any proportionality-preserving tree which branches finitely and equally at each level, would do as well.) A decenary tree is one that branches in ten,

---

[2] This question was put to the author in the summer of 1987 by Bas van Fraassen.

1   2   3   4   5   . . .

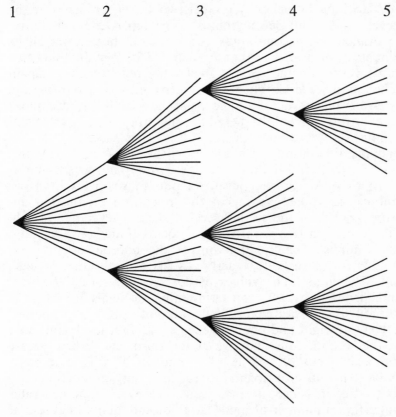

FIG. 4.2

and all of whose branches branch in ten, at each of a denumerable infinity of discrete nodes. The branching is along the time-axis. Imagine a decenary tree that branches in ten once a second forever. The shape of such a tree is shown in Fig. 4.2. Not all the branches are drawn in, but the idea is clear. A decenary tree with a denumerable infinity of levels at which it branches will contain a nondenumerable infinity of branches, namely $10^{\aleph_0}$.

Despite the large number of branches it contains, it is not difficult to demonstrate that there exist exact numerical proportions among sets of branches of a decenary tree. Suppose for example it is desired to construct a tree, each of whose branches is either a 'plus' branch or a 'minus' branch. (Recall the photon passing through the polarizer.) Suppose further that the proportion of 'plus' branches must be exactly $\frac{1}{4}$, and the proportion of 'minus' branches $\frac{3}{4}$. Such a tree may be constructed in the following way, by labelling branches. We label two of the branches at the first level 'plus', and seven 'minus'. The labels are hereditary: once a branch has become 'plus' (or 'minus'), then it and all its descendants (the branches that branch off it) remain 'plus' (or 'minus') forever. At the second level, only ten branches remain unlabelled, and we label five 'plus' and five 'minus'. Counting up, twenty-five branches out of a hundred at the second level are 'plus', and seventy-five 'minus'. By heredity, this proportion is preserved throughout the remainder of the tree, hence exactly $\frac{1}{4}$ of the non-denumerable infinity of branches will be 'plus', and $\frac{3}{4}$ 'minus'.

Suppose now that a decenary tree is desired with a plus–minus ratio of $\frac{2}{3}$ to $\frac{1}{3}$. At the first level, we put 'plus' on six branches and 'minus' on three. We repeat this operation on the ten unlabelled branches of the second level, the third level, the fourth, etc. The result is a tree with 0.6666... of its branches 'plus', and 0.3333... 'minus'. Finally, thinking of the photon, suppose we are looking for a tree with exactly $\cos^2 20°$ of its branches 'plus', and $\sin^2 20°$ 'minus'. The values of $\cos^2 20°$ and $\sin^2 20°$ are real numbers denoted by the non-terminating decimals 0.884023... and 0.115976... respectively. Well, we put 'plus' and 'minus' respectively on eight and one branches at the first level, eight and one at the second, four and five at the third, etc. When we are finished, exactly $\cos^2 20°$ will be 'plus', and $\sin^2 20°$ will be 'minus'.[3]

The construction of a decenary tree (or of any proportionality-preserving tree in which, at any given level, every node generates the same (finite) number of immediate descendants) allows sets of branches of any desired proportion to be singled out. Conversely, from a decenary tree with a fixed proportion of its branches of a certain kind, the exact value of the proportion can be read off, decimal place by decimal place. In infinite sets which are struc-

---

[3] There is a problem here with non-terminating decimals, as was pointed out to me by Christopher Hitchcock. For example, in the decenary tree with $\frac{2}{3}$ of its branches 'plus' and $\frac{1}{3}$ 'minus', at every one of the denumerable infinity of levels there is an unlabelled branch. At what level does this branch get to be 'decided'? Surely it should be decided at *some* level, since every photon which enters a two-channel analyser must emerge in one or the other of the two channels. But it cannot be decided at any *finite* level, since if it were the proportion of 'plus' branches would not be exactly $\frac{2}{3}$. Hitchcock suggests that the one undecided branch might be decided in a *second* decenary tree above the first (it being indifferent for the $\frac{2}{3} - \frac{1}{3}$ ratio whether it was decided 'plus' or 'minus') and this is probably the best solution.

Incidentally, the fact that in a decenary tree the branches corresponding to 0.6666... when added to the branches corresponding to 0.3333... yield only 0.9999..., not 1.0000..., shows that the structure of branches in a decenary tree is not quite isomorphic to the real numbers between 0 and 1. Cf. below, Ch. 5 n. 34.

tured in the way the branches of a decenary tree are structured, well-defined proportionalities exist. In this way, Cantor's difficulty is resolved.

Recall now the vertically polarized photon, about to enter the ϕ^+ analyser. It has a $\cos^2 20°$ chance of emerging ϕ^+, and a $\sin^2 20°$ chance of emerging ϕ^-. These probabilities are built into a decenary tree, but the photon does not have the luxury of being able to wait an infinite length of time to discover what the probabilities are. The probabilistic result must be manifested instantaneously, or almost instantaneously, as the beam of photons splits. In the branched model, therefore, the decenary tree is shrunk down so as to occupy an arbitrarily short (but non-null) interval of time. Suppose the photon enters the polarizer at time t, and that by time $t + \Delta t$ the result of the experiment is decided and the photon is either in the ϕ^+ channel or in the ϕ^- channel. In the branched model the first node of the decenary tree is at t, the second at $t + \frac{1}{2}\Delta t$, the third at $t + \frac{3}{4}\Delta t$, etc. By $t + \Delta t$, the entire decenary tree has been run through, and one and only one of its branches remains. Since branch selection is random at each node, the chance of that branch being a ϕ^+ branch is $\cos^2\phi$, as shown in Fig. 4.3.

The model can now be made less artificial, and more realistic. Since each branch on the decenary tree is a four-dimensional manifold in which quantum and other physical events actually occur, we can drop the fiction of the labels 'plus' and 'minus'. A branch which is 'plus' will be a branch on which the photon enters the ϕ^+ channel, and a 'minus' branch one on which it enters the ϕ^- channel. All this happens in an arbitrarily short time. I shall call a decenary tree which yields probabilities for the photon in the analyser a '$\phi\pm$ prism for the photon'.

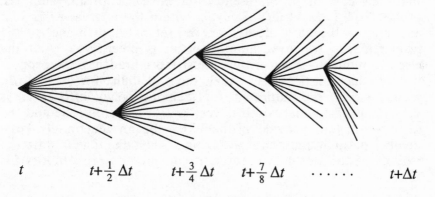

t $t+\frac{1}{2}\Delta t$ $t+\frac{3}{4}\Delta t$ $t+\frac{7}{8}\Delta t$ $\cdot\ \cdot\ \cdot\ \cdot\ \cdot\ \cdot$ $t+\Delta t$

FIG. 4.3

Let us consider in detail what happens when a vertically polarized photon enters an analyser inclined at an angle ϕ, so that the first branch point of the universe is located at the base of a $\phi\pm$ prism for the photon. To take a concrete example, let ϕ be 20°. As the branches drop off, within the arbitrarily short space of time Δt the photon has got to find itself on a ϕ^+ branch with probability 0.884023..., or on a ϕ^- branch with probability 0.115976... At $t + \frac{1}{2}\Delta t$, all but one of the ten branch segments at the first branch point will have dropped off, and one of three possibilities will have obtained. Either (i) the surviving branch is a ϕ^+ branch, in which case one of the branches in the prism which belongs to the set with proportionality 0.884023... will have been selected, or (ii) the surviving branch is a ϕ^- branch, in which case a branch belonging to the set with proportionality 0.115976... will have been selected, or (iii) the surviving branch is neither a ϕ^+ nor a ϕ^- branch, in which case the decision whether the photon will be measured ϕ^+ or ϕ^- is postponed to the interval between $t + \frac{1}{2}\Delta t$ and $t + \Delta t$. Once the process of branch attrition has worked its way through all the node levels between t and $t + \Delta t$, the one surviving branch will be either a ϕ^+ branch, with probability 0.884023..., or a ϕ^- branch, with probability 0.115976.... At the end of the prism, at most one 'undecided' branch remains, and the probability for the photon is established.[4]

The passage of the photon through the polarizer is not the only quantum event that is occurring at time t. On Venus, a radioactive atom has a chance of decaying, and on Alpha Centauri an electron in a hydrogen atom may or may not jump to a new energy level. These quantum events have probabilities too, and the wealth of information capable of being stored within a decenary tree means that they can all be represented on the same prism. Thus the probability of the atom's decaying within the interval Δt is x, if and only if the atom decays on the set of those branches, the proportion of which within the prism is precisely x. And the electron jumps to a new orbit on just those branches, the proportion of which on the prism equals the probability for the electron to make the jump. Although the temporal height of a prism is very small, its spatial width is very great. Therefore one and the same prism can provide probabilities for an indefinitely large number of quantum events occurring at the same time in different regions of the world. Moreover, it can provide probabilities for

---

[4] I am indebted to Jerry Massey for urging me to spell out the process of branch attrition within a prism in this way.

the results of an experiment in which the measured quantity has a continuous spectrum of possible values.

The words 'same time' occurring three lines above bring to mind a feature of the model that was discussed in Chapters 1 and 2 and should be mentioned again. By the special theory of relativity, there is no single preferred simultaneity class of events—no one spacelike hyperplane which brings together all point events simultaneous with 'here-now'. Instead there are many such hyperplanes, each of which constitutes the first branch point, or more properly the first branch surface, of the model in a given frame of reference. The shape of the model, as was pointed out in Chapter 1, is relative not only to a time but also to a frame of reference, and the precise way in which the model branches is a 'perspectival', not a Lorentz-invariant, feature of it. (A Lorentz-invariant description of the branched universe is given in Appendix 2.) The fact that the precise shape of the universe is intimately tied to the coordinate frame in which it is viewed or described might in the minds of some be considered a deficiency, indicating that the model is not wholly objective or 'real'. But such is not the case. In fact, as we shall see in the next section, it is precisely this frame-time-dependency, or hyperplane dependency, of the model which allows us to explain the distant correlations of measurement results in the EPR experiment. The hyperplane dependency of the model, therefore, plays an important role in the interpretation of quantum mechanics.

To sum up, quantum probabilities are interpreted in the model as proportions of sets of branches in prisms, and they and their values are built into the branching structure. They are permanent features of it. In the branched model, quantum mechanics (and hence physics itself) is inherently and irremediably probabilistic, and its probabilistic character is grounded in the treelike topology of space-time. In the branched model, each prism of temporal height Δt has another prism at the tip of every one of its branches, so that the entire structure is a branching stack of prisms. As will be seen in Chapter 5, not only in a single prism, but also in a stack, the probability of an event, relative to the initial conditions prevailing at the base node, is given by the proportion of branches on which that event occurs. Therefore for any event, relative to given initial conditions, the branched model provides an objective single-case probability value.

A noteworthy feature of the model is this: probability values reside in the branching pattern of the *future*, never in the present. When a photon enters a polarizer at time t the probability of its being measured ϕ^+ or ϕ^- is given by patterns of branches later than t. It is, however, a fact about the branching structure of the world, so far as we have been able to observe it, that initial conditions of a specific kind (e.g. a vertically polarized photon entering a $\phi\pm$ analyser) always seem to be correlated with subsequent branching of a specific kind (e.g. ϕ^+ and ϕ^- branches in proportions $\cos^2\phi$ and $\sin^2\phi$ respectively). Correlations of this type between initial conditions and subsequent branching cannot be accidental, yet one might be reluctant to think that they give evidence of some variety of 'pre-established harmony' in the structure.[5] On p. 138 below, and in Chapter 7, p. 217, I outline a hypothesis which may account for them.

(ii) *Quantum non-locality and the EPR experiment*

When two particles are emitted at the same time from a source in what is known as the singlet spin state, and, moving off in different directions, are subjected to different measurements, their behaviour violates 'locality', in the following sense. Suppose that the particles are sufficiently far apart so that the two space-time regions in which the measurements are performed are spacelike separated, i.e. unconnectible by a light ray or any other causal influence. Although seemingly causally independent, the results of the two measurements are statistically correlated in a way that cries out for explanation. Yet there can be no causal influence proceeding from one to the other if the special theory of relativity is correct. Furthermore, as will be seen, the possibility of explaining the correlation by appealing to a common cause, or to a set of local hidden variables which dictate the behaviour of the two particles under all conceivable experimental conditions, is ruled out by Bell's theorem.[6] If the two-particle system is to be regarded as 'real', i.e. existing in space and time, then it seems impossible to avoid the conclusion that its behaviour violates the locality principle that what happens at one place must be unaffected by what happens at some other place, neither event being in the light cone of the other.

---

[5] I owe the use of the phrase 'pre-established harmony' in this context to conversations with John Earman.

[6] Non-technical discussions of Bell's inequality, and its experimental violation, may be found in Clauser and Shimony (1978); d'Espagnat (1979); Bell (1981); Mermin (1981 and 1985); van Fraassen (1982 and 1985); Rae (1986); Hughes (1989: 170–2, 238–48); and the papers contained in Cushing and McMullin (1989).

Can the correlations between the results of performing simultaneous spin measurements upon two distant particles in the singlet state be explained (without invoking hidden variables or a common cause) in some way that is consistent with relativity theory? I shall argue that they can. Furthermore, the explanation need involve no abandonment of the principle of realism; on the contrary it involves an affirmation of realism. Frequently, the challenge posed by Bell's theorem, and by the experimental results of the EPR experiment, has been stated in such a way as to require a choice between locality and realism. The matter is expressed by Clauser and Shimony as follows:

Realism is a philosophical view, according to which external reality is assumed to exist and have definite properties whether or not they are observed by someone. So entrenched is this viewpoint in modern thinking that many scientists and philosophers have sought to devise conceptual foundations for quantum mechanics that are clearly consistent with it. . . . [However, Bell's] theorem proves that all realistic theories, satisfying a very simple and natural condition called locality, may be tested with a simple experiment against quantum mechanics. These two alternatives necessarily lead to significantly different predictions. The theorem has thus inspired various experiments, most of which have yielded results in excellent agreement with quantum mechanics, but in disagreement with the family of local realistic theories. Consequently, it can now be asserted with reasonable confidence that either the thesis of realism or that of locality must be abandoned. Either choice will dramatically change our concepts of reality and of space-time.[7]

The task of this section will be to show that the results of the experiments which test Bell's theorem can indeed be explained by abandoning locality and affirming realism. Moreover, the abandonment of locality to be described below is consistent with, and in the spirit of, special relativity, so that the explanation is both realistic and relativistic. Because the proposed explanatory model for the observed results of the EPR experiment is (i) spatio-temporal, (ii) branched, and (iii) dynamic, it does require, in the words of Clauser and Shimony, that we 'dramatically change our concepts of reality and of space-time'.

In one version of the Bell–EPR experiment (see Fig. 4.4), a pair of correlated photons is emitted from a source S by positronium decay.[8] The angle of plane polarization of each photon is

---

[7] Clauser and Shimony (1978: 1883).

[8] Photons emitted by positronium decay have anti-correlated polarizations, so that if one is observed to be vertically polarized, the other will be horizontally polarized. A pair of photons emitted by a $J = 0 \rightarrow 1 \rightarrow 0$ atomic cascade, by contrast, will have parallel linear polarizations. See Ballentine (1990: 447–9).

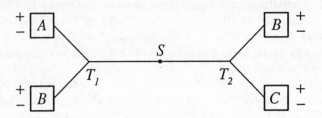

measured in one of two ways: A or B on the left, and B or C on the right.[9]

The measuring devices A, B, and C are two-channel polarization analysers, set at different angles in the plane orthogonal to the photons' line of flight. The measurement axes of the two analysers B on left and right are parallel. (In the experiment designed by Alain Aspect the photons are directed to their respective measuring devices by ultrasonic switches T_1 and T_2 which are set randomly and which operate so rapidly that their settings are not determined until after the photons have left the source.) Each observed result consists of a pair of measured outcomes, e.g. 'A^+ on the left and B^+ on the right', or 'B^+ on the left and C^- on the right', denoted by (A^+, B^+) and (B^+, C^-) respectively. Quantum theory predicts, and experiment confirms, the existence of a precise degree of correlation between the members of any such pair.

If the normalized vectors $|L \updownarrow\rangle$ and $|L \leftrightarrow\rangle$ denote the state of the left-hand photon as vertically polarized and as horizontally polarized respectively, and similarly for $|R \leftrightarrow\rangle$ and $|R \leftrightarrow\rangle$, then the two-photon system is in the following entangled superposition:

$$\Psi = 1/\sqrt{2}(|L \updownarrow\rangle \, |R \leftrightarrow\rangle + |L \leftrightarrow\rangle \, |R \updownarrow\rangle).$$

Furthermore, this state is rotationally symmetric about the axis of propagation of the photons, so that it can also be represented as follows:

$$\Psi = 1/\sqrt{2}(|L \nearrow\rangle \, |R \searrow\rangle + |L \searrow\rangle \, |R \nearrow\rangle),$$

where $\nearrow$ and $\searrow$ are any two orthogonal directions. What this means is that if the photons are measured by any pair of polarization analysers whose axes are parallel, one photon will always pass + and the other −. Hence the joint outcomes (B^+, B^+) and (B^-, B^-) are never observed, no matter what angle of orientation the B-analyser assumes.

[9] Cf. Sudbery (1986: 201).

The second type of correlation exhibited by the outcomes is this: if the angle between the measurement axes of the analysers A and B is θ, then in a sufficiently long run of trials the relative proportion of similar outcomes (A^+, B^+) and (A^-, B^-) will be $\sin^2\theta$, while the relative proportion of dissimilar outcomes (A^+, B^-) and (A^-, B^+) will be $\cos^2\theta$. (If $\theta = 0$ the outcomes never agree, since $\sin 0 = 0$.) Expressing this correlation in terms of probabilities, we write $p(A^+, B^+) = p(A^-, B^-) = \frac{1}{2}\sin^2\theta$, and similarly for the other combinations. The question to be addressed is, what accounts for these precise mathematical correlations between the left-hand and right-hand observations, given that the measurements performed on the two photons may be separated by a considerable distance in space?

A possible explanation is this. As the photons leave the source and travel outwards, they are accompanied by a hidden variable, or set of hidden variables, which guides them through their respective measuring devices and creates the desired degree of correlations between the outcomes. This type of explanation, with a plausible restriction known as *Einstein locality*, is called a *local hidden variables theory*, and rests upon the assumption that an observation made upon one of the two photons should in no way disturb or affect an observation made upon the other. More precisely, a local hidden variables theory asserts the following:[10]

(1) A two-photon system is characterized by a (hidden) state λ which is more complete than the system's quantum-mechanical state Ψ. The domain of all such states λ is denoted by Λ.

(2) Each state λ provides an 'instruction set' for each photon determining its measurement outcome $+$ or $-$ for each of the different polarization tests that may be performed upon it.

(3) The probabilistic, unpredictable character of the measurement outcomes is due to the hidden, uncontrollable character of the states λ. Let pr be a probability measure on Λ. Then $pr(\lambda)$ denotes the probability that a given system is characterized by λ.

(4) The instruction set for the left photon does not depend upon what test is performed on the right photon or on the outcome of that test, and the instruction set for the right photon does not depend upon what test is performed on the left photon or on the outcome of that test.

A theory which conforms to these conditions is a local hidden variables theory. A 'non-local' theory fails to satisfy condition (4). What will be shown is that (i) any local hidden variables theory

[10] Cf. Wigner (1970); Clauser and Shimony (1978: 1888); Shimony (1989b: 385); Ballentine (1990: 445).

entails the satisfaction of an inequality known as the Bell–Wigner inequality, but that (ii) experiment shows that the inequality is in fact violated, hence (iii) local hidden variables cannot be invoked to explain the correlations observed in the Bell–EPR experiment.

In the experiment shown diagramatically in Fig. 4.4, there are three polarization analysers A, B, and C, oriented at different angles. Therefore a local hidden variables theory will indicate how each photon is to react to a particular measuring instrument. For example the instruction set $(++-, --+)$ prescribes that the left photon will be measured $+$ by an A-analyser, $+$ by a B-analyser, and $-$ by a C-analyser, while the right photon will be measured $-$ by an A-analyser, $-$ by a B-analyser, and $+$ by a C-analyser. Because of the requirement that the left and right photons must always take opposite polarization values when measured by analysers with the same orientation, there are, for the three analysers, eight and only eight possible instruction sets provided for any given pair of photons. (For example, the set $(++-, -++)$ is not a possible one.) The eight sets are the following:

1. $(+++, ---)$ 5. $(-++, +--)$
2. $(++-, --+)$ 6. $(-+-, +-+)$
3. $(+-+, -+-)$ 7. $(--+, ++-)$
4. $(+--, -++)$ 8. $(---, +++)$

If these sets are attached to photon-pairs in the correct proportions, as specified by the probability measure pr, their guiding force can account for many of the observed correlations in the EPR experiment. For example, if the angle $\widehat{AB}$ between the A and B analysers is 45°, if $\widehat{BC}$ is 45°, and if $\widehat{AC}$ is 90°, then instruction sets which attach to emitted photon-pairs in the following proportions will explain the observed correlations:

$$pr(1) = pr(+++, \ ---) = 0 \qquad pr(5) = pr(-++, \ +--) = \tfrac{1}{4}$$
$$pr(2) = pr(++-, \ --+) = \tfrac{1}{4} \qquad pr(6) = pr(-+-, \ +-+) = 0$$
$$pr(3) = pr(+-+, \ -+-) = 0 \qquad pr(7) = pr(--+, \ ++-) = \tfrac{1}{4}$$
$$pr(4) = pr(+--, \ -++) = \tfrac{1}{4} \qquad pr(8) = pr(---, \ +++) = 0^{11}$$

However, a local hidden variables theory cannot account for every correlation which quantum theory predicts and which experiment confirms. We see from the above table that

$$p(A^+, B^+) = pr(3) + pr(4)$$
$$p(B^+, C^+) = pr(2) + pr(6)$$
$$p(A^+, C^+) = pr(2) + pr(4)$$

[11] By inspection we see that $p(A^+, B^+) = \tfrac{1}{2}\sin^2 45° = \tfrac{1}{4} = pr(3) + pr(4)$; $p(A^-, C^-) = \tfrac{1}{2}\sin^2 90° = \tfrac{1}{2} = pr(5) + pr(7)$; $p(A^-, C^+) = \tfrac{1}{2}\cos^2 90° = 0 = pr(6) + pr(8)$; etc. as required.

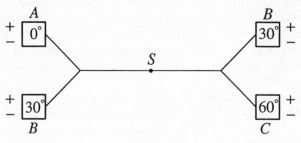

FIG. 4.5

Since all these probability measures *pr* are non-negative, we obtain

$$p(A^+, B^+) + p(B^+, C^+) \geq p(A^+, C^+)$$

which is the Bell–Wigner inequality.[12] It is not difficult to see that this inequality (henceforth referred to as 'Bell's inequality') conflicts with quantum mechanics. Its violation in any sufficiently long run of trials, for suitable choice of orientation of the analysers A, B, and C, is predicted by quantum theory. For example, select $\widehat{AB} = 30°$, $\widehat{BC} = 30°$, $\widehat{AC} = 60°$, as in the experimental arrangement shown in Fig. 4.5. For the appropriate choice of analysers, $p(A^+, B^+) = \frac{1}{2}\sin^2 30° = \frac{1}{8}$, $p(B^+, C^+) = \frac{1}{2}\sin^2 30° = \frac{1}{8}$, $p(A^+, C^+) = \frac{1}{2}\sin^2 60° = \frac{3}{8}$, so that Bell's inequality fails. It follows that no local hidden variables explanation can be given of the correlations observed in the Bell–EPR experiment.

We have arrived at the following state of affairs. Any local hidden variables or 'common cause' type of explanation of the EPR correlations, in which each particle is accompanied by an instruction set or 'program' which dictates its behaviour with respect to arbitrarily oriented measuring devices, appears to be ruled out by the experimental violation of Bell's inequality. On the other hand non-local hidden variables, which could transmit information instantaneously from one measuring device to another, are ruled out by relativity theory. We may be tempted to conclude that no explanation exists at all, and that the distant correlations of the EPR measurement results must simply be accepted as peculiar unexplained facts.[13] But this, I shall argue,

[12] The derivation of the inequality given here comes from Hughes (1989: 170–1), based on Wigner (1970). Similar derivations may be found in d'Espagnat (1979) and Rae (1986: 37–40).

[13] Arthur Fine, in an article entitled 'Do Correlations Need to be Explained?' (1989), takes the view that the stable distant correlations of measurement outcomes observed in the EPR experiment are in no need of explanation, but must simply be accepted as they stand. The view of this chapter, however, is that they *can* be explained, and that the branched model provides an explanation of them.

would be premature. The structure of the branched model, and in particular its branching along hyperplanes in prisms of small temporal height but unlimited spatial breadth, combined with branch attrition, is precisely the mechanism needed to explain EPR's distant correlations. That, at least, is what I shall attempt to demonstrate.

An Aspect-style realization of the experimental set-up in Fig. 4.5, in which the photons pass through switches and have a random chance of encountering either apparatus A or apparatus B on the left, and either B or C on the right, corresponds to the structure in the branched model shown in Fig. 4.6. In this diagram each branch represents a set of branches on the branched model. The four possible random settings (0,30), (30,60), (0,60), and (30,30) occur at the lower branch level. Above them are the different joint outcomes, each with their respective proportions/probabilities. Since $p(0,30) = \frac{1}{4}$, and since, given that the setting (0,30) has been selected, $p(++)$ for that setting is $\frac{1}{8}$, we have that $p(0+, 30+) = \frac{1}{32}$. Multiplying probability values up the diagram in this way, we obtain $p(30+, 60+) = \frac{1}{32}$, and $p(0+, 60+) = \frac{3}{32} \cdot \frac{1}{32} + \frac{1}{32} \not\gg \frac{3}{32}$. In even a reasonably short series of trials, therefore, one would expect Bell's inequality to be violated. Its failure is built into the structure of the branched model.

Let us bring the property of non-locality into sharper focus. Consider a simpler experimental arrangement (Fig. 4.7), with only

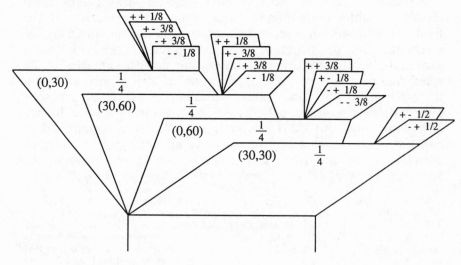

FIG. 4.6

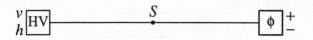

FIG. 4.7

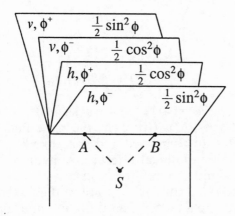

FIG. 4.8

two analysers instead of four, in which the right analyser is set at an angle ϕ to the vertical. Confining ourselves in the first instance to a frame of reference in which the two measurement events A and B at left and right are simultaneous, the branched space-time structure which corresponds to this experimental arrangement is shown in Fig. 4.8.

Except when $\phi = 45°$, in which case $p(v, \phi^+) = p(v, \phi^-) = p(h, \phi^+) = p(h, \phi^-) = \frac{1}{4}$, the fact that the outcomes of the two spatially separated events A and B are not independent of each other is immediately apparent from the branched structure. Designating for the moment the two possible outcomes of A as A^+ and A^- rather than v and h, and similarly for B, the requirement that must be met for A^+ and B^+ to be correlated is that $p(A^+, B^+)$, i.e. $p(A^+ \& B^+)$, should not be equal to $p(A^+) \times p(B^+)$. In Arthur Fine's terminology, $p(A^+, B^+)$ should not be 'factorizable' into $p(A^+)$ and $p(B^+)$.[14] In the two-analyser experiment where $\phi = 30°$ it is obvious that A^+ and B^+ must be correlated, since $\frac{1}{2}\sin^2 30° = \frac{1}{8}$ and $\frac{1}{2}\cos^2 30° = \frac{3}{8}$

---

[14] Fine (1980: 536).

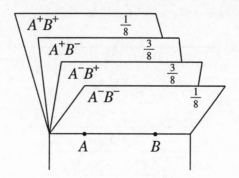

FIG. 4.9

(see Fig. 4.9). Here $p(A^+) = p(A^+, B^+) + p(A^+, B^-) = \frac{1}{2}$, $p(B^+) = \frac{1}{2}$, $p(A^+, B^+) \neq p(A^+) \times p(B^+)$. However, the mere fact that two given events A^+ and B^+ are correlated and spacelike separated does not in itself mean that non-local effects are involved, for A^+ and B^+ might have a common cause.[15] Only in cases where common-cause explanations are ruled out, as they are by Bell's theorem, need we have recourse to explanations based upon a non-local mechanism, and these the branched model is able to supply.

The non-local character of the two-analyser experiment emerges strikingly if we modify the set-up so that one measurement occurs before the other. Suppose that the left analyser is located at **A** a

---

[15] Pick a card at random out of a deck, cut it in half, and mail one half to Vancouver, the other half to Halifax (example from Jeffrey Bub). Let A and B be the two events of opening the letters; let A^+ and B^+ be the finding of a red half-card, and A^- and B^- the finding of a black half-card. Then $p(A^+) = \frac{1}{2}$, $p(B^+) = \frac{1}{2}$, but $p(A^+, B^+) = \frac{1}{2}$, not $\frac{1}{4}$. If I open the Halifax letter and find a red card, then my 'knowledge wave packet' collapses instantaneously and I know immediately the colour of the card in Vancouver. Here there is nothing mysterious: the distant correlation of A^+ and B^+ involves no non-locality, but stems from a common cause.

Another example: out of four cards, two red and two black, pick two at random with your eyes closed, one with your left hand and the other with your right. Let $A^+(A^-)$ be finding a red (black) card in your left hand, etc. Again A^+ and B^+ are statistically correlated:

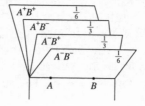

short distance away from the source on earth, and the right analyser at **B** on the moon. A pair of entangled photons is emitted, and the left photon exits in the 'vertical' channel of the polarizer at **A**. As soon as this result is observed, the experimenter on earth can with perfect confidence predict that, on the moon, the right photon will be observed to be polarized horizontally. If the analyser at **B** is set to the angle ϕ, the right photon will be measured ϕ^+ with probability $\sin^2\phi$: in particular if $\phi = 0$ the right photon will invariably exit through the horizontal channel.

In connection with these observed results, three questions arise.[16]

(1) Did the right photon have the property of being horizontally polarized *prior to* the measurement of the left photon at **A**? The answer to this question seems to be no. If the analyser at **A** had registered the left photon as being polarized horizontally instead of vertically, then the right photon would not have had the property in question. Instead, it would have been polarized vertically. If at the last moment the analyser at **A** is removed, so that no measurement is performed on the left photon, then tests of the right photon's polarization at **B** reveal no consistent angle of polarization at all: the photon passes + or − through analysers at any angle with equal frequency. At the level of theory, the right photon is not in an eigenstate of any spin observable. All the evidence suggests that, prior to the measurement at **A**, the right photon did not have the property of horizontal polarization.

(2) Did the right photon have the property of horizontal polarization *after* the measurement at **A**? The answer here is yes. The photon is now in a pure quantum state. Repeated observation shows that if the left photon emerges in the 'vertical' channel, the

---

$p(A^+, B^+) \neq p(A^+) \times p(B^+)$, and once more the explanation lies in a common cause. But not all probability distributions where A^+ and B^+ are correlated are susceptible to common cause explanations. The EPR distribution of $(\frac{1}{8}, \frac{3}{8}, \frac{3}{8}, \frac{1}{8})$ is not, and the empirical discovery of a set of initial conditions in the world which in repeated tests generated the following very different outcome probabilities would prove equally if not more baffling:

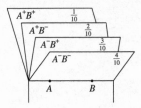

right photon is invariably observed to be horizontally polarized. This is the case whether it is passed through an HV analyser, or through an analyser inclined at an angle ϕ to the vertical. In the latter case it will be measured ϕ^+ in $\sin^2\phi$ of the instances: the mark of a horizontally polarized photon.

(3) How did the right photon acquire the property of being horizontally polarized at the moment of the measurement at **A**, given that it was then on its way to the moon? *Before* the measurement at **A** it was not horizontally polarized; *afterwards* it was. How did the measurement at **A** affect it, given that it was receding from **A** at the speed of light? These questions could be answered if local hidden variables or superluminal transmission of information could be appealed to, but local hidden variable theories are inconsistent with the failure of Bell's inequality, and superluminal transmission of information conflicts with relativity theory. The only way of answering the questions that I know of is to observe what happens to the photons in the branched interpretation.

The two measurement events, at **A** on the earth and at **B** on the moon, lie outside each other's light cones and cannot be joined by a light ray. There will therefore, according to special relativity, be (i) coordinate frames in which the two events occur simultaneously, (ii) frames in which the event at **A** occurs before the event at **B**, and (iii) frames in which the event at **B** occurs before the event at **A**. Let A be the event of the left photon entering its analyser at **A**, and B the corresponding event at **B**. We have

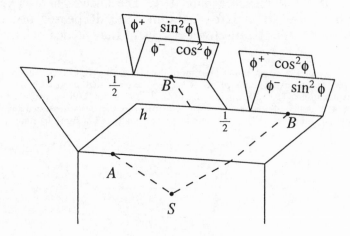

FIG. 4.10

already seen, in Fig. 4.8 above, what the branched model looks like for the case where *A* and *B* are simultaneous. Let us now see, in Fig. 4.10, what its structure is when *A* occurs before *B*.

In this coordinate frame, the left photon has a 50–50 chance of being measured *v* or *h*. When the moment of measurement arrives, and the photon emerges in the 'vertical' channel, then branch attrition in the model ensures that all branches above the hyperplane containing *A* vanish save one, and this branch is a '*v*' branch. In it the left photon is vertically polarized. But the model is structured in such a way that every branch in which the left photon is polarized vertically is a branch on which the right photon is polarized horizontally. Therefore, as soon as the left photon emerges in the 'vertical' channel, and all branches have dropped off except one which shows this, then *ipso facto* the remaining branch shows the right photon as being polarized horizontally. This is no trick, or optical illusion; on that branch, the right photon *really is* polarized horizontally. If, instead, the left photon had emerged in the 'horizontal' channel, then on the surviving branch the right photon would have been polarized vertically.

Whichever result happens to come about—whether the left photon passes '*v*' or '*h*'—branch attrition ensures that the right photon acquires its angle of polarization instantly, no matter how far away it is from the left photon. This is how, in the model, what happens at one place can instantaneously affect what happens at another place. The fact that distant events can occur together on the same branch, plus the fact that branch attrition takes place along spacelike hyperplanes, conspire together to produce non-local effects. These non-local effects are natural consequences of the model's structure.[17]

It might be asked at this point whether the branched model *explains* non-locality—explains why events at one place can seemingly instantaneously influence events at another place—or whether it only *redescribes* non-local phenomena in an equally mysterious way. The same question arises concerning distant correlations: does the model *explain* them, or does it just redescribe or represent

---

[17] In their original paper (1935), Einstein, Podolsky, and Rosen strongly oppose the possibility of non-local effects. For them the angle of plane polarization of the right photon, which following the left measurement can be predicted with certainty, would be an 'element of physical reality'. To allow it to 'depend upon the process of measurement carried out on the first system, which does not disturb the second system in any way' is something which 'no reasonable definition of reality could be expected to permit' (1935: 780). Nevertheless, the branched model shows that 'reality' could be structured in such a way as to make non-local effects possible.

them in some new idiom? I shall argue, in both cases, that it does the former.

Considered in itself and by itself, the branched model explains nothing. It is just a huge spatio-temporal structure. What is *explanatory* is the hypothesis that the world really is like the branched model: that the model models it accurately. If the world really *were* branched in the way the model is, then non-locality, and the distant correlations of the EPR experiment, would not merely be redescribed, they would be explained. They would be explained by the hypothesis that the world has a certain dynamic shape, in the way that Copernicus explained the complex apparent motions of the planets by hypothesizing that the solar system had a certain dynamic shape, with the sun at the centre. (To be sure Cardinal Bellarmine argued persuasively that the Copernican system provided only an alternative description of what had already been described by Ptolemy, but Galileo was not convinced.)

Returning to the perspectival character of the model, Fig. 4.10 pictures the model in a frame of reference in which the earth measurement at **A** occurs before the moon measurement at **B**. There will also, however, be a frame in which the order of these measurements is reversed. Relative to that frame, in which the distant moon measurement can interestingly enough affect the measurement made a few feet away from the source on earth, the model has the form depicted in Fig. 4.11.

A different picture of the same experiment, on a scale which shows better how the results of the moon measurement can

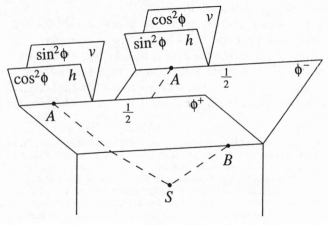

FIG. 4.11

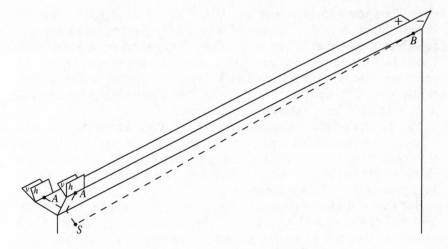

FIG. 4.12

violate locality and affect the earth measurement, is shown in Fig. 4.12.[18]

In this model, the selection of a ϕ^+ branch at the measuring device B on the moon instantaneously gives the left photon the probability $\sin^2\phi$ of being measured v on earth. If the moon measurement had been ϕ^- a ϕ^- branch would have been selected, on which the left photon would have been in the pure state ϕ^+, giving it probability $\cos^2\phi$ of being measured v.

The four diagrams, in Figs. 4.8, 4.10, 4.11, and 4.12, are relativistic variants of one another. Despite their different structures, they are all pictures of the same physical reality, seen from different perspectives, and they all share a number of frame-invariant properties, notably the probability values they assign to the joint outcomes $(v, +)$, $(v, -)$, $(h, +)$ and $(h, -)$. An interesting property of non-local influence that this analysis reveals is its *reciprocal* character.

If a frame of reference is chosen in which the left measurement of the EPR experiment takes place before the right, then the

[18] Fig. 4.12 illustrates the difference between (i) genuine signals, such as light rays, which can be used to send information, and (ii) non-local effects, such as the EPR correlations, which convey no information. The former are represented by space-time trajectories in the branched model, along which information can flow. But the latter are purely the result of branch attrition along spacelike hyperplanes, and bear no relation to moving bodies or to motion of any kind. The model makes a sharp distinction between non-local action-at-a-distance and genuine signals, and explains why EPR-type correlations cannot be used to convey information.

former appears to influence the latter. But if a different frame is chosen the direction of influence is reversed, and the latter appears to affect the former. It may be asked, what kind of influence can there be, the direction of which depends upon the choice of reference frame? If X influences Y in one reference frame, and Y influences X in another, then plainly the relationship between X and Y cannot be a standard causal one.[19]

The reader's first reaction may be, that no candidate for the title of 'causal influence' can be genuine, if its direction of influence can be altered by changing reference frames. An influence that could be 'transformed away' by such a change would be an artefact, not something real.

But the non-local influences which the EPR experiment reveals, and which the branched model explains, are not artefacts. Their *direction* depends upon the choice of reference frame, but their *existence* does not. Given the framework of the branched model, they are real enough. They satisfy, for example, the criterion of counterfactual dependence: if the left photon had not been measured *v*, the right photon would not have been horizontally polarized. But because their direction is in a certain sense arbitrary, there being no more reason to say that X non-locally influences Y than to say that Y non-locally influences X, it is appropriate to describe them as reciprocal.

To say that non-local effects are reciprocal is immediately to set them apart from ordinary causal influences. Ordinary causation is asymmetric—if X causes Y, Y does not cause X. But non-local influences are symmetric. More significantly still, they connect events which are in spacelike separation and lie outside each other's light cones. To summarize the differences between non-local and normal causal effects, the former have the following characteristics:

1. reciprocal, not asymmetric;
2. hold outside the light cone, between spacelike separated events;
3. instantaneous and non-local;
4. not explicable by a common cause;
5. direction of influence dependent on choice of coordinate frame.

Plainly, non-local effects differ radically from ordinary causation.

[19] Abner Shimony states that the disturbance exercised by the measurement of the linear polarization of one photon on the polarization of the other is 'in *some sense* a causal process', although in this case we are presented with 'a kind of causal connection which is generically different from anything that could be characterised classically, since the causal connection cannot be unequivocally analysed into a cause and an effect'. The correlated probabilities of the outcomes on left and right are 'due to reciprocal influence, without singling out one event as the cause and one as the effect' (1989b: 387–8).

Those who are sceptical about the physical reality of non-local connections will focus on the state vector of the two-particle correlated system. Quantum theory itself, as distinct from its interpretation, has no difficulty in accounting for the instantaneous change in state of the right photon in the EPR experiment, consequent upon a measurement performed on the left photon. In Hilbert space the vector Ψ, describing the system composed of the two particles in an entangled state, is a linear combination of two other vectors Ψ₁ and Ψ₂, and when the left measurement is performed the state vector instantly reduces, or 'jumps', to one of Ψ₁ or Ψ₂.

The vectors Ψ₁ and Ψ₂ can be written in many different ways: in one of them the vector Ψ₁ represents the left photon as $\updownarrow$ (vertically polarized) and the right photon as $\leftrightarrow$ (horizontal), while Ψ₂ represents the left photon as $\leftrightarrow$ and the right as $\updownarrow$. In that case Ψ takes the following form:

$$\Psi = \tfrac{1}{\sqrt{2}}(|\,L \updownarrow\rangle\,|\,R \leftrightarrow\rangle + |\,L \leftrightarrow\rangle\,|\,R \updownarrow\rangle).$$

If the left photon is measured v, the state vector Ψ immediately collapses or reduces to the simpler state $|\,L \updownarrow\rangle\,|\,R \leftrightarrow\rangle$, in which case the right photon is polarized horizontally. At the level of quantum theory, there is no need to postulate any instantaneous causal process extending from one measurement event to the other. 'Collapse' is not a physical process, and the correlations between distant events in the EPR experiment are simply the result of mathematical reduction of the state function.

This is all very well as far as quantum theory is concerned. Those who seek an interpretation, on the other hand, who wish to attribute a physical rather than a purely mathematical significance to the state vector, and who try to construct a spatio-temporal picture consistent with quantum theory's account of the Bell–EPR experiment, will not be satisfied. 'Reduction' or 'collapse' may characterize the behaviour of mathematical entities like the state function, but if the state function describes something in the real world, will there not be something physical that corresponds to 'reduction'? In the branched interpretation, the answer is affirmative. What corresponds to 'reduction' is not a physical *process*, because there are no such things as reciprocal instantaneous physical processes. Nevertheless it *is* something physical. What is in question here is not a superluminal signal but something different, something which is part of the physical world and which creates the effect of instantaneous action-at-a-distance, but which does not 'travel' at any speed at all. This is branch attrition. Since branch attrition is a physical effect which involves no motion, and which does not imply that anything travels faster than light, no conflict with relativity arises.

There are two possible conflicts which come to the fore in the EPR experiment. The first of these, discussed by Shimony (1978), and Redhead (1983) and (1987) under the heading of whether there can be 'peaceful coexistence' between quantum mechanics and special relativity, revolves around the question of instantaneous transmission of information. Any incompatibility is resolved by noting that the EPR experiment does not permit faster-than-light signalling, but exhibits at most what Shimony calls 'uncontrollable non-locality', or 'passion-at-a-distance'.[20] The second possible conflict is different. It arises out of whether, in the EPR experiment, the left measurement A occurs earlier than, simultaneous with, or later than, the right measurement B. Since the two measurement events are spacelike separated, it should be a matter of indifference, according to special relativity, whether what is going on in the experiment is described in one or another of three coordinate frames corresponding to $t(A) < t(B)$, $t(A) = t(B)$ and $t(A) > t(B)$.

The difficulty is that, if we appeal to 'collapse' in order to explain how a measurement performed on one particle can affect a measurement performed on another, then it does seem to matter which coordinate frame we choose for the description of the experiment.[21]

Suppose we choose a frame of reference in which the left measurement is made before the right measurement. In that case, the probability of the left photon's being measured v is $\frac{1}{2}$. But if we choose a frame in which the right measurement is made before the left, then, depending on the outcome on the right, the probability of the left photon to be measured v will be either $\sin^2\phi$ or

[20] Jarrett (1984) distinguishes between (i) the requirement that the probability for the outcome of the right measurement be invariant under conditionalization on the *state* (i.e. orientation) of the measuring device on the left, and (ii) the requirement that the probability be invariant under conditionalization on the *outcome* of the measurement on the left. Jarrett calls the first requirement, which is satisfied in the EPR experiment, 'locality', and the second, which is violated, 'completeness'. In Ballentine and Jarrett (1987: 698), it is noted that if the requirement of 'locality' were violated it would in principle be possible for an observer on the left, merely by adjusting the angle of the polarizer, to signal to an observer on the right at speeds faster than light. Shimony (1984) calls this 'controllable non-locality', which the EPR experiment does not exhibit. What the experiment does exhibit is what Shimony refers to as 'uncontrollable non-locality', i.e. the violation of Jarrett's outcome independence or 'completeness'. Proof that the experimental set-up does not permit instantaneous signalling is given in Ghirardi, Rimini, and Weber (1980).

[21] See Penrose (1989: 287 and 371).

$\cos^2\phi$. How can the probability for the left photon be $\frac{1}{2}$ in one frame of reference, and either $\sin^2\phi$ or $\cos^2\phi$ in another? In Penrose's words, we get 'a completely different picture of physical reality' if we choose one frame rather than the other. The trick is to find a way of describing what is going on in the EPR experiment which permits these different pictures to be no more than relativistic variants of one another.

Because it is the reduction of the state vector consequent upon the measurement on the right that gives the left photon its definite polarization or probability value, and also because from a different perspective it is the reduction on the left which gives the right photon its probability value, it follows that what is needed is a relativistic account of state vector reduction.[22] In effect, this is what the branched model provides. If a frame of reference is desired in which the two measurement events of the EPR experiment are simultaneous, the picture of the experimental set-up provided by the branched model is that of Fig. 4.8. If a frame is chosen in which the left measurement takes place before the right, the picture is as in Fig. 4.10. Finally, if the right measurement occurs before the left, the model takes the form of Fig. 4.11 or Fig. 4.12. All these pictures are pictures of the same physical reality, and they all share a number of frame-invariant properties. But they are, at the same time, inherently 'perspectival' or hyperplane dependent (see Appendix 2). This is the price we have to pay, in adopting the branched interpretation, for being able to explain non-locality and the EPR correlations in a way that is both consistent with, and in the spirit of, special relativity.

To conclude this section we shall examine the physical interpretation given by the branched model of what is going on in an even more striking example of non-local influence, the Greenberger, Horne, and Zeilinger (GHZ) experiment.[23] In this experiment a system of three spin-$\frac{1}{2}$ particles, e.g. electrons or protons, is emitted from a central source in an entangled state consisting of the following superposition:

$$(3) \quad \Phi = \tfrac{1}{\sqrt{2}}\,(\,|\,1, 1, 1\,\rangle - |\,{-1}, -1, -1\,\rangle).$$

Here the vector $|\,1, 1, 1\,\rangle$ represents the state in which each of the three particles is spin-up in its direction of propagation (along its

---

[22] See Fleming (1988). The branched model furnishes an ontological picture of Fleming's 'hyperplane dependency'.

[23] See Greenberger, Horne, and Zeilinger (1989); Greenberger, Horne, Shimony, and Zeilinger (1990); Clifton, Redhead, and Butterfield (1991); Mermin (1990a; 1990b); Stapp (1993).

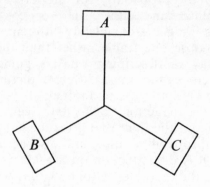

FIG. 4.13

z-axis), and the vector $|-1,-1,-1\rangle$ denotes spin-down for the three particles along their z-axis (Mermin 1990a). The three particles move off in a plane along different directions and pass through Stern–Gerlach devices which can be set to measure their spin orientation perpendicular to their motion in either the x or the y directions (see Fig. 4.13).

If all three detectors are set to measure x spin then quantum mechanics predicts that the product of the three spin measurements will be -1. Consequently, if the measurement at apparatus A is 1 (spin-up) then the outcomes of B and C must be different, whereas if A yields the result -1 (spin-down) then the outcomes at B and C must be the same (both spin-up or both spin-down). Since the experiment can be arranged so that all three measurement events lie outside each other's light cones, the question arises how the information about the outcome at A can be communicated to the other devices in time to bring about the desired correlation of their results. As in the case of the EPR experiment, relativity theory permits no answer to this question since such information transfer would require faster-than-light signalling (Stapp 1993).

The branched model provides the following interpretation of what is going on in the GHZ experiment.[24] If the three apparatuses are all set to measure x-spin, then the only permitted joint results for the three particles are $(1, 1, -1)$, $(1, -1, 1)$, $(-1, 1, 1)$, or $(-1, -1, -1)$. In the model in which all three measurement events are simultaneous there are only four different kinds of branch

<hr />

[24] The experiment is still only a thought-experiment, but Greenberger *et al.* (1990) offer suggestions about ways in which it could one day be performed.

above the relevant hyperplane. That is, there are no branches containing the joint outcomes $(1, 1, 1)$, $(1, -1, -1)$, $(-1, 1, -1)$, or $(-1, -1, 1)$. Therefore if the outcome at A is 1, the outcomes at B and C must differ; if the outcome at A had been -1 and the outcome at B had been 1, the outcome at C would have been 1, etc. The truth-maker for all these conditionals and counterfactuals is the *absence* of certain kinds of branch in the branched model.

As in the case of the EPR experiment, one can also consider frames of reference in which the A measurement occurs before the B and C measurements, or in which A and B are simultaneous followed by C, etc. The situation is symmetric in A, B, and C, so that similar results obtain if frames are selected in which B or C come first. In every instance, the phenomenon of a measurement outcome apparently bestowing a definite spin orientation upon a particle a long distance away will be observed.

For example, the branched model in a frame of reference in which the A measurement occurs first, followed by B and C at the same time, has the following shape. From the structure of the model it is evident why if the outcome of A is positive (spin-up), the outcomes of B and C must differ, while if A is negative they must be the same (see Fig. 4.14).

Unlike theories which explain the correlated results of the GHZ experiment in terms of a purely mathematical reduction of the state vector of the three-particle system, the branched model gives a physical, space-time explanation of what actually occurs when the particles encounter their respective Stern–Gerlach apparatuses. The explanation, briefly, is branch attrition along hyperplanes.[25]

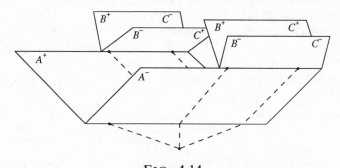

FIG. 4.14

[25] In constructing an indeterministic theory to interpret the results of the GHZ experiment, Henry Stapp says: 'We suppose that scattered throughout the positive time region are special 'break points'. At each of these break points, there is a rupturing of space-time, and the interior of the future cone with apex at that point has two sheets. In these two

To sum up, the branched interpretation succeeds in explaining the results of both the EPR and GHZ experiments without appealing to hidden variables, and does so in terms of a spatio-temporal model that is inherently non-local.[26]

(iii) *The Measurement Problem*

The measurement problem in quantum mechanics arises in the following way. Let Q be an object or *measured system*, and let M be an apparatus or *measuring system*. Suppose we wish to measure some observable or dynamical variable O belonging to Q and to this end couple Q to M. If the initial state of Q, represented in quantum mechanics by the vector Ψ, is an eigenstate of O, then the measuring system M will respond by moving into some definite measurement state, such as a pointer reading on a dial. But if Q is not in an eigenstate of O, as in general it will not be, but is in a linear combination or superposition of such eigenstates, then (if we consider only Schrödinger evolution) the apparatus M will not move into a definite measurement state. Instead quantum theory predicts that, in accordance with the time-dependent Schrödinger equation, O's superposition will spread to the system $Q + M$, with the result that the entire system of object-plus-apparatus ends up in a superposition of states. The measurement problem arises when we try to reconcile this prediction with the fact that the apparatus is always observed to be in a definite state, never in a macroscopic superposition of different pointer readings.

Von Neumann deals with the problem by invoking the *projection postulate*, which in his axiomatization of quantum theory is equal in status to the Schrödinger equation and which describes a very different type of change.[27] In Hilbert space, the effect of Schrödinger evolution consists in the state vector's being acted on by a unitary operator which changes its orientation but not its length.

sheets, the dispositions of things are different. The break points are rare enough so that only one of them is located in any unit interval of time. Nature's process is such that, if there is a break point in the region between t and $t + 1$, then nature will select, at the step that fixes everything in that region, *one* of these two sheets. But there is absolutely nothing in the entire region lying earlier than this break point that determines which of the two sheets will be selected' (Stapp 1993: 851). Stapp's 'break points' are similar to the 'choice points' of App. 2.

[26] App. 3 contains the branched interpretation of a striking version of the EPR experiment devised by David Mermin.

[27] von Neumann (1932, 1955). Opinion nowadays is divided as to whether quantum mechanics ought consistently to allow processes which do not conform to the Schrödinger equation. See e.g. Stein (1970: 102); Putnam (1981: 250); Sudbery (1986: 186).

Under the operator's influence, the state vector evolves smoothly and continuously in time. In contrast, application of the projection postulate results in the state vector's being transformed abruptly and unpredictably into an eigenvector of the dynamical variable being measured. In von Neumann's system, quantum mechanics admits of two incompatible processes: Schrödinger evolution (U) and measurement (R), the latter being based on the projection postulate. When the projection postulate is invoked, the state vector of the combined system $Q + M$ jumps into an eigenvector of the measured observable, and the pointer needle of the apparatus gives a definite reading. The question which von Neumann left unanswered is, at what point in the measurement process does Schrödinger evolution cease and the projection postulate take over?

In the literature on quantum mechanics, many answers to this problem of the 'collapse of the wave packet' have been proposed. None is entirely satisfactory. In what follows I survey three of them, ending up with the 'branched' account of measurement.

London, Bauer, and Wigner

In their 1939 work entitled 'La théorie de l'observation en mécanique quantique' London and Bauer put forward the suggestion that what brings about the collapse of the wave packet is an act of consciousness. Discussing the use of an apparatus to measure the value of an observable on an object, London and Bauer say: 'So far we have only coupled one apparatus with one object. But a coupling, even with a measuring device, is not yet a measurement. A measurement is achieved only when the position of the pointer has been *observed*.'[28] They continue:

At first sight it would appear that in quantum mechanics the concept of scientific objectivity has been strongly shaken. Since the classic period, the idea has become familiar that a physical object is something real, existing outside of the observer, independent of him, and in particular independent of whether or not the object has been subjected to measurement. The situation is not the same in quantum mechanics. Far from it being possible to attribute to a system at every instant its measurable properties, one cannot even claim that to attribute to it so much as a wave function has a well-defined meaning, unless referring explicitly to a definite measurement. Moreover, it looks as if the result of a measurement is intimately linked to the consciousness of the person making it.[29]

---

[28] London and Bauer (1939, 1983: 251).   [29] Ibid. 258.

Wigner, writing on the same theme, asserts:

When the province of physical theory was extended to encompass microscopic phenomena, through the creation of quantum mechanics, the concept of consciousness came to the fore again: it was not possible to formulate the laws of quantum mechanics in a fully consistent way without reference to the consciousness ... [E]ven though the dividing line between the observer, whose consciousness is being affected, and the observed physical object can be shifted towards the one or the other to a considerable degree, it cannot be eliminated.[30]

But to introduce the subjective element of consciousness into the conceptual foundations of physics runs counter to good sense. Are we to believe that before conscious beings evolved, the universe was in a huge macroscopic superposition of states?[31] That the stars, for example, have existed in definite positions only since being observed? Moreover, the consciousness-dependence theory is open to the objection that if an observer$_1$ is needed to reduce the state vector of the object-plus-apparatus system, then a second observer$_2$ may in all consistency be demanded to reduce the vector

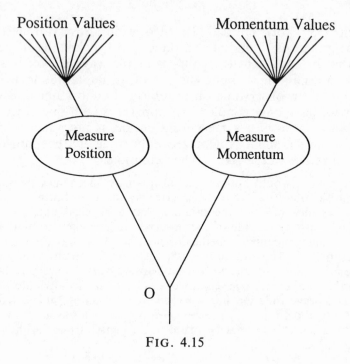

FIG. 4.15

[30] Wigner (1961: 169).   [31] Cf. Bell (1990).

of the object-plus-apparatus-plus-observer$_1$ system, and so on *ad infinitum*. Only if all objective accounts of state vector reduction had been shown to be untenable should we be forced either to conclude that reduction never takes place, or to adopt an observer-based theory of quantum reality.

Although as will be seen the branched model permits a definition of 'measurement' which is independent both of 'observers' and of 'measuring apparatuses', the model cannot and does not exclude the observer from playing an important role in many measurements. For example in the case of incompatible observables such as the x and y components of spin, or a particle's position and momentum, the question of which measurement will be performed may lie within the observer's choice. In this case the model assumes the form in Fig. 4.15.

At the branch point O, the observer chooses whether to measure the position or the momentum of a particle. *How* the observer can do this, for example by making a decision as to which piece of apparatus to use, is discussed in Chapter 9. But London, Bauer, and Wigner are right to this extent, that in the case of incompatible observables the measured properties of quantum systems may depend on an observer's deliberate choice.

Daneri, Loinger, and Prosperi

The aim of these authors is to show that in the case of a medium-sized object like a measuring instrument, the interference effects of coherent superpositions of macroscopic states would be 'washed out' in such a way that being in a superposition would be empirically indistinguishable from being in a mixture of states.[32] (For a macroscopic body to be in a mixture of states A, B, C, ... means that the body is either in A, or in B, or in C, ..., with a certain probability for each.[33]

Suppose for example that the spin of an electron is measured on a Stern–Gerlach apparatus, and that the electron is prepared in such a way that it enters the apparatus in the pure state 'spin up in the z-direction', i.e. $|\uparrow_z\rangle$. The apparatus is oriented so that it measures the electron's x-component of spin, and has a pointer

---

[32] Daneri, Loinger, and Prosperi (1962).

[33] Assuming the 'ignorance interpretation' of mixtures (Putnam 1965: 155); van Fraassen (1972: 328) and (1991: 206); Hughes (1989: 144 and 150). If it should turn out that the ignorance interpretation of mixtures doesn't work, then *a fortiori* Daneri, Loinger, and Prosperi's solution to the measurement problem doesn't work either. As will be seen below, the branched model provides an objective means of distinguishing pure from mixed states, and hence avoids the ignorance interpretation.

with three positions, 'ready', 'up', and 'down'. When the electron has passed through the apparatus, application of the time-dependent Schrödinger equation predicts that the combined system $Q + M$ of electron-plus-apparatus will be in the following linear superposition of states:

$$\tfrac{1}{\sqrt{2}}(|\uparrow_x\rangle \;|\text{pointer reads 'up'}\rangle + |\downarrow_x\rangle \;|\text{pointer reads 'down'}\rangle).$$

However, Daneri, Loinger, and Prosperi (DLP) do not interpret this to mean that the apparatus is in a 'fuzzy' combination of different pointer readings. Making use of an ergodic irreversibility hypothesis which applies to the measuring apparatus in virtue of the large number of elementary particles it contains, DLP demonstrate that the combined system of electron-plus-apparatus behaves exactly as if it were in one of the two states $|\uparrow_x\rangle$ |pointer reads 'up'⟩ or $|\downarrow_x\rangle$ |pointer reads 'down'⟩. The system actually *is* in a superposition, as predicted by the Schrödinger equation, but is *indistinguishable* from what it would be like if it were in a mixture, i.e. in one pure state or the other. In this way DLP save the phenomena, and dispense with any need for the projection postulate.[34]

The DLP solution to the measurement problem has found favour with many, but is open to two serious objections. The first of these is found in Jauch (1968) and Ballentine (1970). DLP rely on the interference effects of a measurement interaction being damped by the size of the measuring apparatus. But, as Ballentine remarks, if the measuring device is a photographic plate, then the essence of measurement may be performed by a single silver-halide complex. The amplification process, which enlarges the measurement state to macroscopic size, does not take place until the plate is developed, which may be months later.[35] Hence the alleged damping effects, if they occur at all, cannot occur at the time of measurement.

The second objection has been made by several authors. It is that although DLP may have shown that the superpositional state of a measuring instrument following a measurement is indistinguishable from a mixture, it is still a superposition. In the words of Nancy Cartwright: 'We want to see a mixture at the conclusion of a measurement. It is no credit to the theory that the superposition which it predicts looks very much like a mixture in a variety of tests.'[36]

---

[34] A more complete description of DLP's solution, together with a criticism, is found in Bub (1988: 142–3).

[35] Ballentine (1970: 371), based on Jauch (1968: 169).   [36] Cartwright (1983: 170).

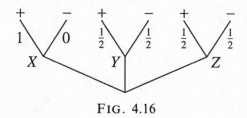

FIG. 4.16

A. J. Leggett takes the view that 'there is an unbridgeable logical chasm between the statements that the system behaves "as if" it were in a definite macroscopic state and that it *is* in such a state.' Leggett's opinion is that relying on our inability in practice to distinguish macroscopic superpositions from mixtures 'totally fails to solve the real problem'.[37] Similar criticisms are found in d'Espagnat (1976: 193–4). We may conclude that DLP fail to provide a satisfactory solution to the measurement problem.

On the subject of mixed states, it is worth noting that the branched interpretation yields a way of distinguishing superpositions from mixtures which is entirely objective and is not based on ignorance. Using electron spin states as an example, let x^+ (x^-) represent the pure state 'spin up (spin down) in the x direction', and similarly for z^+ and z^-. In quantum theory, the pure state represented by the vector x^+ is identical to the superposition $\frac{1}{2}(z^+ + z^-)$, and is quite distinct from the mixed state $\frac{1}{2} Pz^+ + \frac{1}{2} Pz^-$ represented by a weighted sum of projection operators. In the branched model, the pure state corresponds to the configuration in Fig. 4.16. Here X, Y, and Z are apparatuses measuring the x, y, and z components of spin respectively, and the probability of spin up and spin down outcomes are as indicated. As is the case with all pure states, there is one particular measurement (the X-measurement in this instance) which yields a specific measured value with probability one. The mixed state, by contrast, corresponds to the pattern in Fig. 4.17. In this case there is no measurement which yields any outcome with probability one. In the branched model, mixed spin states receive an interpretation which distinguishes them clearly from pure states and is not based on ignorance.

Ghirardi, Rimini, and Weber

In contrast to DLP, Ghirardi, Rimini, and Weber (GRW) take the view that state vector reduction does take place, and that its

---

[37] Leggett (1987: 170). Cf. Bub (1968); Bell (1975).

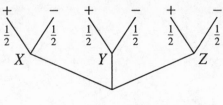

FIG. 4.17

occurrence can be dealt with in the general framework of quantum mechanics not by introducing the projection postulate, but by modifying the Schrödinger equation.[38] Noting that the problem arises because classical Schrödinger dynamics permit superpositions of macroscopically distinguishable states, GRW propose a 'spontaneous localization process', the purpose of which is to collapse these superpositions. This process, the physical origin of which is left unclear, is conceived of as some kind of stochastic background fluctuation, the effect of which is to subject the wave packets of all particles in the universe to random localizations or 'hits'.[39]In the case of a single isolated particle, the average frequency of the hits would be sufficiently low (once in 10^8 or 10^9 years) that no spontaneous collapse of the particle's wave packet would be expected to occur. But if the particle is coupled to a large object having something in the order of 10^{23} constituent particles (as in the case of a measuring apparatus), then the chances are that at least one of the apparatus' particles will suffer a hit every 10^{-7} seconds, and this hit is sufficient to collapse the wave function for the whole object. A macroscopic object consequently has definite, sharp properties. By choosing the parameters of the postulated localization process appropriately, GRW are able to account simultaneously for the 'wavy' behaviour of microscopic quantum systems, and the 'classical' behaviour of macroscopic bodies.[40]

For those who seek a realistic account of the measurement process, the approach of GRW has great appeal. First, it gets rid of macroscopic superpositions. Secondly, it provides a unified dynamics for microscopic particles and macroscopic bodies, permitting definite trajectories to be assigned to the centres of mass of objects like Geiger counters and pointers. At the same time it

---

[38] Ghirardi, Rimini, and Weber (1986; 1988); Pearle (1989; 1990); Ghirardi, Pearle, and Rimini (1989); Pearle and Soucek (1989); Ghirardi, Grassi, and Pearle (1990).

[39] Ghirardi, Rimini, and Weber (1986: 471); (1988: 9, 17); Pearle (1989: 2278); Ghirardi, Pearle, and Rimini (1990: 79); Pearle (1990).

[40] Ghirardi, Rimini, and Weber (1988: 20).

denies well-defined trajectories to individual particles, as is required by the two-slit and delayed choice experiments.[41] Thirdly, GRW permit the Schrödinger wave function to be interpreted, not just as a mathematical or computational device, but as a 'real property of an individual physical system'.[42] As long as the superposition principle permits the mathematical creation of wave functions, a literal reading of which would require a body to be in a macroscopic superposition, a physical interpretation is difficult if not impossible to accept. But on the theory of GRW, inflated wave packets are rapidly and spontaneously collapsed. Therefore a realistic interpretation of the wave function becomes once more an option.

These are considerable advantages. Nevertheless, certain problems remain. As is acknowledged by Philip Pearle, neither in GRW nor in his own extension of GRW do superpositions fully collapse. For a brief period, before they are annihilated by the spontaneous localization process, macroscopic superpositions exist. Even afterward, traces of 'what might have been' linger on.[43] Elimination of these last vestiges of macroscopic superposition would strengthen and add philosophical credibility to the theory. Secondly, there is what Pearle calls the 'trigger problem'.[44] In GRW, no explanation is given of the physical mechanism which causes spontaneous localization. But if state vector reduction is to be an objective physical fact about the world, something which happens independently of being observed, as GRW wish to argue, then some natural mechanism which triggers the reduction should be identified. In the discussion of the 'branched' account of measurement below, a suggestion will be made concerning a suitable triggering mechanism for GRW's spontaneous localization.

The 'branched' account of measurement

The account of measurement which the branched model provides is different from that of any of the theories already considered.

[41] For the delayed choice experiment, see Wheeler and Zurek (1983: 182–4). GRW meet J. S. Bell's requirement for quantum mechanics, namely that it should 'allow electrons to enjoy the cloudiness of waves, while allowing tables and chairs, and ourselves, and black marks on photographs, to be rather definitely in one place rather than another, and to be described in "classical terms" ' (Bell 1987: 190).

[42] Ghirardi, Rimini, and Weber (1988: 25). By contrast, the 'statistical' interpretation of Einstein (1936: 375) and Ballentine (1970) takes the wave function to describe the behaviour, not of an individual system, but of an *ensemble* of similarly prepared systems.

[43] Pearle (1989: 2289 n. 8; 1990: 11). Philosophical objections to this feature of GRW are found in Shimony (1989a: 36) and in Albert and Loewer (1990).

[44] Pearle (1989: 2278). The 'preferred basis problem', which Pearle also identifies, is discussed below on p. 133.

In the branched model, measurement processes **R** are distinguished from Schrödinger or unitary evolution **U** by an objective criterion, having nothing to do with observation or human consciousness. The branched model provides a physical correlate of state vector reduction. Since both the structure of the branching and the process of branch attrition are contingent, not necessary, the question of whether or not superpositions are collapsed, and at what exact point they are collapsed, becomes a matter of empirical fact. Without requiring any modification of the Schrödinger equation, the branched interpretation provides an objective modelling for applications of the projection postulate, so that what takes place in the quantum world, or for that matter in the world of macroscopic objects, is made up of a constant interplay of the processes **U** and **R**. All this comes about in the following way.

On the branched model the difference between **U** and **R** is a difference between prism stacks.[45] Let S be some system, quantum or macroscopic, and let v be the state of S at the base of some prism stack in the model. With regard to some future possible state or property w characterizing S at a later time, there are three alternatives (see Fig. 4.18). If the stack shows S to be in w on the tip of every branch in the stack, or not in w at the tip of every branch, we say that S, with regard to w, is at the base of a **U**-*type prism stack*. If on the other hand the stack shows S to be in w on some branches, and not in w on others, we say that S, with regard to w, is at the base of an **R**-*type prism stack*.

Plainly these three alternatives are exclusive and exhaustive. With regard to any later property w, every system S, on any node of the branched model, either finds itself at the base of a **U**-type prism stack with regard to w, or finds itself at the base of an **R**-type stack. For every system and every property, the difference

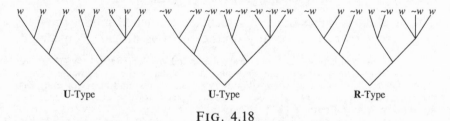

U-Type   U-Type   R-Type

FIG. 4.18

[45] A (finite) prism stack consists of a single prism at its base, on each of the branches of which there sits another prism, and so on up to the top of the stack. We speak of a stack as *extending from* t_1 *to* t_2 if the time of the base node is t_1, and that of the tips of all the branches is t_2 (strictly speaking the open set of instants $< t_2$).

between **U**-type and **R**-type prism stacks is an objective fact about the model.

Given the difference between **U**-type and **R**-type stacks, the difference between unitary evolution and measurement follows. Suppose that, at time t_1, the system S is in state v. This state of affairs is represented on the branched model by S, in state v, being located at the first branch point of the model at time t_1. At t_1, S is at the base of a prism stack which we shall speak of as being in 'privileged position' on the model (i.e. situated directly above the trunk). Suppose further that the stack in question is an **R**-type prism stack with respect to some possible future state or property w of S, and that the stack extends from t_1 to t_2. We shall say that, between t_1 and t_2, S undergoes a *measurement*, and that the measurement in question is a measurement of a certain dynamical variable or observable O if and only if w is an eigenstate of O. If on the other hand the prism stack is a **U**-type prism stack with regard to w (i.e. if w occurs on either all or none of the branches in the interval $t_1 - t_2$) then we shall say that between t_1 and t_2 the change in S with regard to w is an instance of *Schrödinger evolution*.[46]

Why does the difference between the two kinds of prism stack, when in privileged position, make for a difference between two kinds of change or process? Prism stacks, after all, are a *structural* feature of the model, and change is *dynamic*. The reason lies in another property of the model, namely branch attrition, the progressive reduction of the tree to a single branch. In the case of unitary evolution, branch attrition within a **U**-type prism stack with w on every branch ensures that, come what may, the change in the system S will result in its being in the state w at t_2. Schrödinger evolution is smooth, continuous, and deterministic. But in the case of measurement, branch attrition within an **R**-type stack ensures either that S will definitely be in w at t_2, or that S will definitely *not* be in w at t_2. In a measurement process, the state of S jumps suddenly and unpredictably into an eigenstate

---

[46] As defined here, it is a consequence of the difference between Schrödinger evolution and measurement that if at t_1 S is *already* in an eigenstate w of O, then no 'measurement' of O will occur by t_2, since S remains in w on all branches of the stack. For example, an electron which emerges from a beam splitter in the spin state $|\uparrow_z\rangle$ and then enters another similar apparatus oriented in the z direction is not 'measured' by the second, since the electron continues in the state $|\uparrow_z\rangle$ on all branches after passing through the second apparatus. It is also a consequence of the definition that S can undergo a 'measurement' in the sense of an indeterministic change of state between t_1 and t_2, even though the state w of the **R**-type prism stack that S faces at t_1 is not an eigenstate of any known quantum observable O. For example a bridge which at t_1 faces a prism stack with some branches which show it collapsed at t_2, and other branches which show it intact, undergoes a 'measurement' between t_1 and t_2.

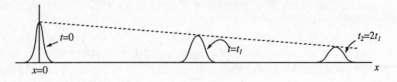

FIG. 4.19

of the measured observable, or into a macroscopic measurement state, in accordance with the projection postulate. Let us examine the difference between **U** and **R** processes in greater detail.

A simple example of Schrödinger evolution is the following: a free electron in a pure state Ψ, represented by a Gaussian wave packet, moves in one dimension down the x-axis, and the wave packet spreads out (as in Fig. 4.19). Here the change in the wave packet is a continuous function of time. Measurement, by contrast, involves discontinuity, and the branched model is able to represent discontinuity. Consider once again the example of an electron which has the x-component of its spin measured by a Stern–Gerlach apparatus.

At t_1, when the electron enters the apparatus, it is in the pure state $|\uparrow_z\rangle$, 'spin up in the z-direction', and also at the same time in the linear combination $\frac{1}{\sqrt{2}}(|\uparrow_x\rangle + |\downarrow_x\rangle)$. The apparatus is in the 'ready' state. On the branched model, electron and apparatus face an **R**-type prism stack in privileged position, on every branch of which, at some later time t_2, the combined system is in one or the other of the two states $|\uparrow_x\rangle$ |pointer reads 'up'\rangle and $|\downarrow_x\rangle$ |pointer reads 'down'\rangle.

At some moment t between t_1 and t_2, branch attrition results in a single prism out of the stack being brought into privileged position, at the base node of which the electron is in the superposition $\frac{1}{\sqrt{2}}(|\uparrow_x\rangle + |\downarrow_x\rangle)$, and on every branch of which the electron is either in the state $|\uparrow_x\rangle$ or in the state $|\downarrow_x\rangle$. The temporal height of this prism is Δt, i.e. arbitrarily short but non-null. At some instant within Δt the superposition collapses, and the state of the electron jumps in a discontinuous way from $\frac{1}{\sqrt{2}}(|\uparrow_x\rangle + |\downarrow_x\rangle)$ to either $|\uparrow_x\rangle$ or $|\downarrow_x\rangle$. This is the moment, t', at which the x-component of the electron's spin is measured, and an instantaneous reduction of the state vector takes place.[47]

[47] More explicitly, if the branching is upper cut (see App. 1 below), the top of the model's trunk is constituted by an open set of hyperplanes at times less than t', where t' lies between t and $t + \Delta t$. In many of these hyperplanes the state of the electron is $\frac{1}{\sqrt{2}}(|\uparrow_x\rangle + |\downarrow_x\rangle)$, but it is either $|\uparrow_x\rangle$ or $|\downarrow_x\rangle$ on each branch at t' and above. State vector reduction takes place at t'.

What about the measuring instrument? Plainly it cannot jump instantaneously, at the time at which the state vector of the electron is reduced, into one of the two states |pointer reads 'up'⟩ or |pointer reads 'down'⟩. But neither need it be in a superposition. Before t', the apparatus is in the 'ready' state. At t', when the electron's state vector is reduced, the instrument changes either to a state which eventually leads to |pointer reads 'up'⟩, or to a state which eventually leads to |pointer reads 'down'⟩. The combined system of electron-plus-apparatus, therefore, is in one or the other of the two states $|\uparrow_x\rangle$|state leading to pointer reads 'up'⟩ and $|\downarrow_x\rangle$|state leading to pointer reads 'down'⟩. At no time is it in a superposition of these two states.

In branch attrition, the branched model provides a triggering mechanism for state vector reduction and spontaneous localization in the theory of GRW, as well as in other theories which recognize 'collapse' as an objective process.[48] In the branched model, reduction is complete and instantaneous, so that the problem of macroscopic superpositions of a measuring instrument existing for a brief moment before being reduced is eliminated. Furthermore, branch attrition within an R-type prism stack provides a *physical mechanism* for state vector reduction. This physical mechanism, and in particular the question of precisely where and when it operates, plays no role in the mathematics of quantum *theory*, but is exclusively a feature of the spatio-temporal *interpretation* of the theory. Quantum theory gives us probabilities for what *may* happen: the branched interpretation gives us, in addition to these, a picture of what actually *does* happen. By contrast, under indeterministic experimental conditions quantum theory says nothing about what actually does happen. This will become clearer in the discussion of the many-worlds interpretation below.

A further feature of the branch-attritional theory of collapse is that besides being complete and instantaneous it is also non-local. This ability to simultaneously influence events which are spatially separated was noted earlier, and may create the impression that such a process contradicts relativity theory because it involves faster-than-light causal influences.[49] It was, in fact, the instantaneous character of the 'collapse of the wave packet', and its consequent ability to make things happen simultaneously in spatially separated regions, that provoked sceptical and critical comment

[48] As noted above on p. 111, the branched model also provides a *relativistic* account of state vector reduction. On the need for this, see Ghirardi, Grassi, and Pearle (1990).
[49] Stapp (1989: 154–9) contains a discussion of 'collapse ontologies' and the conceptual difficulties attending them, including the difficulty concerning relativity theory.

from Einstein at an early stage in the development of quantum theory.[50] However, as was pointed out in section (ii), nothing travels between two events which influence each other non-locally, and the unique variety of reciprocal linkage between them is provided entirely by branch loss along hyperplanes. There is, therefore, no conflict with relativity theory.

Before concluding this section, an argument put forward in Wigner (1963) against any objective theory of measurement must be considered. The argument is based on the possibility of 'measurement reversal', and proceeds as follows.

If a beam of electrons, all of which are in the pure state x^+, 'spin up in the x direction', is directed through a Stern–Gerlach apparatus oriented in the z direction, then the beam is split into two beams, z^+ and z^-, of equal intensity as in Fig. 4.20. For each individual electron, there is a probability $\frac{1}{2}$ of entering the 'z^+' channel, and an equal probability of entering the 'z^-' channel. *Prima facie*, this experiment appears to be a prototypal example of an **R**-type prism stack, and hence of what on the branched model would be a 'measurement'. Each electron faces an **R**-type prism, with equal numbers of 'z^+' and 'z^-' branches. But, Wigner argues, the situation is in fact more complex than this. If the two beams are recombined and passed carefully, without disturbing their phase, through a second apparatus of the same kind, but in

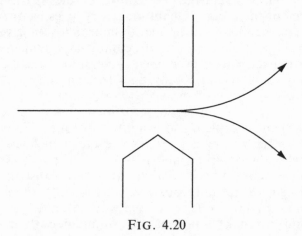

FIG. 4.20

[50] At the fifth Solvay Conference in 1927, Einstein remarked that, if the state function were interpreted as representing an individual system, not a statistical ensemble, then the collapse of the wave packet would represent a peculiar action-at-a-distance. Since the wave function (so interpreted) is distributed continously over some region of space, its instantaneous collapse would appear to violate relativity. See Fine (1986: 28).

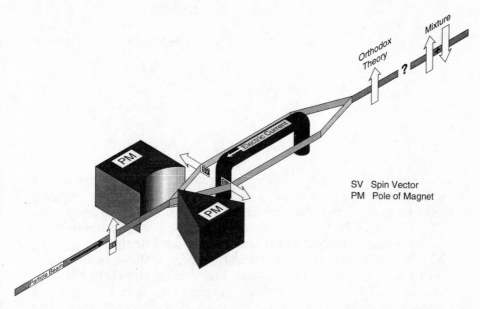

FIG. 4.21

the reverse direction, then quantum theory predicts that every electron in the resultant beam is once more in the pure state 'spin up in the x direction', as in Fig. 4.21.[51]

Wigner's 'reversed measurement' experiment shows that an electron emerging from the first apparatus cannot be in a *mixture* of the spin states z^+ and z^-, but must be in a *superposition*. If it were in a mixture then it would have a probability of $\frac{1}{2}$ of emerging from the second apparatus in each of the two states x^+ and x^-. But this is not what is observed. Therefore, passage of an electron through the first Stern–Gerlach apparatus cannot constitute a true 'measurement', since no particle emerges definitely in the z^+ channel or the z^- channel. Instead, every particle is in a superposition of both these states.

Wigner's thought-experiment is a challenge to any 'objective' theory of measurement which, as in the case of the branched theory, attempts to find some physical or empirical difference between measurements and non-measurements. Just as the two-slit experiment shows that it is a mistake to think of any individual particle as definitely passing through one slit rather than the other, so Wigner's experiment shows that what looks at first sight

[51] Diagram after Wigner, in Wheeler and Zurek (1983: 331).

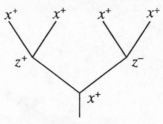

FIG. 4.22

like a measurement, i.e. a definite separation of the two beams, may not be a measurement at all. For Wigner, nothing in the physical world distinguishes a measurement from a non-measurement; the difference is to be found in the consciousness of the observer.

How is the branched interpretation, which attempts to give an objective definition of measurement, to deal with Wigner's objection? There are two possibilities. The first is to retain the idea that the initial apparatus divides the electrons into z^+ and z^-, as represented in the interpretation by an **R**-type prism stack. The effect of the second apparatus is then represented by a **U**-type prism stack which sits on top of every branch of the first stack, and which restores the electrons to their original state as 'spin up in the x direction', x^+, on every one of its branches (shown schematically in Fig. 4.22).

The second alternative is, in the spirit of Bohr, to maintain that quantum phenomena require full specification of the experimental conditions which serve to define them, and therefore that the placing of the reversed apparatus next to the first constitutes a whole new set of experimental conditions. A new apparatus has been created, and it is not to be understood simply as the adding together or juxtaposition of the two original ones. In this vein, the operation of the reversed measurement experiment would be represented (Fig. 4.23) in the branched interpretation by the presence of a single **U**-type stack, not as the conjunction of a **U**-type and an **R**-type.

In the second alternative, the superposition $\frac{1}{\sqrt{2}}(z^+ + z^-)$ differs only notationally from the state x^+. For various reasons, the second branched interpretation of Wigner's experiment seems preferable to the first.[52] Just as in the two-slit experiment it is untrue to say that the electron went through the top slit rather

---

[52] I am indebted to comments on this point from Frank Arntzenius and Christopher Hitchcock.

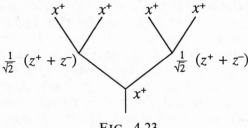

FIG. 4.23

than the bottom unless a detector registers its passage (in which case the interference pattern is destroyed), so in Wigner's experiment it is untrue to say that an individual electron emerges in the z^+ rather than the z^- channel of the first apparatus unless so registered by a detector (in which case the second apparatus will not reconstitute the pure x^+ beam). Unless a Stern–Gerlach apparatus is provided with detectors with which the exiting electrons can either interact or fail to interact, it does not correspond in the branched model to an **R**-type prism stack and is consequently not a measuring device. (The exiting particles may travel through empty space for a year: if at that time one of them collides with another particle, then relative to a year-high **R**-type prism stack the Stern–Gerlach apparatus *is* a measuring device.)

To sum up, whether we conclude that Wigner's 'reversed measurement' experiment is to be regarded as pure **U**, or as involving an element of **R**, the difference between measurement and non-measurement processes in the branched interpretation has not become blurred. Between the two, a gulf remains fixed. It is therefore unnecessary to become sceptical about measurement and to conclude that in principle, on the pattern of Wigner's experiment, *any measurement at all* can be cancelled out by returning the system to its pre-measurement state. This alarming idea is put forward by Alastair Rae.

In Wigner's experiment, an electron is first 'measured' and then 'unmeasured', the result being exactly as if no measurement had been performed at all. But measuring instruments themselves are collections of electrons, protons, and other particles, and Rae's suggestion is that if an electron can be 'unmeasured' and restored to its original state, nothing in principle makes it impossible that exactly the same thing should be done to a Stern–Gerlach apparatus, or to any other instrument.[53] It is, Rae says, always possible

[53] Rae (1986: 57–9).

to envisage a mechanism that would restore the apparatus to its 'ready' state, before the measurement was made, in the same way as Wigner's experiment restored the electron. Such a mechanism would be incredibly complex, but in theory constructible. Since all instruments which record 'measurements', including cameras, Geiger counters, and the human brain, could in principle be restored to their pre-measurement state, Rae's conclusion is that there exists no physical definition of a 'measurement'.

But surely this conclusion is too extreme. The mere possibility of a system's returning to one of its earlier states says nothing of its history in the mean time.[54] On the branched model, the fact that R-type stacks can have U-type stacks on top of them, which restore systems to earlier states, does not annul the difference between R and U. In the model, every process is either a 'measurement' or an instance of Schrödinger evolution, and the two are distributed throughout the branched structure. Nothing rules out the possibility of U and R processes combining so as to re-create earlier states. The moral to be drawn from all this is not that reversed measurement experiments blur the dividing line between measurements and non-measurements, but that they bring out the differences between them in new and interesting ways.

(iv) Comparison with the Many-Worlds Interpretation

In the 1950s the interests of a group of theoretical physicists led by J. A. Wheeler at Princeton were focused on the problem of giving a quantum formulation of general relativity. To this end it seemed desirable to eliminate the distinction between 'unitary evolution' and 'measurement' which von Neumann had introduced, for measurement implies that some line be drawn between 'object' and 'measuring system', and if the 'object' with which general relativity is concerned is the universe as a whole, it is difficult to see what 'measuring system' could possibly be used to measure it. If the idea of a single state vector ψ for the entire universe is to play a role in a unified theory of general relativity and quantum mechanics, or in cosmology, then the problem must be faced of how in the absence of a convenient *deus ex machina* such a universal state vector or wave function could be reduced.

---

[54] Suppose that in the year 2500 the world miraculously returned to exactly the same state it was in 1490. Would we say that history had been wiped out, and that Columbus never discovered America? No: we would say instead that history was cyclical, or in any case permitted repetition of identical world-states.

In 1957 Hugh Everett made the bold suggestion that the state ψ is never reduced. Instead, the definiteness of the macroscopic properties of the world at all times is preserved by the fact that every measurement splits the universe into as many copies as there are distinct measurement outcomes, with each outcome being included in a separate copy or branch.[55] In Everett's solution to the measurement problem macroscopic superpositions are completely eliminated, and quantum mechanics is simplified by doing away with the projection postulate. The enormous number of different branches into which the universe is constantly splitting constitutes the 'many-worlds interpretation' of quantum mechanics, to which the 'branched' interpretation of this book bears obvious analogies.

Nevertheless, despite the analogies, there are important differences between the many-worlds and the branched interpretations. To begin with, the many-worlds interpretation implies that measurements or observations *create* the splitting or branching of the world, and hence that the different copies or branches *come into being* with successive observations.[56] In the branched interpretation, on the other hand, the branches are not created as time goes by; they have all been in existence since the beginning of the universe, and their number progressively decreases rather than increases.

Secondly, the many-worlds interpretation implies that the present state of the universe is multiple; that right now, there exist many different copies of oneself in different branches.[57] By contrast, in the branched interpretation the present is *unique*. All branching lies exclusively in the future, not in the present or past.

These two differences between the many-worlds and the branched interpretations serve to emphasize the different conceptions of temporal process that the two interpretations employ. In the many-worlds interpretation the branching or splitting of the world is an ongoing process, one that has been taking place since the

[55] Everett (1957); DeWitt (1970 and 1971).

[56] 'Thus with each succeeding observation (or interaction), the observer state "branches" into a number of different states' (Everett 1957: 146 of DeWitt and Graham 1973). 'Every quantum transition taking place on every star, in every galaxy, in every remote corner of the universe is splitting our local world into myriads of copies of itself' (DeWitt (1970), in DeWitt and Graham 1973: 161.) Even if the interpretation should not be understood as implying a literal branching or splitting of space-time continua, or of physical systems, as is suggested by Healey (1984: 596), but instead as implying a splitting of *states*, nevertheless the universe has more states after the splitting than it had beforehand.

[57] The many-worlds interpretation envisages the idea of '10^{100+} slightly imperfect copies of oneself all constantly splitting into further copies' (DeWitt, *loc. cit.*).

beginning of time and has by now resulted in the simultaneous but separate existence of a large number of spatio-temporally isolated branches or worlds. In the branched interpretation, on the other hand, the branching is a fixed and permanent feature of the universe, not created by measurement. The only thing that *happens* is branch attrition.

A third difference between the two interpretations is this. The many-worlds interpretation does not attempt to give a physical or ontological interpretation of quantum probabilities in terms of numerical proportions of sets of branches. If probabilities entered into the many-worlds interpretation, they would presumably do so as probability weightings to be attached to the different outcomes of a split. This contrasts with the branched interpretation of probability as branch proportionality.

Fourthly, there exists another fundamental difference in the two interpretations concerning probability. In the branched interpretation, but not in the many-worlds, the probability of a future event is the probability of that event's being or becoming *actual*. In the idiom of the branched interpretation, this is the probability that an event of a certain type X will be instantiated on the single branch which survives branch attrition. But in the many-worlds interpretation, no meaning can be attached to the idea of a unique 'actual' world, or of 'the probability of becoming actual'. If I am in the process of observing a sample of uranium which may or may not decay, it makes no sense to say that the probability of its decaying in the next ten minutes is p, since according to the many-worlds interpretation the world will split and on some branches I shall observe its decay, and on other branches I shall not observe its decay. The concept of probability plays no role here, since what will happen in the future is certain.[58] In the many-worlds interpretation, every eventuality gets realized in some world or other. In the branched interpretation by contrast only some eventualities get realized, and probability is probability of realization.

Next, the branched interpretation does not attempt to eliminate the use of the projection postulate in favour of unitary evolution. Instead, it attempts to show (i) that an objective, ontological difference exists between the two processes **U** and **R**, built directly into the branched structure, and (ii) that the entire history of the world consists of a constant interplay of these two processes.

A consequence of this difference between the two interpretations is that in quantum cosmology, where the idea of a wave function

[58] Cf. Albert and Loewer (1988: 201).

for the entire universe plays an important role, the branched interpretation provides a mechanism by which the universal wave function may be collapsed.[59] In Everett's interpretation, since there is no apparatus or observer outside the universe to collapse it, the universal wave function never reduces but 'splits' into its separate eigenfunctions. The branched interpretation, on the other hand, provides for the random selection of a single branch as actual, and thus avoids the consequence of many coexisting universes or universe states. To cosmologists who make use of the universal wave function but who value the idea of there being a unique present state of the universe, the branched interpretation's reduction mechanism may be of interest.

In addition to those mentioned there exists a further significant difference between the two interpretations, one that bears on an objection that has from time to time been made to the many-worlds interpretation and which has not been satisfactorily answered. Moreover, the same objection could in principle be made to the branched interpretation, and therefore an answer should if possible be given. The objection is the 'preferred basis' objection, clear statements of which are found in Hellman (1984) and in Shimony (1986).

The difficulty is this. Imagine that a supporter of branching or splitting is asked to specify precisely *how* the universe branches, i.e. into what kinds of branches it divides. The natural answer would be, that on the occasion of an A-measurement—measurement of the value of the dynamical variable A on some system S—the universe divides into branches which correspond to the set of A-eigenstates. There are two problems with this answer. One is that the notion of 'branching' is made to depend on an undefined notion of 'measurement', and the other is that no reason is given why the universe should branch into eigenstates of an A-measurement, rather than eigenstates of a B-measurement or a C-measurement. The very concept of branching seems to hang on the idea of there being, on the occasion of a split, a preferred basis or orthonormal set of eigenstates into which the world divides.[60]

This is a serious objection, and if the friends of branching cannot come up with an answer they will have to withdraw their proposal. On the first point, that branching depends on a prior notion of 'measurement', the many-worlds interpretation does seem to be vulnerable. Supporters of the interpretation normally

[59] The wave function of the universe is discussed in Kolb and Turner (1990: 448 ff.).
[60] Hellman (1984: 564); Shimony (1986: 201).

speak as if a measurement *creates* splitting, or at the very least maintain that if there were no measurements, there would be no splitting. But at the same time no physical definition of 'measurement' is provided, so that there is no objective way of telling when the universe splits and when it does not. This objection, however, does not apply to the branched interpretation. In that interpretation, the universe splits whether a measurement is performed or not. Even if all systems underwent only unitary evolution between t_1 and t_2, the branched model would still split. Furthermore, there exists in the branched interpretation a precise and purely physical definition of 'measurement', as opposed to 'unitary evolution', in terms of the difference between R-type and U-type prisms. Hence branching, in the model, does not depend on some mysterious, undefined concept of 'measurement'.[61]

The second point, that branching requires a preferred basis of eigenvectors in order to be a branching of one kind rather than another kind, is more subtle. If a system S is in a superposition of eigenstates of two different incompatible observables A and B, e.g. the x and the y components of an electron's spin, then does the universe split into eigenstates of A or into eigenstates of B? The choice of one or the other seems completely arbitrary, and critics are surely right to protest that there are no preferred bases in quantum theory. However, in spite of this, the branched interpretation does provide an objective answer to the question of how, at every node, the universe branches. Not in quantum *theory*, but in the branched *interpretation*, where the state of any system S at any point on any branch is fixed and determinate, can it be established in a perfectly objective way what kind of branching is in question at any node.

Suppose for example that a system S is at a node n, and suppose that S is in a superposition of eigenstates of the two incompatible observables A and B. Do the branches above n show S in one or another of the eigenstates of A, or do they show S in one or another of the eigenstates of B? In other words, is n a node of A-branching for S, or of B-branching? It cannot be both. Alternatively, do the branches show S in states which are neither eigenstates of A nor eigenstates of B? To all these questions the branched model gives an unequivocal answer. Therefore the mode of branching at n is not at all arbitrary, or vague, or dependent

---

[61] Cf. the objections of J. S. Bell to the lack of any exact definition of measurement in quantum mechanics. These are given explicit expression in two papers written shortly before his death: '*Towards an Exact Quantum Mechanics*' (1989) and '*Against "Measurement"*' (1990).

on some 'preferred' basis. The way the universe branches at n is an empirical matter of fact, dependent entirely upon the systems located within the four-dimensional model, and on the states characterizing these systems on the different branches. In particular the question of whether a given node represents an A-measurement or a B-measurement, or no measurement at all, is one that can be answered in an objective way.

Finally, the branched interpretation differs from the many-worlds interpretation in its treatment of the contrast between *actuality* on the one hand, and *possibility* or *potentiality* on the other. In a footnote to his 1957 paper, Everett says that he rejected the idea of incorporating the distinction in question into his theory on the grounds that it represented an unnecessary complication:

some correspondents have raised the question of the 'transition from possible to actual', arguing that in 'reality' there is—as our experience testifies—no such splitting of observer states, so that only one branch can ever actually exist. Since this point may occur to other readers the following is offered in explanation.

The whole issue of the transition from 'possible' to 'actual' is taken care of in the theory in a very simple way—there is no such transition, nor is such a transition necessary for the theory to be in accord with our experience. From the viewpoint of the theory *all* elements of a superposition (all 'branches') are 'actual', none any more 'real' than the rest. It is unnecessary to suppose that all but one are somehow destroyed, since all the separate elements of a superposition individually obey the wave equation with complete indifference to the presence or absence ('actuality' or not) of any other elements. This total lack of effect of one branch on another also implies that no observer will ever be aware of any 'splitting' process.[62]

In contrast to its treatment in Everett's interpretation, the transition from possibility to actuality is the cornerstone of the branched interpretation. It is precisely this that is represented by the progressive elimination of all branches but one on the model. In opposition to Everett, on the importance of the possibility/actuality dichotomy, stands Heisenberg. Writing in 1958, Heisenberg identified the transition from potentiality to actuality as the essential new element in quantum physics—one moreover that is brought about by observation:

---

[62] Everett (1957: 146). For Everett, an observer would presumably use the word 'actual' as an indexical to denote whatever branch that observer happened to inhabit. Cf. Lewis (1983: 18).

If we want to describe what happens in an atomic event, we have to realize that the word 'happens' can apply only to the observation, not to the state of affairs between two observations. It applies to the physical, not the psychical act of observation, and we may say that the transition from the 'possible' to the 'actual' takes place as soon as the interaction of the object with the measuring device, and thereby with the rest of the world, has come into play; it is not connected with the act of registration of the result by the mind of the observer.[63]

The difficulty is, that neither Heisenberg nor anyone else has been able to say exactly what this 'transition from possibility to actuality' consists in.[64] Everett, in fact, denies that it exists. If it accomplishes nothing else, the branched interpretation makes it clear that a precise definition can be given of the transition in question. It is this, above all else, that sets it apart from the many-worlds interpretation.

(v) *Indeterminism*

The chapter began with the question, 'What would the world have to be like, in order for quantum mechanics to be true?' A salient feature of the answer so far provided is that the world must be *indeterministic*. What the branched model does is to give a literal picture of indeterminism, a concrete and specific description of a world which is characterized by (i) a multiplicity of possible futures, and (ii) the random selection of one of them as actual. It has been argued that this indeterministic model will serve to interpret quantum mechanics.

More explicitly than any other physical theory, quantum mechanics appears to violate the principle of determinism. An electron, prepared in the pure spin state x^+, has a 50 per cent probability of being measured spin-up by a Stern–Gerlach device with magnetic field parallel to the z-axis, and a 50 per cent probability of being measured spin-down. No deterministic explanation of the observed outcome is at present on the horizon. In this respect quantum mechanics differs from theories such as statistical mechanics which are probabilistic, but in which probabilistic behaviour

[63] Heisenberg (1958: 54–5). The branched interpretation differs from Heisenberg in making the concept of 'transition' independent of that of 'measuring device'.

[64] Howard Stein states that while the notion of such a passage from potentiality to actuality may perhaps be 'the radically new idea introduced into physics by quantum mechanics', such a notion, attractive in itself, should be entertained 'only with the reservation that *we do not yet know of any clear case that can be characterized as such a passage*' (Stein 1982: 576); cf. Shimony (1978).

at the macroscopic level is compatible with deterministic behaviour of individual particles at the microscopic level. The probabilism of statistical mechanics is based on ignorance: perfect information concerning a deterministic micro-level would remove the need for probabilities and restore 100 per cent predictability at the macro-level. But the probabilities of quantum mechanics are not based on ignorance. Quantum mechanics gives probabilities for individual microscopic events, and does not claim that these probabilities are even in principle reducible to the limiting values 0 and 1.

If quantum theory gives probabilities but not deterministic explanations for individual quantum events, the interpretation of quantum theory should do the same. That is, the interpretation should show how the world can mirror theory by behaving probabilistically and indeterministically at the quantum level. At the same time, the interpretation should also show how the indeterminism of individual quantum events is compatible with a high degree of regularity and predictability, albeit of a probabilistic variety. This the branched interpretation is able to do. It gives a picture of how a single electron, entering the inhomogeneous magnetic field of a Stern–Gerlach magnet, has a chance of being measured spin-up, and an equal chance of being measured spin-down. This probability is attached to a *single* event, not to the electron in virtue of being a member of an *ensemble*. At the same time, the similarity of probability prisms throughout the model ensures that all other similarly prepared electrons obey the same probabilistic laws.

Throughout history indeterminism has come to be recognized and understood, to the extent that it has been understood at all, by the use of certain images. From Lucretius we have the image of the swerve of an atom. From Laplace and the classical probabilists we have the image of ignorance: nature itself is not indeterministic, but because of our ignorance of its constitution we speak in terms of probabilities rather than certainties. From a variety of sources, ultimately Aristotle and the Presocratics, we have the notion of indifference: if there is no reason why something should be X rather than Y, or Y rather than X, then whether it is X or Y is undetermined, indifferent.[65] From Leibniz, Ryle, and Popper we have the image of a disposition or a propensity which inclines, but which does not necessitate.

But none of these images fits quantum mechanics. The electron in the magnetic field may swerve, but its swerving is probabilistically

---

[65] Cf. Rescher (1959).

controlled, not arbitrary.[66] Quantum probabilities are not based on ignorance, or on indifference as between outcomes. Propensities come closest, but where does the numerical value of a propensity reside? If quantum mechanics is to be interpreted in terms of a world which behaves indeterministically, a new paradigm of indeterminism is needed.

By now it will be clear that the proposed paradigm is that of a branching space-time structure with branch loss. This structure certainly functions indeterministically, but it is open to a fundamental objection: how can the behaviour of a quantum system, for example an electron or an atom, be interpreted in terms of a four-dimensional model? A quantum system is normally understood as a three-dimensional entity, the state of which evolves through time. How does the probabilistic evolution of a three-dimensional system relate to the interpretation of quantum probabilities in terms of proportions of sets of four-dimensional branches? This is Earman's problem of 'pre-established harmony', mentioned above on p. 94.

A full answer must await Chapter 7, but briefly the solution is this. The descriptions of objects either as three-dimensional entities which evolve through time, or as four-dimensional entities with temporal extension, are equivalent. An electron, entering the field of a Stern–Gerlach magnet, may be considered to be a three-dimensional object with a 60 per cent chance of being detected in the spin-up channel, and a 40 per cent chance of being detected in the spin-down channel. We may say, if we wish, that the three-dimensional electron has a *disposition* or a *capacity* or a *probability*, to do this or that.

Now the three-dimensional electron, if what is said in Chapter 7 is correct, may be equivalently described as a branched four-dimensional object, 60 per cent of whose branches show it as spin-up, and 40 per cent of whose branches show it as spin-down. It is in this way that probability values, defined in terms of branch proportionality, attach to three-dimensional objects. Since the three-dimensional and the four-dimensional descriptions of the electron are equivalent, the four-dimensional object is just the extensional spreading-out in space and time of the three-dimensional disposition or capacity. The spreading-out or branching gives the probability, and the selection of a single branch gives what actually happens.

---

[66] Lucretius would find it difficult to explain why some electrons have a probability of exactly 0.884023. . . of swerving.

Perhaps enough has now been said about how an indeterministic four-dimensional space-time structure can serve to interpret the probabilistic behaviour of individual quantum systems. The justification of the equivalence of the three- and the four-dimensional descriptions of an object, which is crucial to the interpretation, is found in Chapter 7. In the mean time an answer has been given to the question with which we began. The branched interpretation may not represent the *only* way the world could be, if quantum mechanics were true, but it represents *one* way.

5

Probability

MANY writers have remarked on the fact that the concept of probability is two-sided, that an ambivalence lies at its very basis. On the one hand, the concept reflects an uncertainty in our beliefs about the world. When we judge that Mallory probably never reached the summit of Everest in 1924, or that in all probability Goldbach's conjecture, that every even number is the sum of two primes, is correct, then it is to our state of mind, and not to the world or to mathematics, that the word 'probably' applies. On the other hand, probability just as often is taken to reflect an uncertainty in the world itself. If we say that the chance of being dealt a hand consisting entirely of a single suit in bridge is one in 158,753,389,900, or that an atom of plutonium 239 has a probability of .5 of decaying within the next 24,000 years, then it is not to our beliefs, but to events, that the notion of probability or chance applies.

Hume, writing in 1739, was quite clear about probability's ambivalent status: 'Probability is of two kinds, either when the object is really in itself uncertain, and to be determined by chance; or when, though the object be already certain, yet 'tis uncertain to our judgment, which finds a number of proofs on each side of the question.'[1] Ian Hacking, in *The Emergence of Probability*, argues that it was not until the two facets of probability came together in people's minds, about 1660, that the present concept of probability was born.[2] And in 1945 Rudolf Carnap[3] distinguished two fundamentally different meanings of the word, referred to as probability$_1$ and probability$_2$, which though not coextensive with Hume's distinction nevertheless served to reinforce the belief that in dealing with probability we are dealing with a multifaceted if not a genuinely ambiguous notion.

In this book we shall not be much concerned with the epistemic side of probability, which has to do with degrees of belief. We

[1] *Treatise*, II. iii. 9, Selby-Bigge (1888: 444). However, Hume also says in Book I that 'what the vulgar call chance is nothing but a secret and conceal'd cause' (Selby-Bigge 1888: 130), and later on in the *Enquiry* he remarks, 'Though there be no such thing as chance in the world . . .' (sect. vi, Selby-Bigge (1894: 56).)
[2] Hacking (1975: 12).   [3] Carnap (1945).

shall, on the other hand, be much concerned with 'objective' probability, so-called, or objective chance. The entire branched model, in fact, can be regarded as embodying an inexhaustible variety of objective, *de re* probability values. These values, via the concept of proportionality discussed in the previous chapter, are built into the very structure of the world. In this chapter we shall explore the consequences of employing a 'branched' definition of probability, and shall examine the similarities and differences between this approach and other theories of probability which philosophers have proposed and adopted.

(i) *Rival Theories of Probability*

There are four main different conceptions of probability currently accepted by probability theorists: the logical, relative frequency, subjective, and propensity theories. We shall survey each of them briefly, and then compare them with the branched interpretation.

The logical theory of probability was originally put forward by John Maynard Keynes in *A Treatise on Probability* in 1921, and was elaborated and systematized by Carnap in his *Logical Foundations of Probability*. Keynes's *idée maîtresse* was that probability is essentially a relative not an absolute concept. One cannot, for him, meaningfully say that A is probable, any more that one can say that A is equal, or that A is greater than. It must be specified that A is probable *relative to a certain body of information*, or *relative to certain evidence*, just as it must be specified that A is equal to B, or greater than C. If h is any hypothesis, such as the hypothesis that the Liberals will win the next election, then we cannot sensibly speak of $p(h)$, but only of $p(h \mid e)$, the probability that the Liberals will win, given the evidence that the polls are favourable, that the party has not been tainted by scandal, etc. Both Keynes and Carnap stress the analogy, indeed the continuity, between probability and logical implication. Thus the special case where the evidence e logically implies the hypothesis h corresponds to $p(h \mid e) = 1$, and the case where e logically excludes h corresponds to $p(h \mid e) = 0$. The intermediate cases, where h relative to e is more or less certain, or where h is supported or confirmed to some degree by e, correspond to $0 < p(h \mid e) < 1$.

For Keynes and Carnap, probability consists in a logical relation between propositions. Thus the statement 'The probability of rain tomorrow on the evidence of meteorological observations is one-fifth' does not ascribe a probability to tomorrow's rain, but rather to a logical relation between the prediction of rain and the meteorological report.[4] This logical relation is 'objective' in the sense that it is independent of the degree of belief that any given person may have in tomorrow's rain, relative to the weather report. Keynes insists that 'a proposition is not probable because we think it so', and 'probability is logical . . . because it is concerned with the degree of belief which it is *rational* to entertain in given conditions, and not merely with the actual beliefs of particular individuals'.[5] But if this is so, then we are entitled to ask what determines the probability value of a rational belief, relative to evidence. Here both Keynes and Carnap are vulnerable.

Keynes holds that there are some degrees of rational belief to which no numerical value can be given, and that the Principle of Indifference must be used to assign equal probabilities to alternatives in the absence of any reason to prefer one of them to the others.[6] Carnap on the other hand maintains that in every case in which a sentence *h* is supported or confirmed to a certain degree by a sentence *e*, it is possible to assign a precise numerical value $c(h, e)$ to the degree of confirmation.[7] This value, however, depends upon the language used to formulate the state-descriptions on which the calculation of $c(h, e)$ is based, and different choices of language yield different values for $c(h, e)$, hence different probability values. Neither Carnap's inductive logic, therefore, nor Keynes's methods can be said to define unique objective values for the logical probability $p(h \mid e)$.

The approach of the relative frequentists is very different. For Venn, as for von Mises and Reichenbach, probability is an empirical, not a logical concept.[8] Relative to a 'collective' or a 'reference class' consisting of, for example, trials which are throws of a pair of dice, the probability of 'double six' is defined as the limit of the ratio of favourable trials to the set of all trials as the size of the reference class increases without limit. If the dice are thrown 10,000 times, of which 592 result in double six, then a value of .0592 is obtained for the relative frequency. This value represents a first approximation to the probability of throwing double six with those (presumably biased) dice, the probability itself being the limit of the relative frequency 'in the long run'.

There are two main issues over which a relative frequentist will disagree with a logical probabilist. First, the frequentist considers probability values to represent empirical facts about the physical

[4] Carnap (1962: 30).   [5] Keynes (1921: 4).   [6] (1921: 34, 42).
[7] Carnap (1962: 288).   [8] Venn (1866); von Mises (1939); Reichenbach (1949).

world. For the logical probabilist, on the other hand, probability values measure logical relations between propositions or sentences. Secondly, a frequentist does not regard probability values as in any way relative to evidence. Given a reference class, the probability of any attribute within that class is fixed, and evidence can at most influence our *estimate* of that probability.

At first sight, the relative frequentist emerges as a sturdy empiricist, the logical probabilist as a philosopher with his head in semantical clouds. But in fact the frequentist's conception of probability is far from empirical. In real life, infinite sequences of trials do not exist, and it is notorious that any relative frequency of favourable cases in any finite initial sequence is compatible with any limiting frequency in an infinite sequence.[9] There is moreover no guarantee that the limit exists: the relative frequency has some chance of taking any value at all, or may oscillate between different values, as the size of the reference class increases. Such oscillations may occur whether the sequence of trials is actual or hypothetical.[10] A frequentist's probability values, therefore, defined as limiting frequencies, are unverifiable and unfalsifiable in any experimental situation, and can only by courtesy be described as 'empirical'. By contrast, the relative frequency values themselves are thoroughly empirical.

A noteworthy feature of the frequency theory, one for which it has been often criticized, is the fact that it assigns probabilities only to *repeatable* events, not to events that occur only once. Thus von Mises:

When we speak of the 'probability of death', the exact meaning of this expression can be defined in the following way only. We must not think of an individual, but of a certain class as a whole, e.g. 'all insured men forty-one years old living in a given country and not engaged in certain dangerous occupations'. A probability of death is attached to the class of men or to another class that can be defined in a similar way. We can say nothing about the probability of death of an individual even if we know his condition of life and health in detail. The phrase 'probability of death' when it refers to a single person, has no meaning at all for us.[11]

More than anything else, the need for a satisfactory account of single-case probabilities has led philosophers to look beyond the

---

[9] See Ch. 3 n. 54 above.

[10] 'There is no such thing as *the* infinite sequence of outcomes, or *the* limiting frequency of heads, that *would* eventuate if some particular coin-toss were somehow repeated forever. Rather there are countless sequences, and countless frequencies, that *might* eventuate and would have some chance (perhaps infinitesimal) of eventuating' (Lewis 1980: 90; see also Tooley 1987: 144–5).

[11] von Mises (1939, 1957: 11).

logical and the frequency theories. The principal alternatives at present, each of which provides for probabilities of single events, are the subjective and the propensity theories.

The subjective theory defines probability as the degree of confidence, or belief, or 'credence', that an individual person X has in a proposition A. The earliest writer to define it in this way seems to have been Jacob Bernoulli, but it is to F. P. Ramsey that we are indebted for an explicit statement of the subjectivist position. Modern subjectivists include Bruno de Finetti and L. J. Savage.[12]

What distinguishes a subjectivist from a logical probabilist or a frequentist is his denial that there is one and only one real number x which may correctly be said to be the probability of A, or for that matter the probability of A on condition B. On the subjectivist's view, the probability x is not uniquely determined, but may take any value between 0 to 1 depending on the inclination of the person whose degree of belief that probability represents.[13] This arbitrary character of probability values, their total dependence on individual choice and inclination, is provocatively summed up by de Finetti in the subjectivist slogan 'Probability does not exist'. For him, 'The only relevant thing is uncertainty—the extent of our own knowledge and ignorance.'[14]

However, even for a subjectivist there are constraints to be placed on probability values. If the propositions A and B are atomic, then no restrictions are imposed on the probability values $p(A)$ and $p(A \mid B)$.[15] But for non-atomic propositions there are restrictions, aimed at imposing a standard of rationality or *coherence* upon subjective probability values. The restrictions stipulate that the probability values assigned to non-atomic propositions must be such that no set of bets can be made, based upon those values, which result in one of the bettors losing no matter what the outcome. For example, the standard probability calculus requires that if $p(A) = x$, then $p(\sim A) = 1 - x$. A bettor who violated this restriction and assigned a probability of say $\frac{4}{5}$ to A and $\frac{2}{5}$ to not-A (i.e. simultaneously accepted odds of 1 to 4 when betting on A and 3 to 2 when betting against) would be sure to lose whether A turned out to be true or false. Coherence, therefore,

---

[12] de Finetti (1937, 1964); Savage (1954). Kyburg and Smokler's anthology (1964) contains an excellent introduction to the subjective theory, and reprints Ramsey's paper 'Truth and Probability', written in 1926, originally published in Ramsey (1931), and reprinted in Ramsey (1990).

[13] Kyburg and Smokler (1964: 6).   [14] de Finetti (1974: i. pp. x, xi).

[15] This is not quite true. David Lewis requires in addition that $p(B) = 0$, and hence that $p(A \mid B)$ be undefined, only if B is logically false (Lewis 1980: 88).

aims at presenting a criterion of how a person's degrees of belief *ought* to be related in order that his actions based on those beliefs should be rational, and is accepted in one form or another as an integral part of the subjective theory of probability. It turns out that coherence is a necessary and sufficient condition for conformity of degrees of belief to the rules of the probability calculus.[16]

Moving on to the propensity theory, the first philosopher to propose a 'propensity interpretation' of the probability calculus was Popper. In two papers appearing in 1957 and 1959, Popper argues that probabilities cannot be the limiting frequencies of *actual* sequences of trials, though they can be regarded as frequencies in *hypothetical* or *virtual* sequences.[17] Suppose, Popper says, a loaded die is being tossed, with a probability $\frac{1}{4}$ not $\frac{1}{6}$ of landing with six uppermost. For a relative frequentist, this probability will show itself in the relative frequency of sixes over a long series of trials. But suppose now, mixed in with these trials, one or two throws with a *regular* die are made. The fact of occurring within a sequence where the relative frequency of sixes was $\frac{1}{4}$ would not entail that the probability of six with the regular die was $\frac{1}{4}$. But if this is so, the naïve identification of probability with relative frequency in an actual sequence, or with a limiting frequency, is incorrect. What we want to say is that when the regular die was substituted for the biased die the experimental conditions were changed, and that the new conditions did not have the same tendency, or disposition, to produce a given distribution of sixes in a sequence of trials as the old one. For the propensity theorist, the key notion is that of a *tendency*. In Popper's words:

Every experimental arrangement is liable to produce . . . a sequence with frequencies. . . . These virtual frequencies may be called probabilities. . . . They characterize the disposition, or the propensity, of the experimental arrangement to give rise to certain characteristic frequencies when the experiment is often repeated.[18]

In virtue of the thought expressed in this passage, Popper is credited with being the author of the 'hypothetical frequency interpretation' of probability. The probability of a (repeatable) event is the frequency the event *would* have, relative to a reference class, if a sufficiently long sequence of trials were made.[19] But, in

---

[16] Kyburg and Smokler (1964: 11).

[17] Popper (1957 and 1959). Salmon (1979) contains an excellent discussion of the development of the propensity theory.

[18] Popper (1957: 67); emphasis in the original omitted.

[19] Recall, however, Lewis's remark about the variability of hypothetical sequences in n. 10 above.

addition to this thought, Popper also puts forward a second proposal, one that defines probability not as a hypothetical relative
frequency, but as a property of the generating conditions of a
single event. This second view is the origin of what Salmon calls
the 'full-blown single case propensity interpretation' of probability,[20] and constitutes a more radical break with other theories
than the hypothetical frequency interpretation. Popper says:

we now take as fundamental the probability of the result of a single
experiment, . . . rather than the frequency of results in a sequence of
experiments. Admittedly, if we wish to test a probability statement, we
have to test an experimental sequence. But now the probability statement
is not a statement about this sequence: it is a statement about certain
properties of the experimental conditions, of the experimental set-up.[21]

Since the late 1950s a number of probability theorists have
embraced one or another form of the propensity interpretation.
The list includes Hacking, Mellor, Fetzer, Suppes, Giere, Gillies,
and Salmon.[22] But it is not at all easy to say which form of the
propensity interpretation each of these authors supports. The task
is made difficult by the fact that Popper himself does not succeed
in specifying exactly what objects or entities propensities are to
be attributed to, and until the matter is clarified, the propensity
interpretation remains shadowy and imprecise. The difficulty is
brought out in a paper by Henry Kyburg.[23]

Kyburg asks, what are propensities attributed to? Is the propensity of a biased penny to fall heads a property of the penny?
Popper's answer is no. Propensities are not properties inherent in
the penny, they are 'relational properties of the experimental
arrangement—of the conditions we intend to keep constant during
repetition'.[24] But, Kyburg asks, suppose the coin is the only thing
we intend to keep constant? Obviously there are many ways the
coin can be tossed: with the right hand, with the left hand, against
a hard surface, a soft surface, high or low, in air or in a vacuum.
If the propensity of the coin to fall heads is a property of the
'experimental arrangement', then presumably there will be as many
different propensity-values as there are different arrangements.
To be consistent, we should not even exclude the 'arrangement' of
deliberately placing the penny heads-up on the table.

---

[20] Salmon (1979: 197).　　　[21] Popper (1957: 68); emphasis in the original omitted.
[22] See Hacking (1965); Mellor (1971); Fetzer (1971 and 1981); Suppes (1973 and 1974);
Giere (1973); Gillies (1973); Salmon (1988).
[23] Kyburg (1974b).
[24] Popper (1959: 37). On this question see Redhead (1987: 49).

Plainly this is not what Popper and other propensity theorists have in mind. What they perhaps mean by 'experimental arrangements' are *fair* tosses of the coin, tosses where the coin's bias (if it exists) has a chance to show itself. Presumably, in a sufficiently long run of 'fair' tosses of a biased coin, the relative frequency of heads will approach closer and closer to some value like .58962, and this value could be, in fact, the propensity of a *single* toss of that coin to show heads. But if that is so, then would Popper not allow that .58962 measured the propensity *of the coin* to fall heads, and that it had this propensity even if it were never tossed? So that propensities were tendencies or dispositions of material objects, like coins, to behave in certain ways? Or would a propensity theorist still insist that no value can be assigned to any propensity unless the experimental conditions are precisely specified, and that any change in those conditions (such as flipping the coin left-handedly) changes the propensity and hence the probability of heads?

We cannot answer these questions here, although no clear understanding of the propensity theory can be achieved without their being settled. We turn instead to a different theory in which the difficulties do not arise: objective single-case probabilities of events based on branching.

(ii) *The 'Branched' Definition of Probability*

In Chapter 4 it was seen how quantum probabilities, in the branched model, are defined by the proportion of branches on which a certain event occurs, relative to the set of all branches above a given branch point. For example, the probability that a vertically polarized photon will be measured ϕ^+ by a polarization analyser inclined at an angle ϕ to the vertical is given by the proportion of branches in which the photon emerges in the ϕ^+ channel, relative to the set of all branches above the branch point at the instant when the photon enters the polarizer. The branched structure yielding this proportionality, namely a decenary tree of arbitrarily short but non-zero temporal length, was named a *prism*. In prisms, probability reduces to proportionality of branches.

It may be asked, what is the status of quantum probabilities defined in this way? Can other, non-quantum probabilities be defined in the same manner? And does the definition of probability as proportionality in branched structures satisfy all classical probability theses? Does the 'branched' theory constitute an interpretation of the standard probability calculus?

In answer to the first question, it would seem that the branched definition yields a completely objective, empirical, *de re* conception of probability. This holds not only for the general conception of probability, but for individual probability values. Thus the proportion of branches in which a photon's angle of plane polarization takes the measured value ϕ in the example cited above is $\cos^2\phi$, with the consequence that the photon has the probability $\cos^2\phi$ of emerging in the ϕ^+ channel. This number, $\cos^2\phi$, is not a measure of the degree of logical support, provided by evidence that the photon is entering a $\phi\pm$ analyser, for the hypothesis that it will emerge in the ϕ^+ channel. Nor is it a measure of the degree of anyone's belief. Whatever people's beliefs may be, the probability has the value $\cos^2\phi$. Finally, and most importantly, the value of the probability is not defined by, and indeed is quite independent of, any sequence of trials in which vertically polarized photons are passed through $\phi\pm$ analysers. What these trials do is provide rough confirmation that the proportion of ϕ^+ branches in the $\phi\pm$ prism is $\cos^2\phi$. But the exact proportion of ϕ^+ to ϕ^- branches in such a prism is an empirical fact about the branched universe model, a fact expressed by saying that the proportionality is $\cos^2\phi$ to $\sin^2\phi$, and it is not to be expected that any sequence of trials, whether finite or infinite, actual or hypothetical, will yield this exact value. Moreover, the same proportionality of future branches, $\cos^2\phi$ to $\sin^2\phi$, characterizes each individual instance of a single photon passing through the analyser, and gives that photon a fixed probability of emerging in one or another channel. The branched definition, therefore, yields precise, objective, empirical, single-case probability values. The same cannot be said of any of the theories of probability previously discussed.

The branched probability values so far mentioned all derive from a single prism, and as such represent only a small fraction of the full range of probabilistic phenomena. In the branched model as a whole, prisms are stacked on top of prisms, and it must be shown that combining 2, or n, or an infinite number of prisms always results in well-defined probability values.

Let us begin with a three-prism case. Suppose that a vertically polarized photon passes through a $\phi\pm$ analyser. If it emerges in the ϕ^+ channel it passes next through a $\theta\pm$ analyser (i.e. an analyser oriented at an angle θ to the photon's new plane of polarization after emerging in the ϕ^+ channel), and if it emerges in the ϕ^- channel it passes through $\pi\pm$ analyser. The probabilities of passing θ^+ in the ϕ^+ beam and π^+ in the ϕ^- beam are $\cos^2\theta$ and

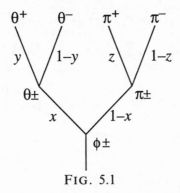

FIG. 5.1

$\cos^2\pi$ respectively. Now, what are the final probabilities for the photon to emerge in each of the four channels θ^+, θ^-, π^+ and π^-? Putting $\cos^2\phi = x$, $\cos^2\theta = y$, and $\cos^2\pi = z$, we have the following branched *probability diagram*. In Fig. 5.1, individual branches represent *sets* of branches in the universe tree, and the number beside each branch represents the *proportionality* of the set of branches in question. The probabilities for the photon to pass through this complex apparatus in each of the four channels are as follows:

$$
\begin{array}{ll}
\theta^+ & xy \\
\theta^- & x(1-y) \\
\pi^+ & z(1-x) \\
\pi^- & (1-x)(1-z)
\end{array}
$$

These four probabilities sum to unity, and the probability of a disjunctive event, such as the probability for the photon to be measured either θ^+ or π^+, will be a partial sum of this total (in this case $xy + z - xz$).

To calculate probabilities in linked systems of prisms, we proceed as follows. We multiply probability values up the branches to obtain the probability of any single outcome, and we add across at any given level to get probabilities for disjunctive events.

An example of a more complex probability diagram is given in Fig. 5.2. Probability diagrams such as these, which are used as a means of computing probabilities of events in the branched model, are not intended to be literal pictures of the model, but are schematic simplifications. Each node on a probability diagram represents the base node of a prism in the full model. Each branch on the diagram represents a set of branches belonging to the full model, and the number beside each branch segment denotes the

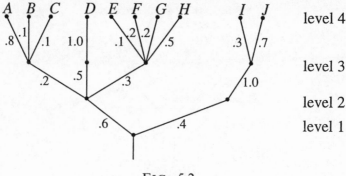

FIG. 5.2

numerical proportion of the corresponding set to the set of all branches above the corresponding node in the full model. A probability diagram is a picture of objective chance.[25] In labelling such diagrams, the sole constraint to be observed is the requirement that, in a branched probability diagram, the probability values of all the branch segments immediately above any given node must sum to unity.

Given this constraint, it is not difficult to show that the probability values for events yielded by probability diagrams obey all the laws of the standard probability calculus, and hence that the 'branched' definition of probability constitutes an interpretation of the calculus. Consider the set of all nodes at any given level. These constitute the set of all possible outcomes (at that level) of the initial conditions prevailing at the base node, and each one has a definite probability value attached to it, computed by multiplying together the numbers on the branch segments which lead up to it. For example, the probability of events occurring at the G-node on the diagram above, relative to events occurring at the base, is $.6 \times .3 \times .2 = .036$. At each level, therefore, there is a *function* whose arguments are nodes and whose values are numbers. To show that this is a probability function we must show that it is additive, non-negative, and sums to unity over its domain. The function is additive in the sense that the probability of any union of disjoint subsets of the domain is the sum of the probabilities of the subsets. It is non-negative because any product of positive numbers is positive. Finally, to show that it sums to unity over its domain requires that it be proven that the sum of

[25] A similar probability diagram, constructed along the same lines as Fig. 5.2, may be found in Lewis (1980: 102).

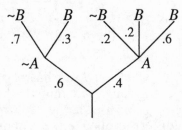

FIG. 5.3

the probabilities of all nodes at a given level is one. This is not difficult.[26]

In addition to yielding objective probabilities for future events, probability diagrams also provide values for *conditional probabilities*, as in Fig. 5.3. Relative to the base node, the probability $p(B)$ of an event of type B is $(.6 \times .3) + (.4 \times .2) + (.4 \times .6)$, i.e. $.5$. The conditional probability $p(B|A)$ of B occurring, given that A occurs, is defined as $p(A \& B)/p(A)$, i.e. the fraction obtained by dividing the proportion of $A \& B$-branches by the proportion of A-branches. In the diagram, $p(A \& B) = (.4 \times .2) + (.4 \times .6) = .08 + .24 = .32$; $p(A) = .4$; and $p(B|A) = .8$. A slightly more complicated example, with B-events occuring above two distinct A-nodes, is given in Fig. 5.4.

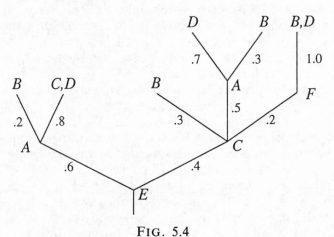

FIG. 5.4

[26] In the example given, the sum of the probabilities of the nodes at level 4 is $.096 + .012 + .012 + .3 + .018 + .036 + .036 + .09 + .12 + .28 = 1$. In general it can be shown that if $B_1, \ldots, B_n$ are the nodes immediately above any node A, then $p(B_1) + \ldots + p(B_n) = p(A)$. The result follows by an induction.

Here

$$p(B \mid A) = p(A \& B)/p(A)$$
$$= (.6 \times .2) + (.4 \times .5 \times .3)/.6 + (.4 \times .5)$$
$$= .12 + .06/.8 = .225$$

This last example, in which $p(B \mid A)$ has a single overall value of .225 for the whole diagram even though the proportion of B branches over the A-node on the left is .2, while on the right it is .3, illustrates the difference between a conditional probability $p(B \mid A)$ having a definite value x on the branched model, and the equation $p(B \mid A) = x$ being a *law of nature*. For $p(B \mid A) = x$ to be a law of nature, the proportion of B-branches would have to be precisely x, no more and no less, above every A-node in the model. This is not so in the example given, thus showing that two universals can be instantiated in such a way that the probability of finding one, conditional upon finding the other, takes a well-defined value, without the universals standing in any lawlike relationship.

The examples of probability diagrams so far given have all been finite. However, there may be cases in which we wish to compute the probabilities of infinitely many different possible outcomes of a single set of initial conditions, or to compute probabilities of events which are infinitely far removed from initial conditions. In order to accommodate such cases, there are two ways in which the restriction of finitude on probability diagrams can be lifted.

To begin with, each consecutive pair of nodes on a probability diagram is separated in time by a small but non-null interval Δt, the temporal height of a single prism. Consequently any finite branch, no matter how long, contains only a finite number of nodes. This property of probability diagrams, known as the *finite branch property*, will have to be removed if we wish to consider probabilities of events lying infinitely far in the future. To remove it, the use of infinite products of numbers between zero and one is required.

The second property of finitude, the *finite fork property*, may also be relaxed. If we wish to allow for a countable infinity of distinct possible outcomes of a given set of initial conditions prevailing at a hyperplane in the branched model, then the number of nodes immediately above a single node on some probability diagrams will have to be countably infinite. In that case the probability function will be *countably additive*. Let us examine what this involves.

Let N denote the set of nodes at a given level of a branched probability diagram, and let S be a sigma-field of subsets of N, meaning (i) S contains N, (ii) S is closed under complementation in N, (iii) S is closed under countable unions of pairwise disjoint sets.[27] To say that the probability function p defined on S is countably additive means that if $\{A_1, A_2, \ldots, A_n, \ldots\}$ is a countable pairwise disjoint set of members of S, then $p(\cup A_i) = \Sigma p(A_i)$. There seems no difficulty in allowing this to be true of the probability function associated with a branched diagram which permits countably infinite branching at a single node. Hence we shall assume that such diagrams yield well-defined probability values for countably infinite different types of outcome.

In order to permit branched diagrams to represent faithfully all the probability distributions found in nature, countable additivity is essential. For example, if a hydrogen atom absorbs and later re-emits a photon, the electron in the atom has a chance of jumping to any one of a countable infinity of higher orbits. In order to assign each one of these possibilities its correct probability, a countable infinity of distinct outcomes is required.

Again, as was seen in the previous chapter, a prism is not merely a prism for a single experiment like sending a photon through a polarization analyser, but is also, at the same time, a prism for an extremely large number of other events occurring in other parts of the world. A prism has a very short temporal height, but every hyperplane in it has a very broad spatial width. Thus one and the same prism serves as a polarization prism for a photon on earth, a time-of-decay prism for a radioactive atom on Betelgeuse, and a probability-of-jump prism for an electron orbiting the nucleus of a hydrogen atom in the Magellanic Cloud, assuming that these events are located on a single spacelike hyperplane. To accommodate all combinations of possible outcomes of all such initial conditions across the face of the universe, a denumerable infinity of different types of branch on a single probability diagram is not too much.

Finally, the calculation of some probability values on branched diagrams may involve even higher infinities. The number of distinct outcomes of a single set of initial conditions may be uncountably infinite, or the probability of a single event may conceivably be different on each of a non-denumerable infinity of separate prisms at the same level. Such cases, which require the property of uncountable additivity for the probability function, are discussed in the next section.

Returning to the question of whether probability defined as branch proportionality constitutes an adequate interpretation of

[27] See e.g. Kyburg (1974a: 13).

the classical probability calculus, one final obstacle needs to be removed. This concerns the notion of inverse probability, derived via Bayes' theorem, and forms the substance of an objection brought by Salmon and Humphreys against the propensity interpretation.[28] The objection is most clearly stated in terms of an example of Salmon's, designed to show that propensities and probabilities are distinct.

Let us imagine two light bulb factories, an old one which produces a high proportion of defective bulbs, and a new one which produces both a lower proportion of defectives and a larger overall number of bulbs. We can speak of the *conditional probability* $p(B \mid A)$ that a bulb, given that it was produced by the old factory, will be defective. Similarly we can speak of the *propensity* of the old factory to produce defective bulbs. Now suppose we hold a defective bulb in our hand. It makes perfect sense to speak of the *inverse probability* $p(A \mid B)$ that this bulb, given that it is defective, was produced by the old factory. Bayes' theorem, as will be seen, provides a way of computing the probability in question. But it makes no sense to speak of the bulb's *propensity to have been made* by the old factory. Salmon and Humphreys conclude that propensities are conceptually unlike probabilities, and do not provide an admissible interpretation of the probability calculus.

It might appear at first sight that the only inverse probability values allowed for by the 'branched' definition of probability are the values zero and one. If I hold up a defective light bulb, and ask what the probability is that it was manufactured by the old factory, then it might seem that *epistemically*, relative to the state of my knowledge, the answer would be a value between zero and one, but that *ontologically*, or *objectively*, only the two extremes of zero and one made sense. Since the bulb was in fact manufactured in either the old or new factory, any uncertainty as to its origin would seem to be confined to my beliefs, and not to characterize the world. On the branched model, a given state of the world can issue in many possible future states, but comes from only one earlier state. Therefore it might seem that inverse probabilities, in the model, were meaningless.

Surprisingly, this is not so. Unlike propensities, probabilities defined as branch proportionalities allow for inverse values. Consider the example shown in Fig. 5.5. At t_2 a coin will be tossed,

---

[28] Salmon (1979: 213); Humphreys (1985). Salmon credits Humphreys as being the originator of the objection.

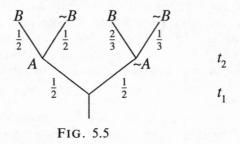

FIG. 5.5

but whether it is a fair coin or a biased one, with twice the chance of heads over tails, will not be decided until t_1. Let A = a fair coin is tossed, B = the result is heads, and let $p(A) = \frac{1}{2}$. Then $p(B \mid A) = p(A \& B)/p(A) = \frac{1}{4} \div \frac{1}{2} = \frac{1}{2}$. Now for the inverse probability $p(A \mid B)$. This is the probability, given the outcome heads, that the coin tossed at t_2 is a fair coin. By Bayes' theorem $p(A \mid B) = [p(B \mid A) \times p(A)]/p(B) = (\frac{1}{2} \times \frac{1}{2}) \div \frac{7}{12} = \frac{3}{7}$, since $p(B) = \frac{1}{4} + \frac{1}{3} = \frac{7}{12}$ on the diagram. Given heads, the probability that the coin that was tossed is a fair one is $\frac{3}{7}$. This probability is computed from the branched structure using Bayes' theorem, and is perfectly objective. Since the example is easily generalized, we may conclude that wherever it permits the definition of conditional probabilities, the branched theory also permits inverse probabilities. This fact marks a difference between it and the propensity theory, and allows the branched theory to serve as an interpretation of the probability calculus.

(iii) *A Probability Value for Every Event*

The object of this section is to show that for every future event, indeed every event, the branched model provides a unique, precise, objective, single-case probability value. It is immaterial whether we succeed in discovering what that value is. Independently of what we know or do not know, the value exists, and supplies a solid foundation in empirical fact for all discourse about probabilities.

Before considering the argument leading to the general conclusion about event-probabilities, the dependence of probabilities upon time must be emphasized. In the branched model, as was seen earlier in discussions of time flow and the laws of nature, time plays an essential role not only in how we see the world, but in what the world *is*. This is especially true of the probabilities of events. As time passes, the probabilities of many future events change, sometimes quite abruptly. The probability of a person's contracting flu, for example, changes with time depending upon the person's resistance, the number of infectious persons he or she comes in contact with, etc. By contrast, the probability of decay of a radioactive atom remains the same from one instant to the next.

A beautiful example of objective probabilities changing with time is given by David Lewis. You enter a labyrinth at 11 a.m., and employ the strategy of tossing a coin whenever you come to a branch point to determine whether you go left or right:

When you enter at 11:00, you may have a 42% chance of reaching the center by noon. But in the first half hour you may stray into a region from which it is hard to reach the center, so that by 11:30 your chance of reaching the center by noon has fallen to 26%. But then you turn lucky; by 11:45 you are not far from the center and your chance of reaching it by noon is 78%. At 11:49 you reach the center; then and forevermore your chance of reaching it by noon is 100%.[29]

Lewis's example illustrates perfectly the way in which probabilities change as the first branch point of the branched model works its way up a stack of prisms. Making your way through a labyrinth with the help of a stochastic decision-maker is very like the random way in which, at every instant, one and only one of the uncountably many branches of a prism at the universe's first branch point is selected, with the consequent effect of this selection on future probabilities. For those who have doubts about whether tossing a coin is a genuinely chance method of selecting

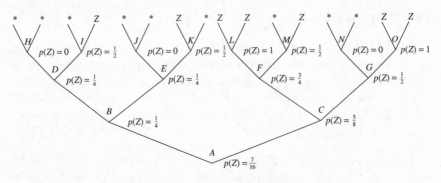

FIG. 5.6

[29] Lewis (1980: 91).

a path, Lewis notes that even air resistance, which affects the fall of the coin, depends partly on the chance making and breaking of chemical bonds between the coin and the molecules it encounters. Although a subjectivist, Lewis in the first half of his paper finds room for objective, physical chance.[30]

A branched probability diagram, which gives the probability at each choice point of reaching the centre of a simple maze by using Lewis's method of tossing a coin, is the following. Let $Z =$ you reach the centre, $* =$ you get eaten by wild beasts. As the branches drop off, the value of $p(Z)$ alters, as shown in Fig. 5.6.

A maze corresponding to this probability diagram is shown in Fig. 5.7. Armed with a coin, we would expect to flip our way through to the centre an average of seven times out of sixteen (another example of von Fraassen's horizontal–vertical problem of Chapter 3).

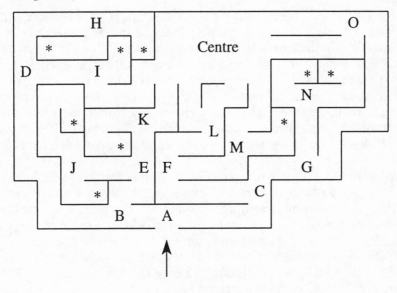

FIG. 5.7

The probability diagram for the maze may be contrasted with one which represents probabilities that do *not* change with time, such as decay probabilities. Suppose that a certain elementary particle has a half-life of one millisecond. The probability diagram which represents the outlook for this particle is shown in Fig. 5.8,

---

[30] However, in the second half of the paper he comes back to the view that chance is 'objectified subjective probability' (1980: 98).

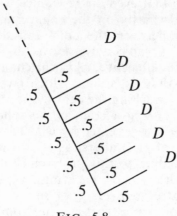

FIG. 5.8

where each space between nodes represents a stack of prisms one millisecond high. D = the particle decays. After a second, the probability of the particle's still existing is $(\frac{1}{2})^{1000} = 2^{-1000}$. On the branched model, the 'fingerprint' of every unstable particle is a prism stack with the structure of the above diagram, where only the temporal distance between nodes distinguishes one type of particle from another. This 'fingerprint' continues to infinity on the branched diagram, and it is the random falling off of branches which determines whether or not the particle decays at any given time.[31]

We proceed now to the main task of this section, which is to show that, without exception, every event has an exact probability. If the event lies in the present or the past, the result is immediate: relative to the state of the world at any moment the probability of each such event is one if it occurred, and zero if it did not. For example, relative to today's branched model, the probability of Thomas E. Dewey's being elected president in 1948 is zero, although in earlier models it was not zero. The interesting cases involve probability values for *future* events.

Let us start with a single prism: a single decenary or other proportionality-preserving tree of temporal height Δt. The number of branches in a prism is of cardinality $2^{\aleph_0}$, and it was seen in Chapter 4 how it is possible for a prism to have a proportion r of its branches be of one type, say A, while $1 - r$ of its branches are not A, for any real number r between zero and one. The

[31] This way of representing decay probabilities avoids the problems discussed in van Fraassen (1989: 109–10).

demonstration of this fact rested on the possibility of in effect copying the number r, in decimal form, step by step onto the branches of a decenary tree, in such a way that at the end of the copying process exactly r of the branches were A-branches, and exactly $1 - r$ were not A-branches. Given that this can be done, it follows conversely that from many although not all prisms, the branches of which are either A or not A, the decimal r which gives the proportion of A-branches can be recovered by a procedure which, though infinite in length, is finitary at each step (since a prism branches finitely at each node). In these cases, the proportion of A-branches in a prism is always a fixed and definite real number r, where r gives the probability that any randomly selected branch will be an A-branch. In other cases, as we shall see, the value of r may have to be a 'hyperreal' number.

For any real $r > 0$, the set of A-branches in a prism whose proportion is r will be non-denumerably infinite in number. (If the decimal expansion of r is not a uniform string of zeros, a non-zero decimal will occur at some finite position, necessitating at least one A-node at the corresponding level in the prism. The A-branches which are the descendants of that A-node will be non-denumerable.) But what if the number of A-branches in a prism is finite? Or denumerably infinite? What if a prism contains just one A-branch? In classical measure theory, the measure of a single point on the real line, or of a denumerable set of points, is zero, and it might be thought, by analogy, that the proportionality of a denumerable set of branches in a prism should also be zero. But to affirm this brings with it undesirable consequences, as is shown by Brian Skyrms, and the introduction of infinitesimal proportionality values leads to a more elegant solution.

Suppose just one branch in a prism is an A-branch, and we decide to assign the proportion zero to the set consisting of that one branch. A consequence of this is that we assign probability one to a proposition (namely that A will not occur) that may possibly turn out to be false.[32] Again, let 1000 branches in a prism be A-branches, of which 500 are also B-branches, and let there be no other B-branches. Since, by classical measure theory, $p(A \& B) = p(\sim A \& B) = 0$, a classicist would presumably be prepared to accept as fair a bet that an $A \& B$-branch will be selected rather than a $\sim A \& B$-branch. This is a bet which he can lose but which he cannot possibly win, thus showing that something is wrong with any probability measure which makes the bet a fair one.

[32] Skyrms (1980: 74).

These problems can be solved by introducing probability values that are infinitesimally small. The existence of infinitesimal numbers, smaller than any real number, was demonstrated in the 1960s by Abraham Robinson, and is now accepted and used by an increasing number of philosophers and mathematicians.[33] In a prism, the chance that the actual branch will be one of a finite or denumerable subset of the set of all branches is non-null but infinitesimally small, and there are enough infinitesimals of different sizes to guarantee the subset property: if the set of A-branches is a proper subset of the set of B-branches, then $p(A) < p(B)$. With the help of infinitesimals, every measurable set of branches in a prism can be assigned an appropriate proportionality.[34]

Moving now from a single prism to a stack of prisms, we wish to show that every future event E, represented by the set of branches on which the event-type E is instantiated, has a fixed objective probability. (The event in question may be *temporally definite*, e.g. a sea battle occurs at Lepanto 3 p.m. 16 September 2055, or *indefinite*, e.g. a sea battle occurs at Lepanto some time in the twenty-first century, or some time in the future. Either way, the method of showing that it has a well-defined probability value is the same.)

The value of $p(E)$, as it is defined here, is always relative to a given branched structure, and may vary over time as branch attrition within the structure takes place. But at each instant it

[33] Let **AN** be any first-order theory of analysis, and let the language of **AN** contain a name for every real number. Add to **AN** the set **S** of all open sentences $v > r_i$ for some free variable v and for every r_i which is a name for a real. Any finite subset **S′** of **AN** + **S** has a model in which v is assigned some finite real number, hence by the compactness theorem for first-order theories **AN** + **S** must have a model. In this model the variable v cannot be assigned any finite number, but must denote an 'infinite' element which is greater than every real. Because of this infinite element, the model for **AN** + **S** cannot be the *intended* or *standard* model of analysis, but nevertheless is still *a* model of analysis, which satisfies every true first-order statement. Among these statements is the statement that every number has a reciprocal, and also that if $x > y$, then $1/x < 1/y$. Consequently, if v is greater than any real number, then the reciprocal of v is less than any real. We call it an *infinitesimal* number: a *hyperreal* number is one that differs from a real by an infinitesimal. For this, and for the proof that there exist infinitely many infinitesimals, and hence infinitely many hyperreals which are 'infinitely close' to every real, see Skyrms (1980: 73–6 and 177–87), based on Robinson (1966).

[34] That the reals are insufficient to capture all the nuances of proportion among sets of branches in a prism follows from a similar but different consideration. In a decenary tree, a set of branches with proportionality denoted by a terminating decimal such as 0.148000 . . . differs from a set with proportionality 0.147999. . ., even though, by convention, the decimals 0.148000. . . and 0.147999. . . are taken to denote the same real number. If we designate the first set of branches A, and the second B, then B is a proper subset of A, having one fewer branch. Therefore $p(A)$ and $p(B)$ must differ by an infinitesimal, even though 0.148000. . . = 0.147999. . . I owe this observation to Christopher Hitchcock.

has a definite value. To show this, it suffices to consider the worst case, namely the case where $p(E)$ is the sum of uncountably many infinitesimals.

This worst case can come about in the following way. Suppose we have a two-tier prism stack, namely a stack consisting of a single prism at the base, and of a prism sitting at the tip of each of the branches of the first prism. These prisms are non-denumerably infinite in number. Suppose further that E occurs on some set of branches on each of the upper row of prisms, but that, for each of these prisms, the value of $p(E)$ is different. What is the value of $p(E)$ for the whole stack? To compute $p(E)$ requires taking an uncountably infinite sum. Where j is the infinitesimal number giving the proportionality of the unit set of each individual branch of the bottom prism, and $p_i(E)$ is the proportion of E-branches in the ith prism of the upper row, then

$$p(E) = \sum_i j \cdot p_i(E).$$

The existence of this uncountably infinite sum, and the fact of its lying between zero and one, are guaranteed by the structure of the two-tier prism stack. To compute it may be beyond the power of any earthly computer, but that is beside the point. The probability functions generated at each level by a stack of prisms will be, in the most general case, uncountably additive, although this does not prevent them from being well-defined.

This concludes the task of this section. Even though the probability of an event in a two-tier stack may be an uncountably infinite sum of infinitesimals, it still takes a fixed and definite value between zero and one. Since any prism stack of finite height consists of a finite number of tiers, the process of computing probability values in a two-tier stack need be iterated only a finite number of times. If the stack is infinitely high, a denumerable infinity of prisms of height Δt still suffices, and the process of computation will be iterated at most a denumerable number of times. In every case, the calculations required follow the pattern of those for the simple two-, three-, and four-tier probability diagrams given earlier in this chapter.

As a consequence, a significant simplification in the construction of probability diagrams becomes available. For example, a hundred years from now, either Canada will be a distinct political entity or it will not. Since a definite numerical value at present exists for the probability that it will be, a simple probability diagram can be constructed (Fig. 5.9). This probability diagram

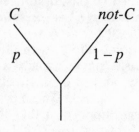

FIG. 5.9

sums up all the information that the universe tree presently
contains concerning the future existence of Canada. In similar
fashion, an analogous diagram can be constructed for any other
future event. We come therefore to the desired conclusion: on the
branched model, every future event has a precise, objective probability value.

6

Conditionals

IF Hitler had invaded England in 1940, he would have won the war. Is this conditional true or false? Much has been written on the problem of determining truth-conditions for counterfactuals, such as the conditional about Hitler.[1] It will be argued that the branched model provides, in theory if not in practice, a clear and definite answer to the question of truth-conditions not only for counterfactuals, but for all conditionals with contingent antecedent and consequent, e.g. 'If we keep on like this, we'll capsize.' As will be seen, some conditionals of this kind have *truth-values*, some have only *probability values*, and some have both. The situation is complicated, but I hope it can be laid out clearly.

(i) *Truth-Values and Probability Values*

In Chapter 5 it was established that, on the branched model, every event has a precise, objective probability value. Furthermore, if the event in question is in the future, located on the branches of the model rather than on the trunk, then its probability value may vary with time. As branch attrition takes its toll, the proportion of branches on which a future event is located may change substantially.[2] As is demonstrated by the probability diagram earlier shown in Fig. 5.4, at time t_1 the probability value $p(B)$ of an event B may be $(.6 \times .2) + (.4 \times .3) + (.4 \times .5 \times .3) + (.4 \times .2 \times 1.0) = .12 + .12 + .06 + .08 = .38$, while at t_2 it may be either .65 or .2, depending on whether branch attrition wipes out the left or the right part of the tree (see Fig. 6.1).

In exactly the same way, the probability value of the conditional $A \to B$ may also vary with time. Defining $p(A \to B)$ as $p(B \mid A)$, as is permissible on branched diagrams where probability values

---

[1] All discussion of the problem goes back to two papers published in the late 1940s: Chisholm (1946) and Goodman (1947), reprinted as the first chapter of Goodman (1955).

[2] Exceptions to this rule are laws of nature. Relative to the base node of the branched model at any time, in any frame of reference, the proportion of branches on which a vertically polarized photon is measured ϕ^+ by a suitably inclined analyser is always $\cos^2\phi$. Similarly, at every instant there is a constant proportion of branches on which an unstable particle or radioactive atom decays in a given interval of time.

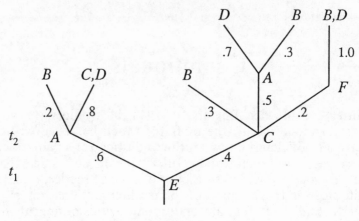

Fig. 6.1

are defined objectively in terms of branch proportionality,[3] we have
that at t_1 $p(A \to B) = p(A \& B)/p(A) = (.6 \times .2) + (.4 \times .5 \times .3)/.6 +$
$(.4 \times .5) = 0.255$, while at t_2 $p(A \to B)$ is either 0.2 or 0.3 according to where branch attrition happens to strike.

So much for probability values. But in addition to these, the
supervenience conception of truth allows propositions about the
future to have truth-values as well. The probability that I stoop
and tie my shoelace at noon tomorrow may be, say, 0.00017,
reflected in the number of branches on which I do tie my shoelace
at that time, but above and beyond having a probability value
the proposition that I stoop and tie my shoelace at noon tomorrow may also be *true*. Whether it is true or false depends on what
happens tomorrow, i.e. on what branches drop off, and there is
of course no way of knowing beforehand which of the values
'true' or 'false' is possessed by the proposition in question. But,
known or not, the proposition *has* a truth-value, and this value
is quite independent of its probability value. On the branched
model, probability values depend on branch proportionality and
truth-values on branch attrition.

However, not all propositions have both probability values and
truth-values. As we shall see, there exist conditionals which have
only the former, and others which have only the latter. Consider a
typical 'causal' or 'empirical' conditional with contingent antecedent

---

[3] On the branched model, the probability of the conditional is the conditional probability,
and the alleged trivialization consequences of identifying the two are avoided. See Lewis
(1976) and Appendix 4 below.

and consequent, such as 'If we keep on like this, we'll capsize.' Suppose we *do* keep on. Then if it turns out that we capsize, the conditional is (and always was) true, whereas if we don't capsize, the conditional is (and always was) false. Contingent conditionals with true antecedents have truth-values. But if we *don't* keep on, then (whether or not we end up capsizing) the conditional has no truth-value. We can argue about its truth or falsehood, perhaps even fight over it, but in reality there neither exists nor ever will exist anything in the world which makes the conditional true or false. There does, however, exist something in the world which gives it a *probability value.* After it's all over we can say 'If we'd kept on like that, we would have capsized,' and the counterfactual we utter has the same probability value as the corresponding present-tense conditional had earlier. Contingent conditionals with false antecedents have probability values,[4] but they don't have truth-values.

In addition to conditionals which have probability values but no truth-values, there exist also conditionals with truth-values but without probability values. Examples of the latter are: 'If Shakespeare didn't write *Hamlet,* someone else did', and 'If Alain is nervous, he doesn't show it', which belong to a category named 'type **B**' conditionals below. Type **A** conditionals such as 'If we keep on like this, we'll capsize' differ from type **B** conditionals in having probability values, and are in turn subdivided into those whose antecedents are true and which consequently have truth-values as well as probability values, and those which, like counterfactuals, have no truth-values.

These distinctions will emerge in what follows. Before setting out the differences between type **A** and type **B** conditionals, I shall give a brief review of various alternative approaches to counterfactual semantics, focusing on Stalnaker's and Lewis's conception of comparative similarity amongst possible worlds as a basis for assigning truth-values. This was an approach that I followed myself at one time, and I shall try to explain why I now favour a different strategy.

(ii) *'Inferential', 'Degree of Belief', and 'Possible World' Analyses of Counterfactuals*

Nelson Goodman, setting out the framework of the problem of counterfactuals that has occupied logicians and philosophers of science since 1947, held that when we say

---

[4] Provided that they are so-called type **A** conditionals. See below.

(1) If that match had been scratched, it would have lighted.

We mean that there exist 'relevant conditions' (e.g. the dryness of the match, the presence of oxygen) such that given their presence, and given the existence of physical or causal laws concerning such things as the heat generated by friction and the ignition point of the chemicals in the match, we can infer 'that match lights' from 'that match is scratched'. In his words: 'A counterfactual is true if and only if the antecedent joined with relevant true statements about the attendant circumstances leads by way of a true general principle to the consequent.'[5]

Goodman's truth-conditions for $A \rightarrow C$ constitute what I shall call the 'inferential' theory of conditionals. A conditional is true, according to this theory, if there exist suitable true statements which, conjoined with the antecedent, enable us to infer the consequent. Sometimes the inferential theory is phrased in such a way that conditionals turn out not to be propositions which are either true or false, but instead become 'inference tickets' which in a variety of ways are either 'acceptable' or 'unacceptable'.[6] For Goodman, counterfactuals *are* either true or false, but the problem of specifying exactly which relevant conditions or supplementary statements are to be conjoined with the antecedent A in order to infer the consequent C is as yet unsolved, for the following reasons.

We have a fairly clear idea of what true statements and physical laws suffice, if joined to the antecedent of (1), to logically imply the consequent. But consider the following:

(2) If match m had been scratched, it would not have been dry.

Statement (2) seems intuitively unacceptable, but on the inferential theory it is as true as (1). Since the set of true statements in the context of which the counterfactuals (1) and (2) are being examined includes both 'match m is not scratched', and 'match m does not light', we can join the second of these statements to the antecedent of (2), together with other truths, to obtain:

(3) Match m is scratched. It does not light. It is well made. Oxygen enough is present . . . etc.

From (3), together with relevant causal laws, we can infer that match m is not dry. Hence the inferential theory yields the unwelcome result that both (1) and (2) are true.

---

[5] Goodman (1955: 8 and 37).
[6] For the notion of an inference ticket, see Ryle (1950: 308–9). Ryle, however, held that conditionals were assertions with truth-values.

Goodman's suggestion for dealing with this problem is to restrict the class of relevant conditions which, together with the antecedent A of a true counterfactual, imply the consequent, to conditions *cotenable* with A. Condition B is 'cotenable' with A if B and A are not only mutually compatible, but if it is not the case that, if A were true, B would be false. This restriction is not met by the statement 'match m does not light' in (3), for given things as they are, if the antecedent of (2) were true, 'match m does not light' would be false. Hence 'match m does not light' is not cotenable (given all the other circumstances that obtain) with 'match m is scratched'. But, as Goodman points out, the notion of cotenability will not solve our problem, since cotenability itself is defined in terms of counterfactuals. To define truth and falsehood for counterfactuals using the notion of cotenability is circular; to proceed without it appears impossible. Goodman sees no way out of this dilemma.

The problem of finding exactly the right set of relevant conditions and physical laws S which, joined to the antecedent A, enable us to infer the consequent C of a true counterfactual, is analogous to the *frame problem* in artificial intelligence. Taking an example from Dan Dennett,[7] suppose RDD is a robot whose only task, in Dennett's words, is to 'fend for itself'. Let RDD be equipped with the most sophisticated data banks, information-retrieval systems, and reasoning abilities that money can buy. One day RDD learns that its precious spare battery is on a wagon in a locked, burning room. How to save it? In a trice RDD locates the key, and figures out that turning it clockwise in the lock will release the catch and that the door opens outward. Now the wagon. It has a handle, and RDD reasons that a certain action it calls PULLOUT (WAGON, ROOM) will result in the wagon being removed from the room. It also deduces from the knowledge available to it that pulling the wagon out of the room will not change the size, the shape, or the colour of the battery. Yet it does nothing, and the battery is melted. Why? Although it knew useful things like ON (BATTERY, WAGON), and IN (WAGON, ROOM), RDD did not know that if it pulled the wagon out of the room, the battery would come with it. Lack of a crucial piece of information prevented RDD from inferring 'the battery is saved' from:

(4) The battery is on the wagon. The wagon is in the room. The room is burning. RDD pulls the wagon out of the room.

---

[7] Dennett (1984*a*).

The problem for RDD's designers is how to ensure that RDD mobilizes exactly the right amount of relevant information to derive the conditionals it needs in order to fend for itself and survive. Suppose RDD learns that ON (BATTERY, WAGON), and PULLOUT (WAGON, ROOM) together imply PULLOUT (BATTERY, ROOM). Then, on the inferential theory, it knows the truth of the counterfactual

(5) If I had pulled the wagon out of the room, I would have pulled the battery out of the room.

Knowing this counterfactual, in turn, will enable it to save the battery in similar circumstances in the future. It follows that a solution to the frame problem would provide the supplementary premisses demanded by the inferential theory of counterfactuals, and conversely, resolving the problems of the inferential theory would be a major step forward in finding an answer to the frame problem. The difficulty is that the frame problem is insoluble, at least in its current form.

To see why this is so, suppose that RDD manages to marshal enough information to conclude that if it pulls the wagon out, it pulls the battery out. The next time, instead of being on a wagon, the battery is on the floor, or on loan, or on charge . . . . Plainly no general method can be devised of providing a robot with the right information to determine, in every situation, which conditionals are relevant to its needs and which are not. From time to time the robot may be lucky, and hit on the vital supplementary premisses which enable it derive the consequent of a true and useful conditional, but its search methods are invariably *ad hoc*. Even the most intelligent machine can always be stumped by a question like the following:

(6) If the wagon had been made of spaghetti, would it have taken the paint off the wall when it was pulled out of the room?

Such questions can be answered by 5-year-olds, but they cannot be answered by RDD. Dennett sees no solution in sight for the frame problem, which means, if I am right, that there's no solution either for the supplementary premiss problem in the inferential theory of counterfactuals.[8]

What characterized the inferential theory was the search for an appropriate set of supplementary premisses for an inference. No

---

[8] The most recent attempt to defend a version of the inferential theory is Kvart (1986). But to supply the requisite supplementary premisses, Kvart appeals to world-histories and to divergent world processes, i.e. to items from the area of possible world interpretations.

principled, non-*ad hoc* way of generating such a set was found. A different approach is to forget about premises and inferences, and to appeal directly to a device which, given the antecedent of a counterfactual, produces (or fails to produce) the consequent directly. Two species of such devices are *minds and their beliefs*, and *possible worlds*.

The originator of the appeal to minds and beliefs was F. P. Ramsey. Suppose you are considering whether or not to accept the conditional $A \to C$. Ramsey's test as formulated by Stalnaker is this:

> Add the antecedent A hypothetically to your beliefs, and then consider whether or not you accept the consequent C as true. If you do, then you believe $A \to C$. If you do not, then you don't believe $A \to C$.[9]

For Ramsey, the acceptability or 'degree of credence' you attach to $A \to C$ reduces to the degree of credence you attach to C, given that you accept A hypothetically. In Ramsey's words, you are 'fixing your degree of belief in C given A'. Of course, if A is a belief-contravening hypothesis, i.e. if you initially believed A to be false, then the thought-experiment of fixing your degree of belief in C given A will involve adjustments to what you believe. Ramsey's test, however, does not in any way specify *how* such adjustments are to be made. For Ramsey's test to work, and for it to produce results without getting entangled in the difficulties of the inferential theory, the mind performing Ramsey's thought-experiment must be regarded as a black box, which produces a degree of credence in C without revealing any details of how this is accomplished.

The strength of Ramsey's method for determining the degree of acceptability of conditionals $A \to C$ is that, unlike the inferential theory, it always produces an answer. Its weakness is that it produces, not truth-conditions, but belief-conditions for conditionals. Furthermore, the actual degree of credence accorded to one and the same conditional may vary widely from individual to individual. Worse still, for some conditionals the results of applying the Ramsey test go directly against intuition, to such a degree that we can only conclude, for these conditionals at least, that Ramsey's is the wrong test.

Imagine that we are Shakespeare scholars, interested in the question of authorship. We perform Ramsey's test, and add

---

[9] Stalnaker (1968), in Sosa (1975: 169); Ramsey (1990: 155). Ramsey's test is found in an essay originally written in 1928.

'Shakespeare did not write *Hamlet*' hypothetically to our system of beliefs, making whatever adjustments are necessary to avoid contradictions. We then ask ourselves whether we accept the proposition 'Some other author wrote *Hamlet*'. The answer is almost certainly yes. But, despite this, we do not accept the truth of the counterfactual 'If Shakespeare had not written *Hamlet*, some other author would have.'[10] The Ramsey test, therefore, is not the right one for counterfactuals. This still leaves open the possibility that it is the right test for the so-called indicative conditional 'If Shakespeare did not write *Hamlet*, someone else did.' We shall return to the difference between counterfactual and indicative conditionals below, in section (iv). But for the time being let us say goodbye to the Ramsey test and the 'degree of belief' analysis of counterfactuals, and turn to the 'possible worlds' theory.

Robert Stalnaker, in his paper 'A Theory of Conditionals',[11] states clearly the difference between 'belief-conditions', i.e. conditions under which belief in a proposition is justified or appropriate, and 'truth-conditions'. Belief-conditions have to do with pragmatic considerations pertaining to conditionals; truth-conditions with their semantics. Stalnaker maintains that conditionals can be studied from the point of view both of their pragmatics and of their semantics, and he is right. We may be interested in the conditions under which a given conditional is *acceptable*, or worthy of belief, and we may also be interested in the conditions under which the same conditional is either *true* or *probable* in some non-subjective sense. In this book, we are interested in the latter. In making the transition from belief-conditions to truth-conditions, Stalnaker asserts that what is needed is the ontological analogue of a stock of hypothetical beliefs, namely the concept of a *possible world*.

It would, I think, be fair to say that for many years those who studied counterfactuals wandered in the wilderness, until at last Stalnaker and Lewis showed us the promised land. This was the land of possible world semantics, investigations into which during the late 1960s and 1970s transformed the study of counterfactuals from a realm of speculation and intuition into something approximating a science. Something *approximating* a science, to be sure, but not quite a science. The Stalnaker–Lewis semantics are based upon the concept of *closeness* or *similarity* of possible

---

[10] The example comes from Bennett (1988: 523). The same point concerning the failure of the Ramsey test is made in Lewis (1973: 71).

[11] Stalnaker (1968).

worlds, and the vagueness inherent in this concept prevents the study of counterfactuals from being truly scientific. Not until the notion of similarity is replaced by something more precise will truth- or probability-conditions which admit of no ambiguousness be attainable. Nevertheless, truth-conditions based on possible worlds and their similarities constituted a giant step forward, and if we can now think more clearly about counterfactuals and other conditionals it is because of the work of Stalnaker and Lewis.

For Stalnaker, a conditional $A \rightarrow C$ is true in our world if and only if C is true in the closest A-world, i.e. the world (i) in which A is true, and (ii) which differs minimally from the actual world. (If there is no possible world in which A is true, i.e. if the antecedent of $A \rightarrow C$ is impossible, then $A \rightarrow C$ is true, but vacuously.[12]) Lewis's semantics are like Stalnaker's, but are more general, more powerful, and more flexible. In Lewis's semantics, there may be no single A-world which is closest to the actual world. Instead there may be ties for closest, or there may be an open infinite sequence of closer and closer A-worlds, but no closest. To allow for these possibilities, Lewis's truth-conditions state that $A \rightarrow C$ is true in a world w (not necessarily the actual world) if and only if one of the following two conditions is met. Either (i) A is true in no world that is 'accessible' from w, or that falls within any of w's 'spheres of accessibility'. In that case $A \rightarrow C$ is vacuously true. Or (ii) A is true in some world w' accessible from w, and $A \supset C$ holds in every world at least as close to w as w' (equivalently, A is true and C is false in no world at least as close to w as w').[13] Lewis's semantics have been successful in enabling us to understand many interesting and puzzling features of counterfactuals, and a striking instance of their success is discussed in the next section.

(iii) Comparative Similarity

In order to define comparative closeness or similarity, Lewis needs some way in which to compare the degree of similarity of the pair of worlds w_1 and w_2 with the degree of similarity of the pair w_1 and w_3. These degrees of comparative similarity are represented in his semantics by an elegant device: a system of concentric 'spheres of accessibility' centred on w_1. The image is a

---

[12] Stalnaker in Sosa (1975: 171).   [13] Lewis (1973: 16).

geometric one.[14] Worlds correspond to points in space, and the world w_2 is 'closer' or 'more similar' to w_1 than w_3 if w_2 is enclosed in some sphere centred on w_1 that does not enclose w_3. Two distinct worlds, if they each belong to exactly the same set of w_1-spheres, will be equally similar to w_1. Among the set of spheres centred on w_1 which contain A-worlds there may be no smallest, thus permitting infinite sequences of closer and closer A-worlds, without limit. Having set up a measure of comparative similarity on possible worlds through this Ptolemaic system of concentric spheres, Lewis uses it to deal with the following problem.

Suppose that last night we gave a party, and that it was a flop. Someone says, 'If Otto had come, the party would have been lively' $(A \rightarrow L)$. 'Yes,' says someone else, 'but if both Otto and Anna had come, the party would have been dull' $(A\&B \rightarrow \sim L)$. 'I agree,' says the first, 'but if Otto and Anna had brought Waldo, the party *would* have been lively' $(A\&B\&C \rightarrow L)$. Obviously this could go on forever, resulting in a potentially infinite sequence of true counterfactuals, any adjacent pair of which have contradictory consequents. But how is this possible? Normally, if a conditional $A \rightarrow L$ is true, the conditional $A\&B \rightarrow L$ derived from it by strengthening the antecedent is also true, or so the traditional laws of logic dictate. If $A\&B \rightarrow \sim L$ is true instead, it might appear either that $A \rightarrow L$ was false in the first place, or that the conjunction $A\&B$ was impossible and that Otto and Anna could not be simultaneously present at the party.

The great virtue of Lewis's Ptolemaic semantics is that they are able to accommodate the simultaneous truth of $A \rightarrow L$, $A\&B \rightarrow \sim L$, $A\&B\&C \rightarrow L, \ldots$ etc. In fact, Lewis constructed his semantics in order to do just that. The key is relative closeness. In the actual world, the party guests included none of Otto, Anna, Waldo, etc. In a close alternative to the actual world, Otto attends, the party is lively, and in no closer world does Otto attend and the party fail to be lively. Next we move to the slightly less close worlds in which Otto and Anna both attend. In each of these worlds the party is dull. Moving farther away from the actual world, in every $A\&B\&C$-world the party is again lively, and so on. The system of Ptolemaic spheres simultaneously accommodates all the counterfactuals we intuitively hold to be true.

---

[14] Or, more generally, a topological one. Lewis (1973: 14) notes that if the regions of accessibility surrounding a world are not required to be spheres, but are instead arbitrary open sets of points, we obtain a generalized 'neighbourhood semantics' in which there is no way of comparing the relative closeness of any two worlds to a third.

The problem with the concept of comparative similarity is that it is inherently vague. According as the vagueness is resolved in different ways, different truth-values can be attached to one and the same proposition. A famous example is the following:

(1) If Bizet and Verdi had been compatriots, Bizet would have been Italian.

(2) If Bizet and Verdi had been compatriots, Verdi would have been French.

Using Lewis's spheres of accessibility, which of (1) and (2) comes out true? The answer depends on whether a possible world in which Bizet was Italian is, or is not, more similar to the actual world than a world in which Verdi was French. Is there any way to choose between these alternatives? No. Different resolutions of the vagueness associated with comparative similarity simply yield different truth-values. Nevertheless, despite its vagueness, Lewis maintains that the concept of comparative similarity can serve as the basis of counterfactual semantics:

Somehow, we *do* have a familiar notion of comparative overall similarity, even of comparative similarity of big, complicated, variegated things like whole people, whole cities, or even—I think—whole possible worlds. However mysterious that notion may be, if we can analyse counterfactuals by means of it we will be left with one mystery in place of two.[15]

A few years ago, I was convinced that a satisfactory counterfactual semantics could be based on comparative similarity of possible worlds or possible courses of events, and furthermore that the branched model provided an objective measure of such similarity, devoid of vagueness.[16] But difficulties with this approach have made me change my mind, and although I still believe that the branched model yields the correct semantics, I now favour a different analysis. In the remainder of this section I describe my earlier attempt to state truth-conditions for counterfactuals based on a branched model interpretation of comparative similarity, and in the next section I give the approach I now prefer.

If Napoleon had won the battle of Waterloo, he would not have died on St Helena. Assuming this conditional is true, what makes it true? Letting W = Napoleon wins at Waterloo, and H = Napoleon dies on St Helena, a branched model structure verifying $W \rightarrow \sim H$ is shown in Fig. 6.2. The truth of $W \rightarrow \sim H$ consists in the fact that every W-branch in the model is a $\sim H$-branch. Since

---

[15] Lewis (1973: 92).   [16] McCall (1984a).

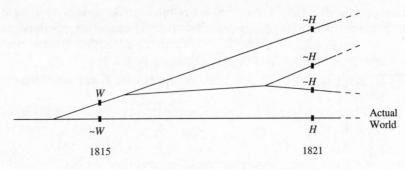

FIG. 6.2

W is false and H is true in the actual world, the branches on which the truth of $W \to {\sim}H$ depends no longer form part of the branched model today. They are not, however, lost to counterfactual analysis, but can be resurrected, since they formed part of past branched models depicted in 'snapshots' of the universe which date back to 1814 and earlier. In Chapter 2, section (iii), these earlier tree structures or states of the branched universe were discussed, and it was concluded that the universe model which is the subject of the book, and which is a dynamic thing, could not be understood in Parmenidean terms as a stack of instantaneous snapshots or universe-states. Nevertheless, these snapshots or earlier universe-states are completely determinate and contain no vagueness or impreciseness. Though past branches do not exist today in the sense of forming part of today's branched model, they used to exist, their past existence being analogous to the past existence of, say, Socrates or Aristotle. This past existence is entirely 'objective', and will be appealed to and made use of for the purpose of evaluating counterfactuals, both in this section and throughout the chapter. A complete semantics for conditionals of all sorts, including counterfactuals, will require as model structures not only the *present* form of the universe tree, in different coordinate frames, but all its *past* forms as well.

Returning to the example of Napoleon, the W-branches, on which he wins at Waterloo, formed part of the branched model in 1814, and appear in full detail in the pictures of past models that we shall assume to have at our disposal. These pictures show the W-branches as forking smoothly off the actual branch, without any hiatuses or discontinuities. Perhaps in some of them Grouchy decided, early on the morning of the battle, to ride back to Waterloo from Ligny with his troops to see how Napoleon was faring. In others Blücher, being an elderly man who had had a fall from his horse the day before, may have chosen to take a day's rest and remain where he was with all his Prussian soldiers. Or in others the Imperial Guard simply broke through the British lines. All these alternative W-branches, on which Napoleon wins, are entirely possible, or rather *were* entirely possible. The truth or falsehood of $W \rightarrow \sim H$ rests on whether each of those W-branches is also a $\sim H$-branch. If it is, then $W \rightarrow \sim H$ is true, if not it is false.

An important requirement, however, has not yet been mentioned. This is (in my 1984 paper) that the W-branches examined in order to see whether, on each of them, Napoleon dies or fails to die on St Helena must be W-branches that branch off the actual world as close as possible to the time of the battle. The reason for this restriction to 'closest' W-branches is that doubtless some branches exist which diverged from our actual history long in the past and which are $W \& H$-branches, not $W \& \sim H$-branches. For instance, Britain might have ceded St Helena to France at the time of the Treaty of Utrecht in 1713, and St Helena might subsequently have become a fashionable retreat for retired French officers. Unlikely, perhaps, but not impossible. Then Napoleon might have won the battle and still died there, thus falsifying $W \rightarrow \sim H$.

It was in order to eliminate such maverick possibilities as these that in my 1984 paper, A-branches used in evaluating a counterfactual $A \rightarrow C$ were restricted to those which branched off the actual world as close as possible to the time of the antecedent A. Of all A-branches, those ones might with some justification be regarded as 'closest' to the actual branch, since they shared the greatest amount of common past with it. The branched model could, in this sense, be regarded as providing an objective measure of *degree of comparative similarity* between branches or 'worlds'. World w_2, on this criterion, would be more similar to w_1 than w_3 if the branch point separating w_1 and w_2 stood higher on the branched model than the branch point separating w_1 and w_3, i.e. if w_1 and w_2 shared more common past than w_1 and w_3. In my 1984 paper, I proposed that this measure of comparative similarity could eliminate the vagueness from Lewis's semantics, and serve as the basis of a new and wholly objective set of truth-conditions for counterfactuals.

Unfortunately this proposal contained a fatal flaw, as is shown by the following example.[17] Suppose I find myself in a traffic jam.

---

[17] The problem was first pointed out to me by Anil Gupta, and is discussed in my 1984 paper on p. 473. The repair I suggested in that paper, however, now strikes me as inadequate.

My son says, 'Dad, if you had turned left, you would have missed this jam.' If we examine those left-turning worlds which branch off closest to the present, then sure enough we find no traffic jam. But a difficulty of a different sort arises. In branches where the left turn is made at the very last moment (which must be examined in order to satisfy the objective criterion of 'closest'), we encounter situations in which I wait until the last possible instant before turning left, then yank the wheel over, step on the gas, and make my turn in the face of oncoming traffic. Such branches verify conditionals like: 'If I had turned left, I would have paint on my bumper from the fenders of oncoming cars.' We may name this difficulty the 'last-minute deviation objection'.

When I wrote the 1984 paper on counterfactuals, I didn't take the last-minute deviation objection very seriously. At the time, I thought it could be dealt with by distinguishing between 'normal' and 'abnormal' ways of realizing the antecedent of a counterfactual—between normal and abnormal ways of turning left. Normal ways involve slowing down, getting in the left lane and signalling, not pulling the wheel over at the last minute, etc. But I now think the last-minute deviation objection *is* a fatal one. For one thing, the concept of normalcy introduces just the element of vagueness and imprecision that the selection of the closest antecedent-branches was supposed to eliminate. For another, the 'last-minute deviation' counter-instances are too numerous and important to ignore. Richard Otte, in a recent note, provides some fine examples, and convinced me that a new approach was needed.[18]

Franz, an incompetent skier, has managed through bribery and deception to enter a World Cup ski race. The course is steep and icy at the top, but flat and easy near the finish. Since Franz is terrified of the difficult early section, he skis slowly and finishes in last place, far behind the other skiers who had little trouble on the slope. Consider Otte's counterfactual:

(3) If Franz had won the race, all of the other skiers would have been ahead of him throughout the race until just short of the finish line.

(We are to imagine a race in which all the skiers start at once, like the old ski races in the 1920s.) Now as Otte points out, (3) is intuitively false. There are various possible worlds branching off from the actual world in which Franz wins the downhill. In some of these all the other racers are disqualified for being late at the start, in some they fall on the steep icy section of the course,

---

[18] Otte (1987).

and in some the others are far ahead of Franz until, just before the finish line, something very improbable happens and none of them actually finishes. If we take the instructions to examine only worlds which share a maximal common past with the actual world literally, we shall, Otte says, discard branches on which Franz's competitors fall on the steep section and retain only branches where disaster overtakes them on the flats just before the finish.[19] But all this is highly implausible, and (3) is surely false.

Consider also

    (4) If Franz had fallen during the race, he would have fallen on the gentle section at the bottom of the course.

If 'closest' means 'sharing maximal amount of common past', then (4) is true. But Otte is surely right in protesting that (4) is false. A new approach is needed, not one that attempts to eliminate the vagueness inherent in the concept of comparative similarity, but one that dispenses with comparative similarity entirely.

(iv) *Two Kinds of Conditionals*

In section (i) it was said that there existed an important difference between conditionals which had probability values, and conditionals which had truth-values but no probability values. It will be argued in this and succeeding sections that all or almost all conditionals fall into one or the other of these two broad groups, which I shall call type **A** and type **B** conditionals. Type **A** conditionals are further subdivided into those which do, and those which do not, have a truth-value in addition to a probability value. Before discussing this second complication, let me first set down the basic difference between type **A** and type **B** conditionals: a difference which has frequently been noticed but not yet, as far as I know, traced back to its semantic roots. With the help of the branched model, the semantic underpinnings of the difference may be laid bare.

A pair of examples will illustrate the line to be drawn between type **A** and type **B** conditionals. The examples were originally given by Ernest Adams, and have been discussed frequently since:[20]

---

[19] In point of fact, I suppose the 'closest' worlds in which Franz wins would be worlds in which Franz finishes last, but in which all the other competitors are later disqualified for missing a gate or for some other reason.

[20] I have modified Adams's conditionals, which appear in Adams (1970), in the way Lewis does (1973: 3).

(1) If Oswald did not shoot Kennedy, then someone else did.

(2) If Oswald had not shot Kennedy, then someone else would have.

The first is in the indicative mood and is true; the second is in the subjunctive and would be regarded by almost everybody as false. Conditional (1) is type **B**; (2) is type **A**. Although the labels 'indicative' and 'subjunctive' have often been used in this connection, I shall avoid marking the distinction between type **B** and type **A** conditionals in these terms.[21] This is not because the difference in grammatical moods seems a superficial one, but because it won't do the job. Indicative conditionals, and to a lesser extent subjunctives, fall on both sides of the line I want to draw.

In order to separate the two classes of conditional I shall first of all attempt to establish a pre-theoretic line between them, based on considerations of grammar and linguistic usage. Subsequently I shall explain how this intuitive, everyday distinction may be grounded in theory. The fact that the distinction can be shown to rest upon a fundamental attribute of the branched model will be additional proof of the latter's explanatory powers.

The pre-theoretic line I shall trace is based on two ordinary language tests, the *tense invariance test* and the *probability test*. The idea for the first of these I got from Brian Ellis, in an article where he adds to Adams's pair of examples a third.[22]

(3) If Oswald does not shoot Kennedy, then someone else will.

Now (3), like (1), is indicative rather than subjunctive, but which of (1) and (2) does it resemble more closely? Ellis opts for (2), on the grounds that (2) and (3) are just different ways of saying the same thing, appropriate to different times. They could be used, for example, as differently tensed assertions by someone who believed there was a conspiracy to kill Kennedy; (3) before the assassination, and (2) after. The truth of (1), on the other hand, is independent of whether or not there was a conspiracy, and (1) has no present or future tense version. I shall say that (2) and (3) exhibit *tense invariance*, in the sense that what they assert is *independent of tense*. One and the same proposition may be asserted by many different tensed forms, each from a different temporal perspective. In the appropriate circumstances, *Brown will*

---

[21] For reasons to be wary of the indicative/subjunctive distinction, see Bennett (1988). The line that Bennett traces between different classes of conditionals, based on V. H. Dudman's 'forward tense shift', is very close, but not identical, to mine, though Bennett's semantics are different.

[22] Ellis (1984).

be out, *Brown is out*, and *Brown was out*, all say the same thing. What they say is tense invariant. On the other hand what (1) says is not tense invariant, but is tied to the past tense. Any attempt to put it in the present or the future tense fails. If what a family of differently tensed sentences says, i.e. the proposition it asserts, is independent of tense, I shall say, loosely, that the members of the family themselves are tense invariant. As we shall see, type **A** conditionals differ from type **B** in that they fall more easily and naturally into tense invariant families.[23]

A different but related example is Grice's, discussed by Stalnaker.[24]

(4) If the butler didn't do it, the gardener did.

This conditional is not tense invariant. But, like (1), it is superficially similar to a tense invariant family:

(5) If the butler hadn't done it, the gardener would have.
(6) If the butler doesn't do it, the gardener will.
(7) If the butler were to fail to do it, the gardener would do it.
(8) If the butler hasn't done it, the gardener has.

For Hercule Poirot, investigating a murder, it is appropriate to utter (4) in circumstances where the list of suspects has been narrowed down to two. But it is not appropriate to utter any of (5)–(8) at any time, either before or after the murder, unless collusion is suspected. What is present in (5)–(8), but absent in (4), is a suggested or implied *link* between antecedent and consequent, a distinguishing feature of type **A** conditionals that will emerge in the semantics. For the moment, the 'tense invariance test' merely separates conditionals which can be arranged *salva veritate* in families differing only in tense or mood from conditionals which cannot, i.e. which resist transformation into a different tense. More examples are given in section (vi).

The second pre-theoretic test, the *probability test*, is this. A conditional 'If *A* then *B*' may or may not permit the insertion of words like 'in all probability', 'there is a chance that', or 'it is unlikely that' immediately before the consequent *B*. If these words can be added without absurdity, without radically altering the sense, or without violating some presupposition that speaker and hearer share, then the conditional passes the test. If not, the conditional fails. Thus adding the words 'in all probability' to (1) creates the suspicion that, at best, the speaker was trying to make

[23] This is not to say that tense invariant families of type **B** conditionals do not exist, but only that they are rare oddities compared to type **A** families. An example of a type **B** family is given on p. 187.
[24] Stalnaker (1975).

a bad joke. If Hercule Poirot were to add them to (4), the listener might justifiably object that he understood the suspects had been limited to *two*. Was Poirot now suggesting there were more? By contrast, the words 'in all probability' can be added to any of (2), (3), (5)–(8) without doing more than relaxing or making more flexible the link between antecedent and consequent. Any parent who makes the transition from 'If you chew gum, you'll lose your allowance' to 'If you chew gum, you'll almost certainly lose your allowance' knows the nature of this relaxation. The conditional still moves in the same direction, but more slowly. Such is not at all the case with a type **B** conditional like 'If Alain is nervous, he doesn't show it.' Adding 'in all probability' here generates nonsense. The same goes for 'If you had bet on Fleetfoot, you would have won', uttered just as Fleetfoot crosses the finish line. More examples are given below.

The tense invariance test and the probability test appear to go hand in hand, picking out the same two classes of conditionals belonging to what I have called type **A**. In the following sections I shall try to show that this is not a coincidence, but results from the fact that tense invariance and probability are related at the semantic level. Section (v) presents semantics for type **A** conditionals based on proportionalities of sets of branches in the branched model. These proportionalities are fixed, permanent features of the model, and are invariant when viewed from different temporal perspectives. The semantics they provide for type **A** conditionals are *probability semantics*, although as will be seen some type **A** conditionals also have truth-values in addition to probability values. Type **B** conditionals, by contrast, have no probability values, and the semantics provided for them by the branched model is a *truth-value semantics*. The basis for this semantics in the model, as will be seen in section (vi), is not branch proportionality but branch attrition. Thus the superficial grammatical differences between the two types of conditional are indicative of, and are underpinned by, a fundamental semantic distinction.

(v) *Semantics for Type A Conditionals*

The semantics for type **A** conditionals presented here differ from traditional counterfactual semantics in at least four respects. (In this section 'conditional' means 'type **A** conditional'.)

(*a*) *Probability values.* The semantics assigns a numerical probability value between zero and one to every conditional, based upon branch proportionalities in the model. In the limiting cases one and zero, conditionals take the values 'true' and 'false'.

(*b*) *Time-dependence*. The probability values possessed by conditionals are sensitive to time, necessitating explicit mention of the time relative to which the probability of the conditional is fixed by the model.

(*c*) *Counterfactuals a special case*. Once allowance is made for time-dependence, probability values assigned to conditionals are invariant under change of tense. Counterfactuals represent one of three kinds of past-tense conditional, the others being *since*-conditionals and what I shall call *neutral past* conditionals.

(*d*) *Truth-values*. In virtue of the 'supervenience' of truth, even conditionals with low probability values can be true. This arises in cases where the conditional's antecedent and consequent both turn out to be true. To distinguish them from conditionals which are 'probabilistically true', i.e. those whose probability value is one, such conditionals will be described as 'superveniently true'. A conditional whose antecedent is false may be probabilistically true, but cannot be either superveniently true or superveniently false.

Each of these features of the semantics will be discussed with the help of examples. Our main example, which illustrates all the points mentioned, derives from the counterfactual stated at the very beginning of the chapter (first used, I think, for philosophical purposes by Strawson). In order to demonstrate clearly the sensitivity of type **A** conditionals to time, I shall narrow down its time interval:

(1) If Hitler had invaded England in August 1940, he would have won the war.

This counterfactual looks back on the events of 1940 from the vantage-point of a later time, and is the past-tense counterfactual version of two present-tense conditionals, each of which could have been, and doubtless was, uttered many times in different languages during 1940:

(2) If Hitler invades England in August 1940, he will win the war,

(3) If Hitler were to invade England in August 1940, he would win the war.

The first of these is in the indicative and the second in the subjunctive, but for our purposes the difference is unimportant; (2) and (3) are the present-tense counterparts of (1), and all three share the same probability value. I shall focus on (2).

No doubt historians will debate for years the question whether (1)–(3) are true or false. If in future they shift their ground a little, and instead ask what probability value (1)–(3) have, then I contend there is an answer, and the answer is to be found in the branched model. I don't mean to say that anyone will ever know enough about the topological shape of the model to be able to find it. But it's there none the less. Let's see how, in principle if not in practice, an exact probability value can be extracted.

Consider the situation in 1939 and the early months of 1940. Britain controlled the surface if not the depths of the seas, and France had a powerful army behind the Maginot line. Any British, French, or German staff officer contemplating (2) would have assigned it a low probability. And in all likelihood he would have been right, in the sense that, on any branched model of say December 1939, the proportion of Hitler-invading-in-August-and-winning branches to Hitler-invading branches was small. I say 'in all likelihood' because we don't *know* the proportion was small. But our knowledge is irrelevant. Whether small or large, the exact proportion, and hence the exact probability, was and still is today an objective fact about December 1939 branched models.

Yet in a few months everything had changed. Surprisingly, the Allies had discounted the possibility of a thrust through the Ardennes, the Germans had broken through, the British army had lost almost all its armour in the evacuation of Dunkirk, and Hitler was assembling an invasion fleet in the French Channel ports. What was the probability value of (2) now? Doubtless much higher than in December. Maybe 25 per cent? 50 per cent? Again the exact value is there in the branched models of June and July, though unknown to us.

What makes the difference between December 1939 and July 1940? Why is the probability value of (2) so different at these different times? The answer lies in the branched structure, and in particular the contingencies of branch attrition. Let (2) be represented for the moment by $A \to C$. Then as of December 1939 there were regions of the tree in which $p(A \to C) = p(C \mid A)$ was low, and regions in which it was high. Had things worked out differently and the first branch point, instead of moving to an area where $p(C \mid A)$ was high, had moved to an area where $p(C \mid A)$ continued to be low—for example the German drive through the Ardennes might have bogged down—then by July 1940 there would have been no better chance of a successful invasion of England in August than there was in December. The difference between 'what might have been' and 'what was' comes from branch attrition within a structure where the numerical proportionality $p(C \mid A)$ is unevenly distributed.

It should now be clear why in the model, probability values vary with time. Moving on to the question of tense invariance, and of how counterfactuals constitute no more than a special case in the overall semantics for type **A** conditionals, suppose that the time has passed beyond 1940. Suppose that someone looking back in 1941, or for that matter in 2000, contemplates the past-tense counterpart of (2). This past-tense counterpart, as it turns out, is the counterfactual (1). But if Hitler *had* invaded England in August 1940, the past-tense version of (2) would have been not (1) but one of two *since*-conditionals:

(4) Since Hitler invaded England in August 1940, he will win the war.
(5) Since Hitler invaded England in August 1940, he won the war.

These conditionals would be appropriately uttered by someone who knew that the invasion had been carried out. But if the speaker is ignorant of whether or not the invasion took place, the appropriate conditional is a 'neutral past' conditional:

(6) If Hitler invaded England in August 1940, he won the war.

Although there are ways of evaluating (4), (5), and (6) on the branched model, we shall not pursue them here. Instead we shall focus on our main concern, the counterfactual (1).

If Hitler had invaded England in August 1940, he would have won the war. Like (2) and (3), this conditional is neither unqualifiedly true nor unqualifiedly false, but has instead a certain probability value. If he had invaded England in August 1940, the probability of Hitler's winning the war would have been p. But what is the value of p? Should p be the low probability value that (2) had in December 1939, or the higher value it had in July 1940? Just as the probability value of (2) was time-dependent, so the probability value of (1) is also time-dependent, and failure to specify the time results in its being impossible to assign any clear numerical value to it.

If historians wish to be precise, therefore, and to aim at estimating or knowing something which, in principle at least, has as exact a value as the melting point of sulphur, they must state a reference point when discussing counterfactuals. There is no single probability value for the counterfactual (1). As of December 1939 the probability of Hitler's winning the war if he had invaded England in August 1940 was p_1; as of May 1940 it was p_2; and as of July it was p_3. In the branched semantics we are proposing, explicit reference to the time, and hence to the precise branched model on which any given counterfactual is evaluated, is essential. To be sure, looking back from 1994, the natural reference time for the probability of (1) which interests us is 1 August 1940, and it would normally be assumed that the sought-for probability value was the value that (2) had as of that date. But without the time reference, no determinate probability value for counterfactuals exists.

Finally, as stated earlier, the fact that type **A** conditionals have probability values does not prevent many of them from having truth-values as well. There are two different ways in which this can come about. First, the probability value possessed by a conditional with contingent antecedent and consequent can be either one or zero. In the former case we say that the conditional in question is *probabilistically true*, or (in the circumstances, given the initial conditions that then existed) *causally necessary*, and in the latter case we say that it is *probabilistically false*, or *causally impossible*.[25] Secondly, a conditional with an antecedent which is contingent, but which through the luck of branch attrition happens to turn out to be true, will have a truth-value as well as a probability value. Furthermore it will have it even though its probability value may be less than one and greater than zero. If the consequent of such a conditional is true we say that the conditional is *superveniently true*, or (perhaps better) *contingently true*; if the consequent is false the conditional is *superveniently* or *contingently false*.

Thus some type **A** conditionals, besides those which are causally necessary or impossible, have truth-values in addition to probability values. But those with false antecedents have probability values only: they are neither true nor false and constitute, in branched semantics, exceptions to the principle of bivalence.

(vi) *Type* **B** *Conditionals*

In the preceding section, semantics based on the branched model were given for type **A** conditionals. These semantics explain the fact that such conditionals (a) are tense-invariant, and (b) pass the 'probability test', permitting the insertion of phrases like 'in

---

[25] As emerged from the discussion above on p. 160, we reserve the probability value zero for future events which occur on no branches at all. Events which occur on sets of branches which are non-empty but have classical measure zero are given infinitesimal probabilities.

all probability' without serious distortion in meaning. The tense invariance of type **A** conditionals, consisting in their falling into families differing only in tense or mood, is explained at the semantic level by the fact that each such family is associated with a specific branched model (a model of the universe at a given instant) and with a unique probability value in that model. The probability value of $A \to C$, given by the proportion of A & C-branches to A-branches, is a permanent feature of the model for that time. As such, it can be looked at from different temporal perspectives, and hence can serve as the ontological basis for the unique probability value of a tense invariant family of conditionals, comprising past-, present-, and future-tense members as well as counterfactuals and *since*-conditionals. At the semantic level, therefore, the tense invariant and the probabilistic character of type **A** conditionals go hand in hand.

Such is not the case with type **B** conditionals, which (with one or two exceptions) neither fall into tense invariant families nor pass the probability test. Some philosophers have held that conditionals like 'If Oswald didn't shoot Kennedy, then someone else did' do not express propositions, and do not possess truth-values. These conditionals, in their view, do not say anything about the world, but only about someone's *beliefs* about the world; they express a person's degree of willingness to accept a belief C, conditional upon acceptance of a belief A. The Oswald/Kennedy conditional, for example, would express a person's willingness to accept that someone else shot Kennedy, given that Oswald did not. This degree of conditional acceptance or conditional credence $p(C \mid A)$, where p is a subjective probability function, allows $A \to C$ to have different levels of acceptability or assertibility. $A \to C$ can be highly acceptable, or highly assertible, or highly justifiable, but it cannot be *true*.[26]

However, although according to this view type **B** conditionals may not express propositions, and may not have truth-values, they are valuable commodities for all that. As Bennett says:

To accept $(A \to C)$ is to become disposed, upon coming to believe A, to come also to believe C. Such a disposition can have a vital role in the intellectual management of states of partial information; it can be communicated from one person to another; it can be acquired or lost in the light of evidence; and all of that is consistent with its not having a truth value.[27]

---

[26] See Ernest Adams (1965; 1975); and Gibbard (1981). In his 1965 paper, Adams says that 'the term "true" has no clear ordinary sense as applied to conditionals' (p. 169).

[27] Bennett (1988: 517). This approach, stemming ultimately from Ramsey, cannot be extended to counterfactuals, for the reasons given above on p. 170.

It is tempting to regard type **B** conditionals in this way, as expressing degrees of conditional acceptability, but to do so requires that we be convinced that a truth-value semantics for them is impossible, and I do not think that this is so. In fact I shall argue that type **B** conditionals do not possess degrees of probability, or degrees of truth, whether subjective or objective, but are straightforwardly either true or false. Analysis in terms of high or low degrees of acceptability or assertibility is therefore inappropriate for them. In section (v) probability semantics were provided for type **A** conditionals, and in this section I shall attempt to give not probability conditions but truth-conditions for type **B** conditionals. These truth-conditions are based not on proportionality of branches, which yields probability values, but on a different feature of the model. This is branch attrition, which out of many pairwise incompatible propositions about the future which *could* be true, leaves one and only one that *is* true, and which serves as the basis for the truth or falsehood of a type **B** conditional. I shall illustrate how this comes about by using examples.

(1) If Alain is nervous, he doesn't show it.

Suppose that Alain is giving an important presentation before a committee, and betrays no sign of nervousness. No doubt there are, or were, other possible worlds or branches in which things went differently and he did show that he was nervous, but as things have turned out he is cool as a cucumber. From this contingent fact, that Alain shows no nervousness, which is true, we infer the type **B** conditional (1), which is also true. It gets its truth-value because it follows from a contingent fact, a fact about branch attrition. Conditional (1) is to be contrasted with the type **A** conditional 'If Alain had been nervous, he wouldn't have shown it.' The latter would typically be based, not on the mere fact of Alain's showing no nervousness, but on some earlier event, such as his having taken a public speaking course or having learned how to control his nerves. This in turn would provide a causal link between antecedent and consequent and give the type **A** counterfactual a probability value, deriving from branch proportionality.

(2) If the butler didn't do it, the gardener did.

It is appropriate for Hercule Poirot to utter (2) when, and only when, the number of suspects has been reduced to two. But if (2) is true, it is true much earlier than this. The truth of (2) is based on the contingent fact that either the butler did it or the gardener did it, which is unlike the fact of Alain's showing no nervousness in being a disjunctive fact, not a fully specific one. But it is a fact none the less, whether Poirot or anyone else is aware of it. And, once the murder has been committed, the corresponding disjunctive proposition is true, not probable. This in turn makes (2) true, not probable: (2) may have the appearance of being an epistemic conditional, but its truth-conditions are as non-epistemic and objective as those of (1).

(3) If Shakespeare didn't write *Hamlet*, someone else did.

The truth of (3) follows from another contingent fact, namely that somebody wrote *Hamlet*. This again is a non-specific fact, but no matter. Perhaps if things had worked out differently in 1599 and Shakespeare had embarked on other projects then nobody would have written *Hamlet*, but as it is (3) follows from a contingent truth.

(4*a*) If Kai isn't out in the boat, someone else is.
(4*b*) If Kai wasn't out in the boat, someone else was.

I look out on the lake as it is getting dark, and utter (4*a*). Later I meet Kai on the path, and exclaim (4*b*). Conditionals (4*a*) and (4*b*) constitute a tense invariant type **B** family, and what they say is either true or false. They are of course very different from the type **A** family 'If Kai isn't out in the boat, someone else will be', 'If Kai hadn't been out in the boat, someone else would have been', etc.

(5) If you thought Tegucigalpa was in Nicaragua, you were mistaken.

It is a fact that Tegucigalpa is in Honduras, and the truth of (5) follows from it.

(6) There are biscuits on the sideboard if you want them.

The truth of (6) follows from there being biscuits on the sideboard, whether you want them or not.

(7) If that man's an electrician, I'll eat my hat.

Fact: that man is not an electrician. Conditional (7) is very different from the type **A** conditional 'If that man's an electrician, he'll fix the fusebox.'

(8) If the dowager had been angry at the abrupt leave of absence he took, she was mightily pleased at his speedy return.[28]
(9) If you ate the whole thing, you were very quiet about it.
(10) I'll just have a look around, if you don't mind.

---

[28] Thackeray, quoted in Dudman (1984: 147).

This normally means that I intend to have a look around, with or without your permission.

(11) If Lachute is bigger than São Paulo, then Lachute is bigger than New York.

Based on the fact that, in about 1985, São Paulo's population overtook New York's. That both antecedent and consequent are false makes no difference to the truth of (11).

The last example, which comes from Jonathan Bennett, is exceptional in that it is a counterfactual.[29] Suppose a fair coin is to be flipped at t_2, and that at t_1 you bet on heads, which is what comes up. Consider the counterfactual

(12) If you had bet on tails at t_1, you would have lost.

Although (12) differs grammatically from (1)–(11), it is still a type **B** conditional, entailed by the contingent proposition that at t_2 the coin fell heads. From this contingent fact follows a host of other type **B** conditionals all of which are true: 'If you had called heads, you would have been right', 'If you had been the captain of a cricket team and called heads, you would have been able to bat first', etc. All these counterfactuals are type **B**. Their truth requires that the antecedent of the conditional and the contingent event on which their truth depends be causally independent. If they are not independent, as in

(13) If you had called tails and substituted your special biased coin for the fair coin, you would not have lost.

the result is a type **A** conditional, not a type **B**. In this case, the substitution of a biased coin for the fair coin causally influences the falling of the coin, while simply calling 'heads', or 'tails', does not. Unlike (12), the truth of (13) does not follow from the coin's falling heads: (13) is a probability conditional, and its probability value is determined by a permanent pattern in the branched model, not by branch attrition.

I hope, with these examples, that a natural line of division has emerged between type **A** and type **B** conditionals. I think the difference is fundamental and deep, not superficial. The linguistic/grammatical criteria I suggest, i.e. the tense invariance and probability tests, give a rough indication of where the line lies, but ultimately the division into two categories rests on a semantic

---

[29] See Slote (1978: 27), Bennett (1984: 76). Slote says that he knows of no theory of counterfactuals that can adequately explain why the utterance of statements like (12) seems natural and correct. Another type **B** counterfactual is 'If you had thought that Tegucigalpa was in Nicaragua, you would have been mistaken.'

basis, explicated in terms of the branched model. Without the branched model, and the fundamental difference between (i) a fan of branches, yielding a probability value, and (ii) a single branch isolated by branch attrition, yielding a truth-value, the difference between the two types of conditional could not be made precise.

In conclusion it may be asked whether the division into type **A** and type **B** conditionals is exhaustive, or whether there is a third type. My view is that the boundary between type **A** and type **B** conditionals is the principal line of division among conditionals, and the two classes it divides are the two most important classes. Analogously, in biology the main line of classification lies between plants and animals, and the plant and animal kingdoms are the two most important divisions. But just as in biology there are other types of organism separate from the first two—e.g. the bacteria and viruses—so in the field of conditionals there may well be a third type, the semantics of which cannot be based upon the branched model. The third variety is the *epistemic conditionals*, which have no objective truth or probability values. Here the degree of acceptability or assertibility is based upon the Ramsey test, i.e. upon what the speaker knows or believes to be the case. Examples of type **C** conditionals are not difficult to find, once the field of search is defined by the limits of the speaker's knowledge:

(14) If Mallory and Leigh actually succeeded in reaching the summit of Everest, they may well have planted a small flag before perishing on the spot.

(15) If the gardener did it, he was probably returning from the scene of the crime when we met him.

(16) If there is a proof of Fermat's last theorem, it must be short.

In the case of (16), we are justified in believing the conditional to the extent that we put faith in Fermat's mathematical acumen. Conditional (16) is not based on fact but on evidence, such as the width of the margin in Fermat's book, and hence differs from both type **A** and type **B** conditionals in not having either a probability value or a truth-value.

Beyond indicating that they probably exist, and that if they do their degree of acceptability is determined by the Ramsey test, I don't want to say any more about 'epistemic' or type **C** conditionals. Our interest in this book lies with conditionals whose truth or probability values derive from the branched model, and which possess these values independently of whether anyone knows them. They are the conditionals belonging to types **A** and **B**.

(vii) *Summary on Conditionals*

This chapter on conditionals has dealt with a number of topics, and an attempt will be made to summarize its conclusions.

In the first section, three ways of analysing counterfactuals were discussed: the 'inferential' approach of Goodman, Ramsey's proposal based on degrees of belief, and the possible world semantics of Stalnaker and Lewis. Lewis's semantics are based upon the notion of comparative similarity or comparative closeness of possible worlds, and different ways of resolving the vagueness inherent in this notion result in different truth-values being assigned to one and the same counterfactual. An earlier proposal by the author was discussed in section (iii), namely to eliminate the lack of precision in the idea of 'degree of similarity between pairs of worlds' by replacing it, on the branched model, by 'amount of shared past between pairs of branches'. This proposal was seen to be unsatisfactory in that its adoption would result in the assignment of the value 'true' to a whole new class of counterfactuals, the 'last-minute deviation' counterfactuals, which intuitively should be rejected.

In Section (iv) a new attempt is made to provide semantics for counterfactuals and other conditionals based on the branched model. To begin with, two different types of conditional are distinguished. At a superficial level, the two categories are distinguished by linguistic or grammatical criteria, but the real difference between them lies in their semantics. Type **A** conditionals, exemplified by 'If Hitler had invaded England in 1940, he would have won the war', are characterized by *probability-semantics* rather than *truth-semantics*. That is to say, the semantics based on the branched model provides them with probability values rather than truth-values. Type **B** conditionals, on the other hand, have truth-values, not probability values.

In Section (v), type **A** conditionals are discussed. Counterfactuals like the one about Hitler's never-realized invasion belong to what are called *tense-invariant families* of conditionals, other members of the invasion family being 'If Hitler invades England in 1940 he will win the war', 'If Hitler were to invade England in 1940 he would win the war', etc. All members of the same tense-invariant family have the same probability value, which is understandable if the value in question, based on branch proportionality, is an objective feature of the branched model at any given time.

It is, however, also a feature of the branched model that the probability value of the invasion conditional varied considerably over time. In December 1939 it was low, but the contingencies of branch attrition brought it to a high after the fall of Dunkirk, where it remained until the autumn gales ruled out Channel crossings. As a consequence, even today the invasion counterfactual cannot be given a probability value unless a time of reference such as December 1939 or July 1940, is specified.

In addition to probability values, type **A** conditionals may also have truth-values. If their probability value takes either of the two limiting values one and zero they will be 'probabilistically' true or false, but short of that they will be 'superveniently' true or false if their antecedents happen to become true via branch attrition. In that case they are true or false (and have been so since the beginning of time) according to the truth-value of their consequent.

Section (vi) deals with type **B** conditionals, exemplified by 'If Shakespeare didn't write *Hamlet*, someone else did.' At the semantic level, type **B** conditionals differ from type **A** conditionals in having only truth-values, not probability values. The feature of the branched model on which their truth-conditions are based is not a permanent one, like the fan of branches on which probabilities rest, but one which is based on the progressive loss of branches. Branch attrition is responsible for contingency. It is responsible for the way in which contingent facts come to be, and, once established, remain. Each true type **B** conditional has as its basis a contingent fact, which implies it. This contingent fact, once it *is* a fact, is true, not probable, and hence the type **B** conditional which it implies is true, not probable. For this reason the subjective 'degree of assertibility' approach, which associates with each conditional a degree of conditional subjective probability that in turn determines the conditional's acceptability or assertibility, is inappropriate for type **B** conditionals. Type **B** conditionals are straightforwardly either true or false, not assertible to a high or low degree.

The dividing-line between type **A** and type **B** conditionals is basic and fundamental. It was first pointed out in 1970 by Ernest Adams, and has since been recognized in one form or another by almost every writer on the topic. These writers may differ over the membership of the two classes of conditionals, but they generally agree that there *is* a basic difference, which they all characterize in their own ways. With the help of the branched model, a basic semantic distinction can be drawn. This distinction, I believe, explains and accounts for the obvious and yet puzzling differences between the two types of conditional.

7

Individuals and Identity

OUR subject of investigation, the branched model, makes at best a modest contribution to the philosophical problem of individuals and their identity. Nevertheless, the model lends itself to the elucidation of three recently debated theses concerning individuals—the theses of *actualism*, of *transworld identity*, and of *essential properties*. Actualism is the view that there is only one possible world, the actual one, and that there exist no individuals other than actual individuals. The branched model is actualist inasmuch as its present and past are unique, but it also supports possibilism in that it admits a multiplicity of possible futures. As will be seen, four quite different ontologies of individuals are consistent with it. Secondly, the model provides a natural and compelling account of the transworld identity of individuals, based upon a thesis to the effect that the three-dimensional and the four-dimensional representations of an individual are equivalent. Finally, the model lends support to the doctrine that certain properties of an individual are essential to it, and gives an indication of what these properties are. The two theses of actualism and full-blooded transworld identity might seem on the face of it to be strange bedfellows, but in the branched model they coexist peacefully and harmoniously.

These questions will be addressed in this chapter and the next. But before we begin, a preliminary question deserves mention. This is the problem of whether there *are* any such things as individuals. Does the universe need them? Could the category of individual be abolished? Here are the answers of three philosophers to these questions.

Aristotle held that the concept of an individual substance was central to our understanding of the world. In the *Categories*, a substance is defined as that of which other things are predicated, but which neither is itself predicated *of* a subject (as 'swiftness' is predicated of an individual horse) nor is *in* a subject (as 'the swiftness of that horse' is in that horse—and in no other horse).[1]

---

[1] *Categories* 1ª20–2ª14. 'The swiftness of that horse' is what has been called an *abstract individual*: such individuals are discussed in McCall (1965).

The most distinctive feature of substances is that one and the same substance can receive contradictory properties. For example an individual man becomes pale at one time and dark at another, or hot and cold, or good and bad. 'Nothing like this', Aristotle remarks, 'is to be seen in any other case' (4ᵃ18–21). During the twenty-three centuries that separate us from Aristotle, philosophers have not ceased trying to understand how this can be; how there can be something (the individual substance) which is the same, and yet different at different times.

A good short and decisive way of dealing with this problem is that of Hume, who held that substances were fictions. Hume did not maintain that the world contained no individuals, but his individuals were impressions and ideas, not continuants like this table and this pen. Although we have the belief that physical objects, animals, and persons exist continuously, in reality they are nothing but collections of momentary impressions, which the mind unites and binds into one:

I may venture to affirm of the rest of mankind, that they are nothing but a bundle or collection of different perceptions, which succeed each other with an inconceivable rapidity, and are in a perpetual flux and movement. ... That action of the imagination, by which we consider the uninterrupted and invariable object, and that by which we reflect on the succession of related objects, are about the same to the feeling, nor is there much more effort of thought required in the latter case than in the former. The relation facilitates the transition of the mind from one object to the other, and renders its passage as smooth as if it contemplated one continued object. This resemblance is the cause of the confusion and mistake, and makes us substitute the notion of identity, instead of that of related objects.[2]

For Hume, the true individuals of this world are things of exceedingly short duration, not long-lasting things like human beings. Russell, in *An Inquiry into Meaning and Truth* goes further than Hume and tries to show that the category of individuals can be abolished entirely. This may seem a vain attempt, since the most general form of assertion might be thought to lie in the ascription of an attribute to a particular or set of particulars, but to Russell must go the credit of almost demonstrating that the notion of an individual can be consistently denied, and hence showing that a world without individuals of any kind is possible. I say 'almost' because I don't think that Russell's attempt is ultimately successful, but let us follow him as far as we can.

[2] *Treatise*, I. iv. 6, Selby-Bigge (1888: 252–4).

Russell proposes to eliminate what he calls 'particulars', and to be content with words for universals like 'red', 'blue', 'hard', 'soft', etc. The most general type of proposition would no longer be of subject–predicate form, e.g. 'Individual x has property P', but of the form 'Universal U is instantiated at place p and time t':

I wish to suggest that 'this is red' is not a subject–predicate proposition, but is of the form 'redness is here'; that 'red' is a name, not a predicate; and that what would commonly be called a 'thing' is nothing but a bundle of coexisting qualities such as redness, hardness, etc.[3]

This does away with particulars only if, as Russell acknowledges, places and times can also be analysed away. If not, then place-times are left as particulars, characterized by the qualities 'red', 'hard', etc. Russell makes a valiant effort to reduce places and times to qualities, appealing to quality-based spatial and temporal coordinate systems. But every coordinate system needs an origin, and he gives no hint of how to construe the origin as a quality. So in the end his proposal to abolish individuals must be counted a failure.

The answer to our question, then, of whether individuals might fail to exist, i.e. whether the category of 'individual' could be abolished, is that its abolition seems inconceivable. The philosophical conception of an individual is complex, interesting, and indispensable. We shall see what light, if any, the branched model can throw on it, beginning with the thesis of actualism, and the difference between actual and possible individuals.

(i) *Actualism and Modal Realism*

There are two very different schools of thought about possible worlds, one of them being the actualist, the other the modal realist point of view. The most gifted and articulate defender of modal realism is David Lewis, who states his position in a paragraph that has been quoted so often that an apology is needed for quoting it again:

I believe that there are possible worlds other than the one we happen to inhabit. If an argument is wanted, it is this. It is uncontroversially true that things might have been otherwise than they are. I believe, and so do you, that things could have been different in countless ways. But what does this mean? Ordinary language permits the paraphrase: there are many ways things could have been besides the way they actually

---

[3] Russell (1940: 97). Cf. also Russell (1948: pt. IV ch. 8).

are. On the face of it, this sentence is an existential quantification. It says that there exist many entities of a certain description, to wit 'ways things could have been'. I believe that things could have been different in countless ways; I believe permissible paraphrases of what I believe; taking the paraphrase at its face value, I therefore believe in the existence of entities that might be called 'ways things could have been'. I prefer to call them 'possible worlds'.[4]

For Lewis, a possible world is a complete four-dimensional universe, each one differing slightly (or substantially) from the others in the events and objects it contains. Every world has its own internal spatio-temporal relations, and none has any spatio-temporal relations with any of the others. Each is isolated.[5] Our universe is one of these, and we call it 'the actual world', but many other such worlds are no doubt inhabited by intelligent beings who call *their* world 'actual'. Who is to say they are wrong? The word 'actual', for Lewis, is an indexical like 'I' or 'here'. So every possible world which contains intelligent beings is rightly called 'actual' by its inhabitants. Lewis has been described as an 'extreme' modal realist, but even those who attack him, or react to his theory with blank stares of incredulity, are forced to admit the clarity and internal coherence of his position.

Opposed to modal realism are the actualists, who maintain that only one world exists. Other possible worlds exist in an abstract sense; not as concrete spatio-temporal universes but in the way mathematical objects exist. They may be regarded as maximal states of affairs (Plantinga), or as maximal consistent sets of propositions (R. M. Adams). A possible world, from an actualist's perspective, is like a *description* of something that does not or may not exist. For example, 'A large hairy ape-like creature living at high altitudes in the Himalayas' is the description of something that does not or may not exist. The actual world has a similar description, to which it answers, but according to the actualist nothing answers to the description of a non-actual possible world. What we are left with is the description itself, which is what the actualist defines a non-actual possible world to be. Let us look more closely at this theory.

Alvin Plantinga reports that he worked out his conception of possible worlds during the years 1968–9.[6] For him, a possible world is constituted by *states of affairs*, e.g. 'Socrates being shorter than Plato', or 'Julius Caesar's living to the age of 75'. Some states of affairs obtain or are *actual*, others are not. A state of affairs is

[4] Lewis (1973: 84).   [5] Lewis (1986b: 2).
[6] Plantinga (1985: 88). His theory appears in Plantinga (1974).

possible if and only if it could have been actual. If *S* and *S** are states of affairs, *S includes S** if it is not possible that *S* be actual and *S** not be actual; *S precludes S** if it is not possible that both *S* and *S** be actual. *S* is *maximal* if for every *S**, *S* includes *S** or *S* precludes *S**. A maximal possible state of affairs is what Plantinga calls a *possible world*.

Possible worlds, for Plantinga, are abstract objects, like properties, propositions, and sets. Nevertheless there really are such things; talk about them is not just a *façon de parler*.[7] This opinion is shared by R. M. Adams, who maintains that possible worlds are 'logically constructed out of the furniture of the actual world'.[8] Adams's possible worlds are defined in terms of what he calls *world-stories*, i.e. maximal consistent sets of propositions, and what distinguishes the actual world from other possible worlds is that all the members of its world-story are true. When Adams says that possible worlds are logically constructed out of the furniture of the actual world, he presumably means that in the actual world there exist language-using intelligent beings with enough logical expertise to be able to construct maximal consistent sets of propositions. If the actual world lacked such beings, non-actual possible worlds would presumably fail to exist, even as abstract objects.[9]

The theory about possible worlds that seems to emerge most naturally from the branched model is identical neither with actualism nor with modal realism, though it has affinities with both. The branched model pictures the universe as one, but branched. Relative to the present, each branch may be regarded as a future possible world. There exist therefore, in the branched model, possible worlds, as the model realist claims, and these possible worlds are not abstract, but concrete. They exist in space and time, and do not depend for their existence on being constructed, or imagined,

---

[7] Confusingly, Plantinga underlines his commitment to the existence of possible worlds by describing himself as a modal realist (Plantinga 1985: 88). But it is best to reserve that description for someone who holds that possible worlds exist concretely in space and time.

[8] R. M. Adams (1974: 224).

[9] This last point might seem debatable. In a world which lacked intelligent beings, there would still exist five and only five perfect solids, and a single prime between 31 and 41, so why not the full panoply of possible worlds? Well, suppose that (i) the actual world *w* contained the sun and the moon, but lacked intelligent beings, and that in it (ii) the distance *d* between the centres of gravity of the sun and the moon at time *t* happened to be 93 million miles. Would there exist, in *w*, an abstract object consisting of a state of affairs in which the distance *d* at time *t* was 93.1 million miles? Without anyone to imagine or construct such a 'contingent' abstract object? Worse still, would there exist in *w* an abstract object consisting of Socrates' being shorter than Plato when neither Socrates nor Plato existed in *w*? This is doubtful. Hence the existence of intelligent beings to construct such objects seems to be required.

or thought of. In the branched model, they are as real as the present or the past.

The possible worlds of the branched model differ, however, from David Lewis's worlds in that they lie entirely in the future, and are not spatio-temporally isolated from the actual world. We can travel to them, or rather to one and only one of them at each branch point, by simply waiting. The transition to whichever one of them becomes actual is smooth and uniform, with no hiatuses or discontinuities. They are not, in Lewis's words, 'ways things could have been', but rather 'ways things could be'. They represent not the might-have-beens, but the may-bes. The events in them have not yet taken place, and most of them will never take place. None the less these possible future worlds, or possible future extensions of the actual world, are as *real* and as *concrete* as the fall of Constantinople, or the evolution of the galaxies. They are real, but they are not *actual*.[10]

The actual world, in the branched model, is restricted to the trunk—the past and the present. The branched picture is an *actualist* one in that it contains no non-actual possible worlds contemporary with the present. Actuality is unique: there are no other worlds in which, at this very moment, a Storrs McCall or a David Lewis or a counterpart of same is using the word 'actual' as an indexical and asserting his world to be the actual one. There is only *the* actual world and its possible continuations, each one of which has a chance of becoming actual. Of not one of these is it true to say that it *might have been* actual, but in fact is not.

It is this feature of the branched model which puts it in the actualist camp, or at least puts one of its feet in the actualist camp. For the actualist denies that non-actual possible worlds

[10] Lewis (1986b: 206–9) says that where branching worlds share a common initial segment, he would prefer to describe this as branching *within* worlds, rather than as branching *of* worlds. Although I shall continue to speak of 'branching future worlds', I have no objection to Lewis's proposal, and readers if they wish may silently translate talk of *worlds* in the branched model into talk of *branches*. The problem of 'transworld identity' would then become the problem of 'transbranch identity'. However, Lewis then goes on to say that he rejects branching in favour of what he calls 'divergence' of worlds, this term applying to worlds that have no common initial segment. Lewis's reason for rejecting branching is that for the inhabitants of a branched world there is no unique future, but instead a multitude of different possible futures. Therefore, he says, it makes no sense to wonder what *the* future will bring. In the branched model, however, wondering what the future will bring does make sense, and consists in wondering which future branches are going to drop off, and which is the single branch that will remain. To wonder, for example, what will happen tomorrow is to wonder, in the branched model, which branch point will become the first branch point of the universe tomorrow.

Lewis's reason for rejecting branching, viz. that it fails to make sense of genuine and perfectly justifiable wonderment about the future, does not apply to the branched model. On this matter see also Belnap (1993: sect. 4).

exist concretely. He denies that, right now, there exists a real, concrete, three-dimensional world almost precisely like the actual world except that in it the book that you, the reader, are now reading has 300 pages instead of 328. The branched model agrees. There *were*, no doubt, branches on the model years ago which had three-dimensional cross-sections of this type; for example branches in which the printer managed to condense the text to save paper, branches in which I decided to shorten some lengthy passages, etc. But these branches have by now dropped off. They no longer form part of the model.

The branched model conforms therefore to the actualist's belief that only the actual world exists concretely, and to his denial of the concrete existence of non-actual counterfactual worlds. Such counterfactual worlds used to exist as part of the branched model, but they do so no longer. Their past existence, of course, may be appealed to for purposes of modal and counterfactual semantics, in the way that snapshots of Sophie aged 6 may be appealed to and compared with Sophie now.[11] But there are no flesh-and-blood counterfactual Sophies in the branched model, only the actual Sophie. In this respect, therefore, the branched model supports actualism at the expense of modal realism.

When we come to the future, however, the situation is different. Here the branched model supports modal realism. Future possible worlds are not abstract objects on the branched model, but are concrete, spatio-temporal continuations of the actual world. They are as real as any other part of the model, and the relationship between them and what is actual is dynamic, not static. What enables the model to have a foot, therefore, in both the actualist and the modal realist camps is the element of *time*, which has hitherto been a neglected factor in discussions concerning actualism and realism. From the perspective of the branched model, the actualist is right concerning the *past* and *present*, while the modal realist is right concerning the *future*. There is only one actual world, and there are no concrete non-actual worlds contemporary with it. There are, however, many possible future worlds, all of which exist concretely, and all of which are spatio-temporally (and causally) continuous with the actual world. The theory of possible worlds which accords most naturally with the branched model, therefore, combines elements from both the actualist and the modal realist schools of thought. It constitutes what Robert Adams would describe as a 'simple property' theory of actuality, representing actuality as a property which the actual world possesses

[11] See above, pp. 9, 11, and 31.

absolutely, rather that in its relation either to God or to the intelligent beings which inhabit it and call it 'actual'.[12] On the branched model, what is actual has the ontological property of uniqueness, and what is possible has the ontological property of multiplicity. Both are real, and between them they provide the basis for a novel theory of individuals.

(ii) *Individuals in the Branched Model*

Leaving aside for the moment possible individuals and future individuals, there are two kinds of individual in the branched model: (i) presently existing individuals (e.g. Wayne Gretsky, Venice) and (ii) individuals that used to exist but do so no longer (Aristotle, the Holy Roman Empire). Apart from their temporal location, there is an important structural difference between these two varieties. The histories of Aristotle and the Holy Roman Empire are closed. Nothing now can alter a jot or tittle of what they did or what was done to them. But the histories of Wayne Gretsky and Venice are still being shaped. Each has many different options open to it, and if we could inspect the branched model we would see the various possibilities reflected in the different world-lines or spatio-temporal volumes associated with these two individuals. On one future branch his world-line shows Gretsky scoring 200 goals in one season—on another he gets tired of hockey and becomes a movie star. On some branches, Venice disappears beneath the Adriatic. Just as the universe model is branched, so presently existing individuals are also branched, with Gretsky as a 200 goal scorer and Gretsky as a movie idol both being spatio-temporally continuous with the world-line of Gretsky in the actual world. By contrast, the world-lines of Aristotle and the Holy Roman Empire are unbranched and static, lying entirely in the trunk of the universe model.

What now of future individuals? Here the situation is less clear. There are, in fact, three possible theories that may be adopted concerning future individuals, namely (1) there are none, (2) there are exactly those future individuals that will come into existence in years to come, and (3) there are all the future individuals that *may* come into existence in years to come, i.e. all possible future individuals. To these three we shall add a fourth, which concerns not just future individuals but all possible ones. This is the theory that (4) there exist all possible individuals, i.e. all those that either

[12] R. M. Adams (1974: 221).

are, were, may be, or might have been, back to the time of the universe's origin. Each of these four theories gives rise to a different conception of what constitutes an individual: (1) and (2) yield different extensions of the class of *actual* individuals, and (3) and (4) yield different extensions of the class of *possible* individuals. As will be seen, (1) and (3) are dynamic conceptions, producing extensions which change with time, while (2) and (4) are static. All four are philosophically respectable theories of individuals, which the branched model helps to define clearly.

The first view, that there are no future individuals, has much to recommend it. It does not hold that there *will be* no future individuals, but only that there *are* none. Membership in the class of individuals, on this view, is confined to past and presently existing individuals, and the size of the class increases monotonically with time. As new individuals come into existence they join the class, membership being a one-way affair, with no old members ever resigning. The class of individuals, on this theory, is restricted to the class of individuals that have by now become actual.

This is an attractive theory, but for it to be made plausible on the branched model it must be reconciled with the model's realism concerning future branches. If future branches exist concretely in space and time, as they do, then why don't the individuals they contain also exist? If we could examine the trunk of the branched model with a magic retro-telescope, which enabled us to see into but not affect the past, then, in principle at least, the whole life of a past individual like Julius Caesar would be open to inspection. We could view Caesar at each stage of his life: as a boy; conquering Gaul; dying at the hands of the conspiracy. That is, we could inspect Caesar's world-line, the four-dimensional volume that represents the changing three-dimensional person. Now if we could do this in the case of Caesar, why could we not do the same for any future individual? Presumably there will be babies born on New Year's Day in the year 2000, or for that matter on New Year's Day in the year 2500 (or so we hope). Why could our magic retro-telescope not be switched into a forward-telescope, and directed away from the trunk and towards the branches of the model?[13] Would it not be able to pick out a multitude of new individuals, none of which now exist, and follow them through their branching histories? Would the fact that this is at least in principle possible not entail that, as far as the branched model was concerned, the existence of future individuals was on a par with past and present ones?

[13] Directed towards branches or possible futures, the telescope resembles David Kaplan's Jules Verne-o-scope. See Kaplan (1979: 93).

It's best to go slowly here. What is implied by the ideas of the previous paragraph is theory (3), the view that the class of individuals includes future possible ones in addition to those past and present. Future possible individuals, on this view, are located on the branches of the model, where they exist concretely in space and time. According to theory (1) the class of individuals increases monotonically with time, but according to theory (3) it decreases monotonically, as branch attrition reduces the number of future possible individuals. The class of individuals, on this view, is a progressively shrinking class. No new members are permitted to join, and at every moment large numbers are forced out of existence. Theories (1) and (3) are dynamic theories of individuals, the first regarding individualhood as analogous to an honour bestowed and never lost, while the second likens it to a property with which many are endowed but which few retain.

There are probably many philosophers who have doubts about both these two dynamic theories, or at least about the way they have been presented. If individualhood is something acquired, who or what acquires it? If lost, then surely what once had it and now lacks it is a former individual, and a former individual, like Aristotle, is still an individual. Many, in all likelihood, would prefer a theory like (2) or (4), in which the class of individuals remains constant throughout time. But before we embrace a theory such as (4), which asserts the existence of an enormous collection of actual, possible, and formerly possible individuals, or (2), which asserts the existence of just those future possibles who will one day become actual, some hard questions concerning the identity of future individuals on the branched model must be faced.

Quine, many years ago, asked the kind of question that must be taken seriously if any respectable theory of possible individuals is to be constructed. Suppose someone, in the style of Wyman, were to invite us to consider the existence of a possible fat man in that doorway.[14] And a possible bald man, in the same doorway. Are they the same possible man, or two possible men? What criteria of identity are appropriate for possible entities? In the branched model, an attempt to answer this question might run as follows.

Possible individuals are branched, and one and the same individual may appear on different branches. Pending our discussion (still to come) of individuals as four-dimensional objects or world-lines, let us identify an individual-as-appearing-on-a-branch with its world-line. Two world-lines on different branches are *connected*

[14] Quine (1953: 4).

if they form part of a single branched world-line. A *possible individual* may then be identified with the topological union of a maximal set of pairwise connected world-lines, each one on a different branch. The topological intersection of all the members of each such maximal set is the individual's root or *origin*, which will be a particular place and time on a particular branch. It follows that two possible individuals are identical, on the branched model, if and only if they have the same origin.[15]

To return to the fat man and the bald man, it would appear at first sight that the branched model could provide an answer to Quine's question. Let x be the possible fat man, and y the possible bald man. Then x and y are the same individual if and only if they have the same origin in the branched model. There are, however, difficulties with this answer, as illustrated by the problem of *Mrs Smith's daughters*.

Mr and Mrs Smith had three sons, and very much wanted a daughter. 'If only we had had a daughter after Jack, say in 1985 or thereabouts,' laments Mrs Smith. 'I would have called her Mary, and she would have been a blessing to us.' Now is there, or was there, a possible individual who for one reason or another never became actual, and who is Mrs Smith's daughter Mary? As far as the branched model is concerned, one might think, why not? Surely there was a branch on the model before 1985 on which Mrs Smith had a daughter. As of 1995, there may *still* be a branch on which she has a daughter, though doubtless by 2010 it will be too late. This is perfectly all right as long as we think in general terms—Mrs Smith on some past or future branch coming home from the hospital with a baby girl, Mr Smith proudly pushing a pram, etc. But the question was not, *could Mrs Smith have had a daughter*, but *is there a specific individual—a daughter—whom Mrs Smith could have had*? When we move from the general to the specific, the answer becomes less clear. Remember Quine's principle—no entity without identity. Is Mrs. Smith's possible but non-actual daughter the one with long curls? Or is she the one with short hair who lost a tooth on her eighth birthday? Are these two the same little girl, or different ones? Failing an answer to questions like these, there will be no possible individual at all who is Mrs Smith's daughter. And in the branched model, there will be infinitely many candidates about whom such questions can be asked.

The problem of proceeding from the general to the specific in discussing possible individuals of the future, or individuals who

<hr/>

[15] The necessity of origin will be discussed in the next chapter.

might have existed but didn't, is the problem raised by C. S. Peirce:

If you and I talk of the great tragedians who have acted in New York within the last ten years, a definite list can be drawn up of them, and each of them has his or her proper name. But suppose we open the question of how far the general influences of the theatrical world at present favour the development of female stars rather than of male stars. In order to discuss that, we have to go beyond our *completed* experience . . . and have to consider the possible or probable stars of the immediate future. We can no longer assign proper names to each. The individual actors to which our discourse now relates become largely merged into general varieties; and their separate identities are partially lost.[16]

Peirce takes the view that attempts to specify, in discourse, a *unique* future individual are doomed to failure. We can talk only about general kinds or classes. This is certainly true, on the branched model, of Mrs Smith's possible daughters. Suppose, for purposes of argument, we managed to describe one of these daughters, X, in the most precise way imaginable, complete with exact date and place of birth, name, atom-by-atom physical development and growth, spatio-temporal location over many years, relations with other individuals, etc. Would we have succeeded in identifying a unique possible individual? Not necessarily. Imagine that just before X's conception, an electron in a hydrogen atom on Betelgeuse happened to fall to a lower orbit, emitting a photon of energy. Then there would be many Xs; Xs conceived on branches where the electron fell, and Xs conceived on branches where it did not. Because of the dense branching in the model, there would in general be non-denumerably many Xs of each kind, each with her own origin on a separate branch.

Very well, someone may say, then there will be indenumerably many possible daughters of Mrs Smith's, one for each possible origin. Despite their indistinguishable looks, indistinguishable physical constitutions, and indistinguishable movements they will not be identical, because they have their origin on branches that are not indistinguishable, but are differentiated by micro-events on Betelgeuse and elsewhere. We shall have as many of Mrs Smith's daughters as there are distinct points of origin for such individuals on the branched model. Alas, even this hope of giving a separate identity to each of Mrs Smith's possible daughters may fail. Nothing on the branched model prohibits there being two or more completely indistinguishable branches, i.e. branches which are

---

[16] Peirce (1931–5: iv. 4.172); quoted in Prior (1957: 113–14).

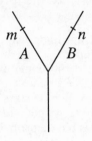

FIG. 7.1

qualitatively identical in every respect (see Fig. 7.1). If branch A and branch B are indistinguishable throughout their entire length, and if one of Mrs Smith's daughters is conceived at point m on branch A and another at the corresponding point on branch B, then by Leibniz's law of the identity of indiscernibles the daughters are identical. But by the principle of having different origins they are not identical. Which principle are we prepared to let go?

At first sight it might seem as if we had a choice in this matter: either abandon Leibniz's law and recognize a multiplicity of indiscernible individuals in the branched model, each with a different origin, or else retain Leibniz's law and use it to reduce the enormous multiplicity to manageable size. For many philosophers, including myself, to contemplate abandoning Leibniz's law would involve a step of fairly heroic proportions. Therefore initially I favoured the second alternative. But subsequent reflection has convinced me that we don't really have a choice after all: that the branched model either satisfies Leibniz's law or it does not.

Quite apart from the question of what constitutes an individual, the branched model may or may not contain two or more branches that are qualitatively identical but numerically distinct. By this is meant two branches that are qualitatively identical throughout their entire history, from the beginning of the universe to the end. The reason why there might be two or more such branches is that they might be required in order for the values of future far-distant probabilities to come out right. Of course, it would be an elegant and beautiful feature of the branched model if all values of future probabilities were exactly represented by branch proportionality, while at the same time no two branches were qualitatively identical throughout the whole of time. This would certainly be an aesthetically pleasing solution. But we can't assume it.

What I am saying is that the question of whether Leibniz's law holds in the branched model is a matter of fact. It may—and this would be the elegant solution—but then again it may not. Since we simply don't know, we have to allow for the possibility of there being many sets of individuals which have all their properties in common but which are nevertheless distinct and non-identical. Theory (4) allows for this possibility: it holds that individuals are defined by their point of origin throughout the model at the time of the beginning of the universe, and that individuals so defined include all might-have-beens and may-bes above and beyond those that are actual. Consequently it allows for the possibility that two numerically distinct individuals may share all their properties. Although theory (4) would also permit individuals with exactly similar or approximately similar properties to be grouped together in equivalence classes, thus reducing their numbers somewhat, plainly the overall collection of individuals envisaged by theory (4) is very large.[17] For those who feel uneasy about its size there is an alternative, namely to return to theories (1) and (2) and recognize, on the branched model, only *actual* individuals.

As was said earlier, theory (1) has much to recommend it. The criterion of identity for actual individuals is unproblematic, at least compared to that for possible individuals. There are, of course, well-known difficulties about criteria of identity for individuals over time and across worlds, and these will be discussed in the next two sections. But the idea of individuals being *created* by the passage of time, an image which on the branched model can be interpreted literally, accords well with intuition and is worth preserving. On the branched model, individuals do not exist atemporally. They are creatures of time. At the big bang, there were no individuals. All that now exist (i.e. past and present individuals) have since that time come into being through branch attrition.[18]

However, before we close the books on the relative merits of theories (1)–(4), supporters of (2) must be heard. Theory (2) is the one which, in addition to past and present individuals, recognizes just those individuals that *will* become actual. The argument for their existence derives from the argument given in Chapter 1,

---

[17] See Black (1952) for a different type of example of individuals violating Leibniz's law. The question of putting possible individuals together in equivalence classes according to the theory of 'propagule essentialism' is taken up below in Ch. 8.

[18] An extension of this line of thought would be to hold that individuals owe to branch attrition not only their original appearance upon the scene, but also their 'formation' or the development of their individuality: what Goethe called *Bildung*. An individual starts off branched and ends up unbranched. In the case of a human being, the end result is what that individual makes of her/himself.

p. 14 above, for the existence of the set of all true propositions about the future. Whether or not we, or God, or anyone, knows the composition of this set of propositions, it exists, and its membership is dictated by future events, i.e. by branch attrition. As was stated in Chapter 1, present truth is supervenient upon future events. In exactly the same way, the composition of the set of individuals (2) also supervenes upon future events. No doubt there are many different possible babies that could be born to Mr and Mrs Smith in the year 2001. But, assuming that the Smiths do have a child in 2001, there is one and only one of these many babies who *will* be born. Which one depends upon a future event, namely which of the branches in 2001 drop off and which survives. Just as the class of true propositions about the future supervenes upon future events, so the class of individuals (2) who are or will become actual also supervenes upon future events, specifically upon branch attrition. For a philosopher, theory (2) has its attractions.

To sum up, the branched model offers four alternative theories (1)–(4) of what constitutes the set of all individuals. Of these, (1) and (2) are theories of actual individuals, while (3) and (4) are theories of possible individuals. Since what is possible includes what is actual, the sets of individuals associated with the four theories can be put in order of increasing size:

$$(1) < (2) < (3) < (4).$$

However, set (1) is getting progressively larger, while (3) is getting progressively smaller. Both are converging on (2). Consequently at the end of time we shall be left with only two classes: the class of all actual individuals (2), and the class of all possible individuals (4).

(iii) *Transtemporal Identity*

A feature of the branched model is that it provides a new and interesting answer to the problem of transworld identity: the problem of how a single individual can exist in two or more possible worlds. In this section not the problem of transworld identity but an analogous problem will be discussed, the problem of transtemporal identity. A solution to this problem will be proposed which argues (i) that to conceive of an individual as a three-dimensional object which endures through time, or alternatively as a four-dimensional object which is extended in time, are *equivalent*, and (ii) that the transtemporal identity of the three-dimensional object from one moment to the next may be based upon the topological connectedness of the corresponding four-dimensional object. The proposed analysis of identity through time will set the stage for the next section's analysis of identity across possible worlds.

As was noted at the beginning of the chapter, Aristotle was aware that the notion of transtemporal identity created difficulties, or might create difficulties if misunderstood. The same man at different times can possess contradictory properties—be pale and then dark. Aristotle's remark to the effect that this is a unique phenomenon ('Nothing like this is to be seen in any other case') alerts us to the possibility of trouble. And indeed, how can the *same man* be pale and dark? If 'same' means 'having all properties in common', then would the man who is pale not be a different man from the man who is dark? Plainly some sophistication in our understanding of what it is to be the same individual at different times is needed.

An obvious move is to have recourse to time-indexed properties. The man who is pale can be identical with the man who is dark if he is pale-at-t_1 and dark-at-t_2. Let identity, i.e. *strict identity*,[19] be understood in the usual way as an equivalence relation which satisfies both limbs of Leibniz's law, namely the identity of indiscernibles and the indiscernibility of identicals.[20] Then if Mengo weighs 92 lb. on his fourteenth birthday, and 104 lb. on his fifteenth, the person who weighs-92 lb.-at-age-14 is strictly identical with the person who weighs-104 lb.-at-age-15. The principle of the indiscernibility of identicals is preserved, since 'weighing 92 lb. at 14' and 'weighing 104 lb. at 15' are distinct non-conflicting properties of the same individual, on a par with 'liking maple syrup'.

This way of dealing with identity through time has much to recommend it. It provides, at first sight at least, a complete answer to the problem of how something at one time can be strictly identical with something at another time. The answer is that the thing in question is identical with itself. There need be no mystery about this. If we want to know how the boy who weighed 92 lb. at age 14 can be identical to the boy who weighed 104 lb. at age 15, we simply ask the boy in question to come and sit down in front of us. We attach two labels to him: 'Boy who weighed 92 lb.

[19] This term is used to distinguish it from 'identity under a sortal K', or 'K-identity'. See below, p. 213.
[20] What was loosely referred to as Leibniz's law in the previous section was only the first limb, the identity of indiscernibles.

at 14', and 'Boy who weighed 104 lb. at 15'. Mengo *is* the boy
who weighed 92 lb. at 14, and he *is* the boy who weighed 104 lb.
at 15. Therefore, by symmetry and transitivity of identity, the boy
who weighed 92 lb. at 14 *is* the boy who weighed 104 lb. at 15.
To assert this is simply to assert Mengo's self-identity.[21] It would
appear that the *diachronic* identity of the 14-year-old with the
15-year-old has been reduced to the *synchronic* identity of Mengo
with himself.

This resolution of the problem of transtemporal identity fits
well with the formal definition of the identity relation. Identity is
very straightforward: it is the smallest relation (smallest set of
ordered pairs) that holds between every object and itself. In David
Lewis's words:

Identity is utterly simple and unproblematic. Everything is identical to
itself; nothing is ever identical to anything except itself. There is never
any problem about what makes something identical to itself; nothing
can ever fail to be. And there is never any problem about what makes
two things identical; two things can never be identical.[22]

This being said, has the problem of identity through time been
resolved? It would be nice to think it has, but difficulties remain.
A danger flag should have been raised by the apparent reduction
in the last example of diachronic identity to synchronic identity.
Is the problem really that simple? Does the identity of *x* in 1982
with *x* in 1989 reduce to the self-identity of *x* in 1989? Yes, but
only if *x* exists in both 1982 and 1989. And this requires the
transtemporal identity of *x*. Recall that when we sat Mengo down
in front of us we attached two labels to him: 'Boy who weighed
92 lb. at 14', and 'Boy who weighed 104 lb. at 15'. The very
possibility of being able to do this presupposes that Mengo is a
transtemporal being: what W. E. Johnson called a 'continuant'.
The law of self-identity, therefore, plus the labelling of a present
individual as one who had some property *P* at some earlier time
t, do not explain identity through time, they simply restate it.

This result should not disappoint us, for restating them in more
illuminating ways is probably all we are ever able to do with the
perennial problems of philosophy, one of which is the problem
of identity through time. (This fact in itself bears witness to the
durability of the problem.) Rarely if ever can we make a philosophical problem disappear, or reduce it or analyse it without

---

[21] This is the way in which van Inwagen (1985) deals with the problem of identity through
time.

[22] Lewis (1986*b*: 192–3).

remainder into something else. The rest of this section will attempt
to shed some light upon the problem of transtemporal identity by
restating it in ways which are not new, but which are worth
reviewing.

Heraclitus could step twice into the same river, but not twice
into the same water, or into what Quine calls a river-stage.[23] For
Quine, a river is a process through time, and a river-stage is
one of its momentary parts. In the same style, a rabbit-stage is not
a rabbit, but a brief temporal segment of a rabbit.[24] The introduction
of river-stages and rabbit-stages takes us from the three-dimensional
to the four-dimensional world. Either a rabbit-stage may be instantaneous, in which case it is a 'rabbit-slice', a four-dimensional
object with zero extension in the time direction, or it may be a
four-dimensional object with small but non-zero temporal extension. In either case it differs from a three-dimensional object, since
even though a rabbit-slice has zero temporal extension it has a
precise temporal *location*. A three-dimensional rabbit, on the other
hand, is not tied to any particular temporal location or date.

Introduction of the notion of a four-dimensional object, and
of such things as rabbit-stages, might well appear to be counter-
productive at this point. Aristotle's problem of the man who was
both pale and dark was dealt with by recognizing the existence
of the idea of a three-dimensional being who persists through time
and who has different properties at different times. But to move
to the idea of a four-dimensional object made up of stages may
seem to be a step backwards. No two rabbit-stages are identical,
and hence transtemporal identity of rabbits becomes problematic
once more. However, this difficulty can be overcome, as will be
seen. Conceiving of objects four-dimensionally has distinct advantages in some contexts, while conceiving of them three-dimensionally has advantages in others. So we shall employ both modes,
at the same time showing them to be equivalent.

The first group of philosophers to contemplate conceiving of
individuals four-dimensionally seems to have been the group at
Cambridge in the early 1900s, which included Russell, McTaggart,
Broad and W. E. Johnson. A classic statement of the four-
dimensional approach is Broad's:

We usually call a flash of lightning or a motor accident an event, and
refuse to apply this name to the history of the cliffs of Dover. Now the
only relevant difference between the flash and the cliffs is that the former

[23] Quine (1953: 65–7).
[24] Quine (1960: 51). See also Quine (1976a; 1976b; and 1981: 10–14).

lasts for a short time and the latter for a long time. And the only relevant difference between the accident and the cliffs is that, if successive slices, each of one second long, be cut in the histories of both, the contents of a pair of adjacent slices may be very different in the first case and will be very similar in the second case. Such merely quantitative differences as these give no good ground for calling one bit of history an event and refusing to call another bit of history by the same name.[25]

Later, in the 1950s, the same theme was taken up by Carnap and Nelson Goodman.[26] The following passage from Quine is a good summary statement of the position:

A physical thing—whether a river or a human body or a stone—is at any one moment a sum of simultaneous momentary stages of partially scattered atoms or other small physical constituents. Now just as the thing at a moment is a sum of these spatially small parts, so we may think of the thing over a period of time as a sum of temporally small parts which are its successive momentary stages. Combining these conceptions, we see the thing as extended in time and in space alike; the thing becomes a sum of momentary stages of particles, or briefly particle-moments, scattered over a stretch of time as well as space.[27]

This way of looking at individuals, as four-dimensional objects with temporal extension, is powerful and elegant, but needs to be handled carefully. How does it cope with transtemporal identity? As mentioned earlier, if four-dimensional objects are made up of momentary slices—if they are sets or aggregates of such slices— then the question is, what makes two slices segments of the *same object*? This question becomes particularly important in the light of Quine's proposal for constructing objects out of arbitrary collections of spatio-temporal volumes. For Quine, a physical object is 'the material content of any portion of space-time, however scattered and discontinuous'.[28] The world's water is one big physical object in all its dispersed raindrop-, lake-, and sea-stages. There is, for Quine, a 'gerrymandered' object, part of which is a momentary stage of a silver dollar at present in Quine's pocket, and the rest of which is a temporal segment of the Eiffel Tower through its third decade.[29] But a liberal ontology of this type, in which momentary slices or stages can be assembled together into 'objects' in purely arbitrary ways, makes nonsense

[25] Broad (1923: 54).   [26] See Goodman (1951: 94); and Carnap (1958: 158).
[27] Quine (1982: 269). I owe this reference to Dau (1986).
[28] Quine (1976a: 859). Cf. Goodman (1951: 46): 'If the Arctic Sea and a speck of dust in the Sahara are individuals, then their sum is an individual.'
[29] Quine (1976a: 859).

of the idea of transtemporal identity. There will never be any definite answer to the question, whether two momentary stages are stages of the same thing, or of different things. In fact, for any two stages whatever, both answers will always be correct. Quine's ontology of objects, therefore, is of no help in giving an account of identity through time.

However, Quine's ontology is not the only one when it comes to four-dimensional objects, nor is it the most natural one. Instead of taking momentary stages, or slices, as our basic particulars, we could instead take whole four-dimensional volumes, corresponding to the entire life or history of an individual. We could abandon, in other words, what Shoemaker describes as

a view about identity that goes back at least to Hume, namely that identity through time, or at any rate what we count as identity through time, consists at least in part in the holding of causal relations of certain kinds between momentary entities—events, or momentary thing-stages (phases, slices)—existing or occurring at different times.[30]

For a living organism like a rabbit, there are two possibilities when we proceed four-dimensionally. (i) We can regard the four-dimensional rabbit as an aggregate of, i.e. as built up out of, rabbit-stages or rabbit-slices. In that case we may posit a relation R, called the 'unity relation', such that two rabbit-stages r_1 and r_2 are stages of the same rabbit just in case r_1 bears R to r_2.[31] Alternatively (ii) we may regard the four-dimensional object itself as the basic entity, and the stages or slices as derivative, i.e. as abstractions in the way that the earth's equator and meridian lines are abstractions. In this case two rabbit-stages are stages of the same rabbit if and only if they belong to the *same connected four-dimensional object*, such objects having a natural shape and not being artificial constructs out of arbitrary portions of space-time.[32] Natural shapes in the four-dimensional world are associated with *sortals*; thus to the sortal 'frog' there corresponds one characteristic shape, to the sortal 'chair' another, etc. Organic shapes exhibit a characteristic temporal developmental pattern; a

---

[30] Shoemaker (1979: 321).   [31] See Perry (1975: 9).

[32] The distinction between a 'natural' and an 'artificial' 4-dimensional shape requires much closer examination than it can be given here, but is intended to correspond roughly to the philosophical difference between a substance and a non-substance. Airlines, for example, count 'passengers' differently from 'human beings': two connecting flights may together carry 60 'passengers', but only 40 'human beings'. (Flight A has e.g. 30 people on it, 10 of whom leave while 10 new people join for flight B.) Passengers can be represented 4-dimensionally as the temporal segments of humans while travelling with an airline. Such segments are not 'natural' shapes, and passengers, unlike humans, are not substances. The example is Gupta's (1980: 23).

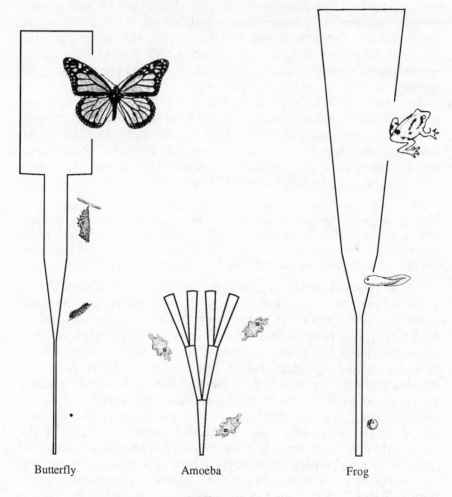

Butterfly   Amoeba   Frog

FIG. 7.2

frog-shape includes successively an egg phase, a tadpole phase, and an adult phase.[33] In addition (see Fig. 7.2) they exhibit a characteristic spatial pattern or morphology at each temporal stage, and a characteristic rhythmic pattern occupying short spatio-temporal intervals, e.g. heart-beats.

[33] The association of 4-dimensional shapes with sortals calls to mind the 'sortal dependency' of identity, or the 'relative identity', of Geach, Wiggins, and others. Sortals like 'tadpole', 'caterpillar', and 'child' are what Wiggins calls 'phase sortals', and, like 'passenger', do not correspond to substances. See Geach (1962), Wiggins (1967; 1980), and Ch. 8 below. Lombard, by contrast (1986: 90–2), thinks that there is no single creature that is first a caterpillar and then a butterfly, or first a tadpole and then a frog.

In the case of an animal or a biological organism, belonging to a sortal or kind K, two K-stages are stages of the same organism if they belong to the same natural four-dimensional volume. Thus the four-dimensional view provides what we might call a principle of *diachronic K-identity*. The principle of *synchronic K-identity*, on the other hand, simply specifies what it is for any arbitrary stage to be a K-stage.[34] In the case not of an organism but of an artefact, synchronic K-identity is relatively unproblematic but diachronic K-identity is complicated and in some cases quite unclear. In Hobbes's example of Theseus' ship, some of the planks in the ship can be taken out and replaced without the ship's ceasing to be the same ship. But if too many such changes are made, the result may be *a ship*, but not *the same ship*.[35] Synchronic K-identity may have been satisfied at each instant of the ship's existence, and yet diachronic K-identity is not satisfied. So far, no one has yet succeeded in proposing a workable principle of diachronic K-identity for artefacts.[36]

Leaving aside such problem cases as Theseus' ship, the advantages of adopting the four-dimensional picture of individual substances lie in the availability of a neat geometrical principle of identity through time. Transtemporal identity derives from the

---

[34] Synchronic K-identity is the ontological correlate of what Dummett calls a 'criterion of application', while diachronic K-identity is what he calls a 'criterion of identity', both relative to a sortal K. See Dummett (1973: 546); also Shoemaker (1979: 333–4); and Noonan (1988: 83).

[35] Quine (1981: 12) says that 'whether we choose to reckon [Theseus' ship] still as the same ship is a question not of "same" but of "ship" '. It is true that synchronic ship-identity may be, as Quine says, a question only of 'ship', but diachronic ship-identity, which involves two or more different times, is a question of 'same ship', i.e. both of 'ship' and of 'same'. Lewis (1986b: 193) takes Quine's position, arguing that many problems which are stated in terms of identity can be restated without identity. It is, e.g., a good question, as Lewis says, 'whether a restaurant is something that can continue to exist through a simultaneous change in ownership and location and name'. This question is not stated in terms of identity. But suppose Louis's restaurant, where Sam works, closes at noon, and a few minutes later you enter a restaurant on a different street named Sam's restaurant, with Louis employed as the chef. Your friend says 'Look, Louis's restaurant continues to exist!' How do you express your disagreement except by replying that what existed before noon, and what exists now, are not the same restaurant? Diachronic identity differs from synchronic identity, but it is still identity.

[36] A workable principle may perhaps be extracted from what is said below about individuals as rooted four-dimensional branched volumes. Theseus' ship e.g. is such a branched 4-dimensional volume, and the fact that on one branch its planks are taken out and replaced by new planks does not change its identity. On other branches, no doubt, the ship just rots. But the whole branched four-dimensional object, rotted or refitted, is still Theseus' ship. Even the fact that the old planks may be reassembled into a second object which is qualitatively indistinguishable from Theseus' ship does not affect the original ship's identity; it entails instead that on one branch a second ship has been created. This second object is again branched and 4-dimensional. See below, p. 238.

topological connectedness of a 'natural' four-dimensional object. To adopt this line of thought is to reduce the *transtemporal* identity of a person or animal or thing to the *strict* identity of the corresponding four-dimensional object: it is to make spatio-temporal continuity the basis of identity.[37] Nevertheless, despite these advantages, there will still be many philosophers who object strongly to the four-dimensional representation of individuals. However 'natural' a four-dimensional object may be, they will say, it is still divisible into temporal parts in the same way as the surface of the earth is divisible by its meridians into spatial parts. But the individuals we are familiar with have no temporal parts. The very idea of a 'temporal part' of an individual is absurd. Or so they will say.[38]

Those who object to temporal parts might agree that *events* have temporal parts, and for that reason a four-dimensional representation of events is possible, even appropriate. A baseball game, for example, is divided into nine innings, and the whole game does not exist until each of its temporal parts have come to exist. But things are not like events. When you hold a rabbit in your hand, you don't just hold a part of a rabbit, you hold the *whole rabbit*. As I sit writing at my table, the table at which I write is *all there* at each moment.[39] If I buy a car, I don't buy a temporal part of the car, I buy the entire car. To buy only a temporal part of a car is called car-rental or hire-purchase, and hire-purchase is quite different from purchase.

These are strong objections. If the four-dimensional conception of individuals is to be acceptable, some way must be found of distinguishing *things* as four-dimensional objects from *events* as four-dimensional objects, and the difference must somehow enable things but not events to be conceived of as being 'all there' at each moment of their existence.

As a first step towards defending the four-dimensional point of view, I shall argue that the description of a thing as a four-

---

[37] Those who would contend, with Locke, that in the case of persons it is consciousness or memory rather than spatio-temporal continuity which determines identity, must find an answer to Butler (1736): 'But though consciousness of what is past does thus ascertain our personal identity to ourselves, yet to say, that it makes personal identity, or is necessary to our being the same persons, is to say, that a person has not existed a single moment, nor done one action, but what he can remember; indeed none but what he reflects upon. And one should really think it self-evident, that consciousness of personal identity presupposes, and therefore cannot constitute, personal identity; any more than knowledge, in any other case, can constitute truth, which it presupposes.' See also p. 224 below.

[38] See e.g. Geach (1965: 329); Chisholm (1976: 138–44); and van Inwagen (1985). For a defence of temporal parts, see Armstrong (1980) and Lewis (1983: 76–7).

[39] Cf. Wiggins (1980: 25), and Stalnaker (1986: 134–5).

dimensional object, and the description of the same thing as a three-dimensional object which persists through time and yet changes, are *equivalent*. Everything that can be said about an object using three-dimensional language can also be said using four-dimensional language, and vice versa. Can a three-dimensional baseball travel from the pitcher to the batter to the left-field bleachers? Yes. In four-dimensional language this journey is described as follows: the world-line of the baseball is contiguous with that of the pitcher's hand at time t_1, with that of the bat at time t_2, and with that of the bleachers at time t_3. Does a four-dimensional object have temporal parts? Yes. In three-dimensional language we would say that the 'life' or 'history' of the object (as distinct from the object itself) is divisible into as many periods as there are spatial intervals on the real line. (Prufrock's life, for example, was measured out by coffee spoons.) What's the four-dimensional equivalent of a caterpillar turning into a butterfly? Well, the four-dimensional object that is the complete insect has an early wiggly section corresponding to the caterpillar crawling about, then it has an intermediate smooth tubular section corresponding to the stationary cocoon (which is spun from the caterpillar and counts as an extension of the body), and lastly it has a pulsating kaleidoscopic section corresponding to the butterfly flying about (cf. Fig. 7.2 above). Finally, if a four-dimensional world-line forks in time within a single branch of the model and divides into two similar world-lines, the three-dimensional equivalent is exemplified by an amoeba or a cell dividing in two, or by particle decay. In every case, the two modes of describing an enduring thing—the three-dimensional mode and the four-dimensional mode—are strictly equivalent and intertranslatable. One mode is useful for some purposes, and the other for other purposes.[40]

If, as I have argued, the description of an individual substance or thing as a four-dimensional object, or as a three-dimensional object which endures through time, are equivalent, then we are half-way towards resolving the problem of temporal parts. Considered as a four-dimensional object, every substance has temporal parts. But we do not have to describe, or conceive of, a substance

[40] In special relativity, e.g., we can say, using the 3-dimensional mode, that an object's length undergoes a 'Fitzgerald contraction' when measured by a rapidly moving observer. But the most elegant way of describing this phenomenon is to shift to 4-dimensional language and show how a 4-dimensional object's apparent spatial and temporal size changes when the object is measured in different coordinate frames. For details of the geometrical treatment of objects as 4-dimensional entities see Mermin (1968: ch. 17).

as four-dimensional. We may, if we choose, describe or conceive of it as persisting and three-dimensional. In the three-dimensional mode, but not in the four-dimensional, the substance or thing can be regarded as *wholly present*, or 'all there', at each moment of its existence. When you hold a rabbit in your hand, therefore, you don't just hold a rabbit-stage. You hold the whole rabbit. Further, the rabbit you hold in your hand today is *strictly identical* with the rabbit you held yesterday, this identity being equivalent to the *topological connectedness* of the four-dimensional rabbit volume. Since the three- and the four-dimensional modes of describing substances are equivalent, there is no need to translate ordinary truths about enduring three-dimensional rabbits into talk of rabbit-stages, though we may do so if we wish.

What remains is to account for the difference between *things*, which we normally regard as having spatial parts but no temporal parts, and *events*, which have temporal parts but no spatial parts. Both are four-dimensional objects. But the difference between them consists in the fact that while things can be redescribed as three-dimensional objects which are wholly present at each moment, events cannot. Events are the sum of their temporal parts, and are never wholly present at any moment. There is no three-dimensional object which is an event—no choice, therefore, about whether to describe an event as a three-dimensional or as a four-dimensional object. An event can be conceived of in only one way, as a four-dimensional object with temporal parts, while a thing can also be conceived of three-dimensionally, with no temporal parts. This is an important difference—perhaps the only important difference—between events and things. Once these facts are recognized, there remains no further obstacle to conceiving of an individual substance or thing *either* as a four-dimensional object, *or* as a three-dimensional object which endures through time.

In Chapter 4 (pp. 94, 138) I referred to a remarkable fact about the branched model and used the phrase 'pre-established harmony' to describe it. It is this. At each branch point, above which there is a probability distribution of branches in a prism which is uniform throughout the model, and consequently represents a lawlike state of affairs, there exists, at the branch point in question, a set of initial conditions that could be appropriately described as the 'categorical ground' (or 'dispositional ground') of the probability distribution above it. For example, initial conditions as a node consisting of a vertically polarized photon entering a $\phi\pm$ analyser, are invariably conjoined with a set of branches of proportion $\cos^2\phi$ above the node in which the photon emerges in the ϕ^+ channel, and a set of branches of proportion $\sin^2\phi$ in which the photon emerges in the ϕ^- channel. Why does this invariable relationship prevail throughout the model (if indeed it does prevail)? Here is a speculative hypothesis which may account for it.

I have argued that the three-dimensional and the four-dimensional descriptions of an object are equivalent. A rabbit can be considered indifferently as a four-dimensional space-time 'worm', or as a three-dimensional object that persists through time. Now suppose we were to extend this idea of equivalence to the universe as a whole. In that case the universe could be regarded either as a large three-dimensional object that persisted through time, or alternatively as an enormous branched four-dimensional object: the branched model in fact. These two ways of conceiving of the universe would be equivalent. In that case, consider the three-dimensional universe at the instant at which the photon enters the polarization analyser. The universe, so regarded, persists through time, and an instant later is characterized (with probability $\cos^2\phi$) by the photon emerging in the ϕ^+ channel, or alternatively is characterized (with probability $\sin^2\phi$) by the photon emerging in the ϕ^- channel. Here the universe is a three-dimensional object enduring through time which changes probabilistically. But there is another, equivalent, way of looking at the universe. This is as a branched four-dimensional object, with $\cos^2\phi$ branches above the photon-entering-analyser node in which the photon is measured ϕ^+, and $\sin^2\phi$ branches in which the photon is measured ϕ^-. The four-dimensional universe, via branch proportionality, gives a numerical value to the probabilities of change for the three-dimensional universe. If the three- and the four-dimensional descriptions of an object such as an electron are truly equivalent, then the latter can be regarded as a picture of that object's potentialities or powers, spread out in branched four-dimensional space-time. In this way the 'pre-established harmony' of the three-dimensional categorical ground and the four-dimensional probability measure would be accounted for.

Let me emphasize that this is a purely speculative hypothesis. I have no very clear idea (at this point) what the implications would be of extending the three dimensional/four dimensional equivalence to the whole universe. But, if it were done, it would certainly explain the remarkable correlation in the branched model between initial conditions of a certain kind and branch proportionality of a certain kind.

(iv) *Transworld Identity*

The subject of this section is the identity of individuals not across times, but across worlds. To the question, 'Can one and the same individual exist in, or be a part of, two or more possible worlds?' the branched model provides an affirmative answer. In the next few pages, we shall see what this answer is based upon.

In the first place, the 'possible worlds' of the branched model are Minkowski worlds, four-dimensional space-time manifolds which lie in the future and in which events and individuals have their various spatio-temporal locations. To assert, therefore, that two individuals in different worlds or branches are the same individual is to assert that the individual in question leads a double life—that it is part of the first world, and also part of the second.[41] Being 'part of' a possible world, on the branched model, or being 'in' it, is not like being 'part of', or 'in', a possible world for philosophers who maintain that possible worlds are abstract states of affairs, or maximal consistent sets of propositions. For those philosophers, something is 'in' a possible world if a proposition asserting that thing's existence is among the set of propositions which constitutes that world.

Plantinga puts it more elegantly: an object x 'exists in' a state of affairs S if and only if necessarily, if S had been actual, x would have existed.[42] But existing concretely, in a spatio-temporal world, is different from existing in an abstract state of affairs, and it is the former not the latter that is at issue in the branched model.

The way in which a single individual can exist in two or more branches of the model is by virtue of being itself branched, i.e. by virtue of being represented as a branched four-dimensional object. As was seen in the previous section, every individual can be represented or conceived of indifferently as a three-dimensional object existing in time, or as a four-dimensional object. The most natural way of picturing transworld identity in the branched model is by means of a four-dimensional object that is itself branched, one which reaches into the model's branches. A good illustration of transworld identity is Kripke's example of throwing dice.[43]

I hold in my hand two dice, and prepare to throw them. If the dice are fair, what is the probability of throwing eleven? To calculate this probability, Kripke says, we work with a set of miniature 'possible worlds'. The thirty-six possible states of the dice are literally thirty-six 'miniworlds', each one possible relative to the initial state of preparing to throw the dice. The probability of eleven

[41] Cf. Lewis (1986*b*: 198).   [42] Plantinga (1985: 89).   [43] Kripke (1980: 16).

is obtained by dividing the number of outcomes which yield eleven by the total number of possible outcomes. Since there are two outcomes which yield eleven, namely (i) die A six, die B five; and (ii) die A five, die B six; the value is $\frac{2}{36} = \frac{1}{18}$. This method of calculating the probability value requires that the two miniworlds (i) and (ii) should be distinct, i.e. that the world in which it is die A (and not B) that is six should be different from the world in which it is die A (and not B) that is five. And this in turn requires that die A, the individual die that I hold in my hand in the actual world, must exist in two different future possible worlds. It is not enough that there be a miniworld in which one die turns up six and the other five. We need *two* miniworlds, which are not identical.[44] Even if the two dice are qualitatively indistinguishable, so that there is no way in practice of telling A from B, there still must be a difference between a world in which A rather than B is six and a world in which B rather than A is six. It seems therefore that even a simple exercise in calculating probabilities requires that we recognize that die A must exist in more than one possible miniworld, i.e. that we recognize the transworld identity of individuals.

A highly simplified picture of a concrete example showing how, on the branched model, an individual can exist in more than one future possible branch is shown in Fig. 7.3. The shaded area

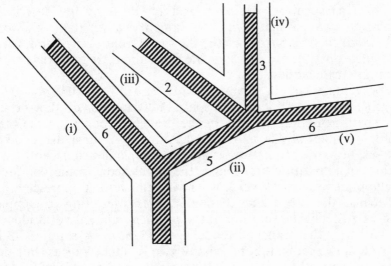

FIG. 7.3

---

[44] Kripke points out that if these two miniworlds are not distinguished, calculation of the probability of rolling eleven will yield the incorrect value $\frac{1}{21}$, not $\frac{2}{36}$ (1980: 17).

represents a branched four-dimensional object, in this case a die. There are five branch segments numbered (i)–(v), and at each branch point the die is thrown. In branch (i) it comes up six, and in branch (ii) five. In branch (ii) it is rolled again, and in each of branches (iii)–(v) it shows a different number. In branch (v), after coming up six, the die is destroyed, and its world-line comes to an end. The diagram shows how, when represented four-dimensionally, an object can exist in different possible worlds, and have different properties in each. Since the three-dimensional and the four-dimensional ways of representing persisting objects are equivalent, the object *need* not be represented in this way, but instead could be represented as an object which is wholly present in each possible world at each moment of time. However, the four-dimensional representation gives the clearest picture of transworld identity in the branched model.

An individual in the branched model exists in two or more possible worlds by being *physically present*, or *spatio-temporally located*, within each of the worlds in which it exists. This is not the way in which Kripke conceives of an individual existing in different possible worlds. For him, the thirty-six miniworlds which are the possible outcomes of throwing two dice are *abstract states of the dice*, not complex physical entities.[45] Die *A* exists in each one of these miniworlds because the abstract states are *given* or *specified* as such. A possible world is 'given by the descriptive conditions we associate with it',[46] and if we stipulate that some individual such as Nixon, or the die that I hold in my hand, forms part of a possible world, then the individual exists in that world without further question.

For Kripke, as for Plantinga and R. M. Adams, non-actual possible worlds do not exist in space and time, nor do they occupy any spatio-temporal volume. In the branched model, on the other hand, possible worlds and the individuals in them are concrete entities, not abstract; and what allows a single individual to exist in many possible future branches is the connectedness and continuity of the branched object which is its four-dimensional representation.

Because the two are so dissimilar, it is interesting to compare in greater detail the solution to the problem of transworld identity provided by the branched model with Plantinga's solution. Both allow a single individual to exist in two or more worlds, but they do so in very different ways. Where the state of affairs *S* lies in the future, Plantinga's definition of what it is for an individual *x*

        [45] Kripke (1980: 17).   [46] Ibid. 44; cf. Kripke (1971: 148).

to 'exist in' S reads: x exists in S if and only if necessarily, if S were to be actual, x would exist. Consider now two future possible states of affairs: (i) Mikhail Gorbachev being elected secretary-general of the UN in the year 2000, (ii) Mikhail Gorbachev becoming president of Harvard in 2000. By Plantinga's criterion, Gorbachev exists in both these states of affairs, since he would exist if they were actual. Hence Gorbachev exists in at least two distinct possible worlds. For Plantinga, this is all there is to the 'problem of transworld identity'. It is a problem with a simple and straightforward solution. In particular, there is no question of 'identifying' Gorbachev in these two possible worlds: examining them with a telescope or with David Kaplan's Jules Verne-o-scope to see whether or not Gorbachev is in them. For Plantinga, possible worlds are not like that; they are not, as Kripke would say, like 'distant planets'. They are, instead, 'given by the descriptive conditions we associate with them'. If Gorbachev's being secretary-general of the UN is one of the descriptive conditions we associate with a world w_1, and if Gorbachev's being president of Harvard is a condition we associate with w_2, then Gorbachev exists in both those possible worlds.

In the branched model, on the other hand, transworld identity has an entirely different basis. Since future possible worlds, or branches, are spatio-temporal entities, they could be peered into by a Jules Verne-o-scope if a Jules Verne-o-scope existed. (What the Verne-o-scope could *not* perceive, of course, is which future branch was going to be actual.) Let us imagine that the Verne-o-scope picks out two individuals, on different branches, who look like Gorbachev. *Are* they Gorbachev? If each of them is linked by a continuous world-line with the author of perestroika, i.e. if they both form part of the same four-dimensional branched individual, then the answer is yes.[47] In the branched model, transworld identity is not an unanalysable notion, but is based on the strict identity of a connected, non-artificial, branched four-dimensional object. When the right conditions are satisfied, two three-dimensional objects in different branches can be identical. This follows from the equivalence of the three- and the four-dimensional representations of an object.

According to the thesis of equivalence, an individual can be regarded as a four-dimensional object, or equivalently as a three-dimensional object which changes and which persists through

[47] In the terminology of Reichenbach, the two Gorbachevs would be linked to the author of perestroika by the relation of *genidentity*, a relation that Reichenbach derived from the psychologist Kurt Lewin. See Reichenbach (1957: 270).

time. If this is correct, and if the same four-dimensional object can extend into and be a spatio-temporal part of two or more future branches, then the corresponding three-dimensional objects in the different branches are *strictly identical*. The situation in the case of transworld identity is exactly the same as in the case of transtemporal identity. Just as the boy who weighed 92 lb. on his fourteenth birthday is strictly identical to the boy who weighed 104 lb. on his fifteenth, so the boy who in one possible future works on economic development is strictly identical with the boy who on another branch becomes a hockey star. Being wholly present on these two different branches, he leads what David Lewis calls a 'double life', although the double life in question is no more problematic than the double life of an object that is *F* at one time and *G* at another. In both cases it's the *same individual*, but with *different properties*.

The leading of double lives, according to Lewis, is what best deserves to be called 'transworld identity'. An object leads a double life if it is part of one possible world *and* part of others as well. Lewis confesses that he cannot name one single philosopher who favours transworld identity, thus understood.[48] Many insist that there must be such a thing as transworld identity, but none including Lewis himself maintains that the same individual can exist as a spatio-temporal part of two or more worlds.[49] This is, however, precisely what is suggested in this section.

Despite its lack of supporters, Lewis believes that the thesis of 'full-blooded' transworld identity, according to which a single individual can be spatio-temporally located in two or more worlds, deserves our attention. The principal objection to it, he finds, is that it implies that the same individual can have different properties in different worlds. This, in the branched model, is indeed the case. In the next federal election, Mr Chrétien can emerge the winner, or he can emerge the loser. These are two very different properties, yet it is the same person who has them on different branches. How, Lewis asks, can this be? How can the same man be both a winner and a loser? Lewis calls this the problem of *accidental intrinsics*. Well, the answer according to the branched model is that the same individual can have different accidental intrinsic properties at different times, by virtue of being conceived of as a persisting or enduring three-dimensional object which changes. That very individual can also have different accidental

---

[48] Lewis (1986*b*: 198).

[49] Lewis's own view (1968) is that while every individual is confined to its own world, it can have counterparts in other worlds.

intrinsic properties on different branches, by virtue of being conceived of as a branched four-dimensional object. Hence Chrétien can be a winner on one branch and a loser on another, or pale on one branch and dark on another, or indeed have any other accidental intrinsic property compatible with his present state.[50] He cannot, of course, cease to have any of his *essential* properties, but the discussion of essential properties must await the next chapter.

(v) *Puzzle Cases of Identity*

The picture of an individual that emerges from the previous section is that of a branched four-dimensional object, rooted in its origin, which on its different branches undergoes every experience, and encounters every fate, that is physically possible for it. At the moment the individual comes into being its four-dimensional representation is a branched structure with no trunk; throughout its life it progressively sheds branches until, like Aristotle and the Holy Roman Empire, it eventually becomes a single unbranched four-dimensional volume located in the past.[51] Its *identity* lies in its fixed position in the branched model. Every separate individual has a separate origin, and is represented by a separate branched (or unbranched) structure.

These considerations lead to the resolution of a number of puzzle cases in personal identity. Locke, who originated the puzzle

---

[50] Lewis (1986*b*: 199–202) makes the discussion of accidental intrinsics more difficult for a proponent of the branched model by raising the possibility that Humphrey could have had six fingers on his left hand. (By this Lewis means a 'proper six-fingered hand', not just one with an extra finger stuck on.) Though it may not be one of his essential properties that he not have six fingers, I imagine we would have to go back to one of Humphrey's very early branches to find him with a proper six-fingered hand. Recall that branching on the branched model is smooth and regular, so that any miracle which sees Humphrey with five fingers one day and six the next is ruled out. This being said, I suppose we might find, in a model of many years ago, a six-finger branch stemming from a branch point at which the genetic material in Humphrey's zygote undergoes some kind of mutation. But unlike Kripke, Plantinga, and Lewis, we can't be assured a priori that some possible world contains a six-fingered Humphrey, nor can we, on the branched model, *stipulate* that there is one. Whether there is one or not is an *empirical* question.

[51] The 4-dimensional representation of some objects as branched and of others as unbranched resolves Geach's problem in Ch. 2 above of how to distinguish between real and Cambridge change. Objects which exist in the present undergo *real change* in the sense that their 4-dimensional branched shape is constantly changing. But objects such as Aristotle which exist wholly in the past, and whose 4-dimensional shape is a single unbranched volume, undergo only *Cambridge change*. Their histories are over and done with and nothing can now affect them, whereas presently-existing objects are continually modified by branch attrition. In this way the model provides for an objective difference between Geach's two varieties of change.

cases, imagined a situation in which the consciousness of one person entered the body of another, and in so doing set the pattern for much discussion about identity:

For should the soul of a prince, carrying with it the consciousness of the prince's past life, enter and inform the body of a cobbler, as soon as deserted by his own soul, every one sees he would be the same *person* with the prince, accountable only for the prince's actions: but who would say it was the same *man*? The body too goes to the making the man, and would, I guess, to everybody determine the man in this case, wherein the soul, with all its princely thoughts about it, would not make another man: but he would be the same cobbler to every one besides himself.[52]

Cases such as these are difficult and confusing, but the concept of an individual within the branched model provides a possible way of resolving them. The following is a modern 'worst-case' version of Locke's example.

Let A, B, and C be three individuals. Suppose that one of A's cerebral hemispheres is transplanted into B, and the other into C, so that B and C have A's memories, perhaps in addition to their own. Which of the resultant individuals is A, which B, and which C? Could one individual A live inside of, or alongside of, another individual B? Could B and C both, in some sense, be A? Could A be shared between B and C?

Phrased in this way, these questions appear to be either insoluble or senseless. But in the branched model they can be given a sense, and can in principle be answered. In the model, A, B, and C are themselves branched individuals. In the example we imagine that on one branch, A's hemispheric lobes are transplanted into B and C. Does this imply that A becomes B, or that B becomes A, or that the identities of A and B are in any way mingled or blurred? No. A remains A, with its origin and memories intact, even though on one of its branches it undergoes a brain transplant. B, on the other hand, experiences some new 'memories' on one of its branches—except that these 'memories' are strictly speaking false, since the events they represent did not happen to B and are not in B's past. On the vast majority of its branches, B experiences no false memories. The brain transplant does not therefore change A into B, or B into A, or give A a new body, or affect the identities of A and B in any way. Still less would it be true to say that the identity of A had been 'shared' or 'divided'

[52] Locke (1689: II. xxvii. 15). Similar examples are found in Quinton (1962: 401); Shoemaker (1963: 23); Wiggins (1967: 53); Williams (1970: 161); Perry (1975: 3); Shoemaker and Swinburne (1984: 109); Parfit (1984: 254).

between B and C: nothing that is done to A on one of its branches changes A's identity.

To sum up, an individual's identity is fixed and forever established by the location and shape of the branched four-dimensional structure which constitutes it at its origin. Therefore the vicissitudes of A's, B's, and C's lives—whether for example they participate in brain transplants on some of their branches—in no way affects their identities. To give an analogy in the three-dimensional world, a forester might link two trees growing in a forest by grafting together two twigs, so that the trees grew together and sent nutrients to some shared leaves. But the grafting operation no more makes them one tree than a brain transplant on one of their branches makes two branched individuals one. To focus on only one branch, as in the single-manifold Minkowski world, may make it appear that in certain circumstances two individuals can fuse their separate identities and become one. But restoring an individual's full complement of branches, as in the model, restores its particularity, and gives it a separate identity it cannot lose.

8

Essential Properties

ESSENTIALISM is the doctrine that some or all individuals have non-trivially essential properties. An essential property is a property that an individual has to have if it exists at all; it is a property that it has at every instant it exists, and in every possible world in which it exists. A non-trivially essential property is a property that an individual has essentially, but in virtue of facts about the world, not in virtue of truths of logic. Examples of trivially essential properties, which individuals have through logical necessity, are such properties as being self-identical, being a male if a bachelor, not being both round and square, etc.[1] The question is, are there any non-trivially essential properties? Since we shall not be concerned with trivially essential properties in what follows, the question becomes simply, are there any essential properties?

(i) *Essential Properties*

Those who have read through up to here will perceive what the answer provided by the branched model must be. An essential property of an individual x is one that x has at every instant, on every branch, at which x exists. The answer to the question of whether there are any such properties lies in the branched structure. In the previous chapter an 'individual' was represented as being a branched, connected four-dimensional object. The shape of this object provides both a diachronic criterion of identity through time, and a transworld criterion of identity across branches, for the individual in question—the very same individual on different branches when represented three-dimensionally. (The equivalence of the three- and four-dimensional representation of individuals is argued above, on pp. 215–17.) Determining which if any properties

---

[1] See Forbes (1986: 4), who also lists existence as a trivially essential property: as a matter of logic, there are no non-existent individuals. For the objection that existence should *not* be a trivially essential property, but should be essential to something only if that thing exists necessarily, see Kripke (1971: n. 11). Kripke's objection would entail redefining the notion of an essential property: such a property would be one (apart from existence) that an individual would have to have if it existed at all.

of an individual x are essential to it becomes therefore a matter of examining branches. If all x's branches show x as having property P, then P is essential to x; if some show x as lacking P, then P is not essential. The question of whether or not essential properties exist becomes on the branched model an empirical question, one that depends on the way the world is.

A caveat must be registered at this point. Individuals which exist in the past, such as Julius Caesar, are no longer branched. If therefore their essential properties are those they have on every branch, all the properties that they actually come to have, such as crossing the Rubicon and being stabbed by Brutus, are essential to them. Although such a view of what constitutes an essential property would not be uncongenial to those who consider an individual to be defined by the set of its properties, and for whom all an individual's properties are therefore essential to it,[2] most philosophers today would reject the idea that it is essential to Caesar that he crossed the Rubicon, or was assassinated on the Ides of March. Such properties of Caesar are contingent, not essential. Caesar's essential properties, therefore, are not those which he actually possesses, i.e. which he possesses once his life has ended, but those which he possesses when his life has just begun, and which apply to him on every branch. These properties are determined by the structure and shape of the four-dimensional object that represents Caesar at the moment of his birth, or better at the moment of his conception. Let us name this enormous branched object, which has not yet undergone branch attrition, *Caesar's life-tree*. Then, on the branched model, the properties that are essential to Caesar are those that apply to him on every branch of his life-tree.

An individual's essential properties are the ones it possesses on all branches of the branched structure which represents it at the moment when it starts to exist. Its other properties, which it has on some branches but not on others, will be contingent or accidental. It is, for example, contingent that I am of the particular height and weight that I am. At other times I weigh more or less than I now do; on other branches I am taller or shorter than I am actually. To say that I *might have been* 6 feet tall is to say that on some branch or branches of my life-tree I *am* 6 feet tall. Are there any essential properties here? Is there, for example, any

---

[2] In a letter to Arnauld, Leibniz says that 'if, in the life of any person, and even in the whole universe, anything went differently from what it has, nothing could prevent us from saying that it was another person or another possible universe which God had chosen. It would then be indeed another individual' (Leibniz 1902: 127–8, cited in Plantinga 1974: 90).

height that I am essentially less than? No doubt there is. Although
the maximum height that I attain on any branch of my life-tree
is unknown, this height is a feature of the branched model at the
time of my conception, and is therefore in principle knowable. I
would imagine it would be, in my own case, a little less than 6
feet 6 inches, and that I reach it on a branch where for some
reason I consume quantities of growth hormone when young. If
this is so, then indeed it will be an objective fact about the
world—an empirical fact—that one of S.McC.'s essential properties is that he is less than 6 feet 6 inches high.

It may be asked, why 6 feet 6 inches? What makes being shorter
than 6 feet 6 an essential property of S.McC., while being shorter
than 8 feet is an essential property of Bona Ring, a Dinka of
Southern Sudan? Each individual, on the branched model, has its
own particular set of essential properties, and what makes being
less than a certain height an essential property of a living creature
is a combination of genetic and environmental factors. The phrase
'genetic factors' refers broadly not only to the genes encoded in
the organism's DNA, but to its whole state at the moment of
conception, i.e. what the organism is like at the root of its life-tree.

'Environmental factors', on the other hand, concern the branches
of the life-tree. On one set of branches the organism is fed adequately, on another it is starved, on others it is afflicted with disease.
What we shall refer to as the *essential physical characteristics* of
a living being, therefore—such things as the greatest possible size,
weight, strength, etc. that the individual is capable of achieving—
depend partly on its heredity and partly on its environment. Since
'environment' includes technological development, the essential
physical characteristics of an organism are to a large extent contingent on the ability of technological skill to either supplement
or overcome genetic factors.[3]

The dependence of essential properties upon time, and upon
such factors as scientific knowledge and technological development, is worth emphasizing. It provides, on the branched model,
yet another example of how the probabilities of future event-types
are not absolute, but are always relative to the state of the universe at a particular moment. Take as an illustration Plantinga's
hypothetical state of affairs *Agnew's swimming the Atlantic Ocean.*
Plantinga intends this to be an example of something that is
possible in a sense that is wider than *causal* or *natural* possibility,

---

[3] e.g. haemophilia is a genetic disease that was normally fatal until modern methods of
managing it were discovered. Hence the essential properties of haemophiliacs today differ
from what they were a hundred years ago.

and narrower than *logical* possibility.[4] Michael Loux, in the introduction to his anthology (1979: 27), uses the term *metaphysical possibility* to describe the intermediate kind of possibility that Plantinga has in mind. Plantinga's example, of Agnew's swimming the Atlantic, is meant as an example of something that is metaphysically possible but causally impossible. An instance of something that is logically possible but metaphysically impossible (in Plantinga's sense) might be Agnew's standing in the middle of the Atlantic and reaching out to touch New York with one hand and Lisbon with the other. Or Agnew's playing football with the moon. Since one of Plantinga's metaphysically possible worlds is a world in which Agnew swims the Atlantic, *not being able to swim the Atlantic* cannot be, for Plantinga, one of Agnew's essential properties. On the branched model, on the other hand, *not being able to swim the Atlantic* is indeed one of Agnew's essential properties, given Agnew's constitution and today's technology. But fifty years from now, who knows? Agnew's grandson, equipped with a body cream that protects him from the cold and a hi-tech way of storing oxygen in his lungs at times of low exertion that enables him to doze in the water, may well swim the Atlantic on some future branches. This would be an example of how the essential physical characteristics of human beings change with time. I have no idea whether there *is* a future branch on which Agnew's grandson swims the Atlantic, but what people know or don't know is irrelevant. Either such a branch exists, in which case what was essential to Agnew is not essential to his grandson, or it does not.

There is, I think, a more general lesson to be learned from the example of Agnew and the Atlantic. This is that the branched model imposes on us, when talking or thinking about possibility, a discipline which curbs the flight of our imaginations. The model offers us, above and beyond logical possibility, a clear criterion of what is empirically possible and what is not. Relative always to a set of initial conditions obtaining at a time, on a hyperplane, what is possible is what is on at least one future branch; what is impossible is what is on none. Possibilities do not come to us out of the blue, they are there on the branched model. Furthermore, between logical possibility, which is governed solely by the law of contradiction, and physical or causal possibility, there is no third alternative. On the branched model, there is no room for Plantinga's concept of 'metaphysical' possibility. This is not to say

[4] Plantinga (1973), in Loux (1979: 146).

that the way of thinking about modality that goes with the branched model is better or worse than Plantinga's, but only that the two are different. Corresponding to the different conceptions of possibility, there will be different conceptions of essential properties.

So far we have discussed only essential physical characteristics of individuals, such as 'being incapable of lifting more than 1,000 lb.'. There are, however, two other extremely interesting kinds of essential property, namely *substance-type* and *origin*. I shall discuss the alleged essentiality of substance-type here, and that of origin in the next section.

Of all the different predicates that a thing can fall under, e.g. *heavy, white, snub-nosed, walking, humourless, tiger*, and *gold*, those which answer the question 'What is it?' make up a special category. Aristotle points out, in the *Categories* and the *Metaphysics*, that it is no answer to the question 'What is x?' to give one of x's qualities; what we want is something that reveals x's substance. For Aristotle this involves giving the species or kind of thing that x is:

Only [species and genera, among predicables] reveal the primary substances [e.g. Socrates]. For if one is to say of the individual man *what he is*, it will be in place to give the species or the genus (though more informative to give man than animal); but to give any other thing would be out of place—for example to say 'white' or 'runs' or anything like that.[5]

For when we say of what quality a thing is, we say that it is good or bad, not that it is three cubits long or that it is a man; but when we say *what* it is, we do not say 'white' or 'hot' or 'three cubits long', but 'a man' or 'a god'.[6]

Let us call a predicate which answers the question 'What is x?' a *sortal*. Among sortals, which specify the sort or kind to which a thing belongs, there are some which apply to an individual at all times that the individual exists, while others apply only during part of its existence. Examples of the first type are *tiger, gold, table, automobile*, and *animal*, while the second type, which Wiggins calls 'phase sortals', includes *child, tadpole, foal*, and *caterpillar*. We shall refer to sortals of the first type as 'substance sortals'. Thus the phase sortal *child* denotes a phase or subsection of the life of an individual falling under the substance sortal *human being*.

By definition, a substance sortal applies to an individual at all times at which that individual exists. If an individual falls under

---

[5] *Categories* 2^b30–7, translated in Ackrill (1963), with italics and square brackets added as in Wiggins (1980: 14).
[6] *Metaphysics* Z 1028^a15–18, tr. W. D. Ross (Oxford, 1928).

a substance sortal, the only way in which the sortal can cease to apply to it is by the individual's ceasing to exist. This holds on all branches. Therefore, a substance sortal which characterizes an individual characterizes that individual essentially.[7]

The two kinds of essential property that have so far been discussed, essential physical characteristics and substance-type, differ significantly from each other. Essential physical characteristics vary from one individual to another within a species, a common type of variation being variation in degree. Thus it is essential to one individual that it cannot run faster than 16 m.p.h., to another that it cannot run faster than 17 m.p.h. But substance-type is all-or-nothing, not a matter of degree, and is common to all members of a given species. If a biologist has on the table in front of him a small white egg, about two millimetres in length, he may speculate about whether it is the egg of *Camponotus*, the carpenter ant, or of *Dorylus*, the African safari ant. But whichever it is, the organism inside the egg belongs to its species essentially. Even if, by some sophisticated operation, the biologist were able to make the young carpenter ant behave and look like a safari ant, all he would have succeeded in doing, on one branch of the ant's life-tree, would be to make a carpenter ant resemble a safari ant. He would have done nothing to change the substance-type of the ant itself, which characterizes the ant on all branches of its life-tree and which the scientist is powerless to alter. In particular, he would not have 'changed a carpenter ant into a safari ant'. The substance-type which an individual is born into is the substance-type that stamps it indelibly, not only throughout its actual career, but throughout every possible career permitted it by the branched model. The model leaves open the possibility that some individuals may belong to no substance-type. But those that do belong, do so essentially.

(ii) *The Necessity of Origin*

Of the three kinds of essential property that individuals may be thought to possess—essential physical characteristics, substance-type, and origin—it is the third, origin, that has been the most

---

[7] Wiggins (1980: 60 ff.) attempts to show that every individual falls under at least one substance sortal. In Aristotelian terms this would involve showing, for any individual x, both that there was an answer to the question 'What is x?' at all times, and also that there was an answer which was the same at all times that x existed. That the latter should be true seems not implausible, at least for the individuals which are familiar to us, but there is no obvious proof.

active subject of debate in recent years.[8] As with many other philosophical debates over the last twenty-five years, the discussion of origins began with Kripke. In *Naming and Necessity* Kripke asks the by now famous question, whether Queen Elizabeth could have been the child of Mr and Mrs Truman? A simple-looking question, but one that admits of no simple answer.

Kripke makes it clear that the question is not whether the Queen could have originated from an ovum transplant, her biological parents still being the Duke of York and Elizabeth Bowes-Lyon. *That* is obviously possible.[9] Nor is the question whether an announcement in the papers might not one day reveal the shocking truth, involving babies being smuggled back and forth across the Atlantic, that Elizabeth was in fact the child of the Trumans, and that her parentage had been successfully concealed up to now. That too is obviously possible. But Kripke's question is a different one: whether this very person, born of the then Duke and Duchess of York in England in 1926, could instead have had a very different origin, being the child of different parents and originating from a different sperm and egg. And this Kripke thinks is not possible.

How could a person originating from different parents, from a totally different sperm and egg, be *this very woman*? One can imagine, *given* the woman, that various things in her life could have changed: that she should have become a pauper; that her royal blood should have been unknown, and so on. One is given, let's say, a previous history of the world up to a certain time, and from that time it diverges considerably from the actual course. This seems to be possible. And so it's possible that even though she were born of these parents she never became queen . . . But what is harder to imagine is her being born of different parents. It seems to me that anything coming from a different origin would not be this object.[10]

Those who have commented on this passage since its appearance in 1972 have been divided on the question of whether an individual's origin is essential to it. John Mackie, Colin McGinn, and Graeme Forbes have defended Kripke's position, or something close to it, while Patricia Johnston, M. S. Price, and Robert Coburn have opposed it. Roughly and broadly, as will be argued in this section, the metaphysical picture of the world provided by the branched model supports origin essentialism.

---

[8] See e.g. Mackie (1974*b*); McGinn (1976); Johnston (1977); Forbes (1980; 1985; 1986); Price (1982); and Coburn (1986).
[9] Kripke (1980: 112).   [10] Kripke (1980: 113).

Kripke asks, how could a person originating from a different sperm and egg from the one she actually came from, *be* that person? The implication is that she could not, and that it is consequently essential, for every person x, that x should originate from the very sperm and egg that x actually originates from. A fertilized ovum is a *zygote*. Colin McGinn argues in support of Kripke that *zygote*, like *child* or *adult*, is a phase sortal of the substance-type *human being*.[11] Just as an adult is transtemporally identical with the child he or she once was, so the child is transtemporally identical with the infant, the infant with the foetus, and the foetus with the zygote. As McGinn says, any attempt to break the obvious biological continuity here would surely be arbitrary. The zygote and the adult are one and the same human being, but at different stages of development.[12] Since things that are identical are necessarily identical, every human being is necessarily identical with the zygote he or she came from, and therefore could have come from no other zygote. This reasoning, if correct, suffices to establish the first part of Kripke's doctrine of origin essentialism.

The second step is to show the necessity of the link between *gametes* (sperm and egg) and zygote. McGinn cannot appeal to phase sortals here, since there is no underlying substance-type of which *zygote, sperm*, and *ovum* are phases. Instead he constructs the following argument. Suppose that I come from some other gametes, say the ones Nixon came from. Imagine a possible world in which this occurs. Now add my actual gametes to this world (since the two sets of gametes do not exclude one another, this is always possible) and let these gametes develop into an adult. Which of the two individuals that result has the stronger title to be me? McGinn says that his intuitions favour the second individual, pointing out that the same conclusion would be reached even if the first set of gametes were genetically similar to mine, and conjectures that it is by reason of a certain sort of spatio-temporal continuity that we prefer the actual gametes of a person as a criterion of identity for that person. McGinn calls the relation of 'coming from', instanced both by my coming from my zygote, and my zygote's coming from its gametes, the relation of *d-continuity*.[13] The adult is d-continuous with its zygote, and the

---

[11] McGinn (1976: 132). What is meant of course is *human zygote*.

[12] The phase sortal *person* is not at issue here. Those working in biomedical ethics are generally agreed that the foetus does not become a person until a minimal degree of autonomy and independent viability has been attained.

[13] McGinn (1976: 132–3); d-continuity bears obvious analogies to the relation of genidentity (Ch. 7 n. 47 above).

zygote is d-continuous with its gametes. Furthermore, the relation is *rigid* in the sense that if x is d-continuous with or comes from y, then x comes from y in every possible world in which x exists. If McGinn's arguments are correct, the rigidity of the relation of d-continuity lies at the basis of origin essentialism.

However, McGinn's arguments may not be correct, and they have been challenged on separate grounds by both Johnston and Forbes. Johnston's objection is that it is easy to imagine circumstances in which, given two individuals in some possible world, the one with the stronger claim to be x is *not* the one that came from x's gametes. If this were so, origin would not be essential.

Suppose that in some possible world Nixon's actual gametes develop into an adult; suppose further that at each stage in the development, this individual has a life the events of which cannot be distinguished from those of the actual Hitler at corresponding stages. Now add Hitler's actual gametes to the world, and suppose them to develop into a person whose life is qualitatively identical with the life of the actual Nixon. Which of these individuals has the greater title to be Hitler? My intuitions are decidedly in favour of the former individual.[14]

Johnston's conclusion is that questions of what is essential and what is not essential to an individual depend on *context*. There is no absolute answer to such questions: in some contexts origin may be the most important consideration in determining identity, but in others it may carry so little weight that drastic alterations in origin are compatible with no change in an individual's identity. What kind of reply could McGinn make to Johnston's objection? Here the branched model may be of use, for it lends itself to a type of essentialism in which answers to questions about what is essential can be given which are *absolute*, not dependent on context. The model supports the necessity of origin, and can supply reasons why, in Johnston's example, it is Nixon, not Hitler who should be identified with the person coming from Nixon's gametes. Johnston invites us to consider a possible world in which the person coming from Nixon's gametes leads a life indistinguishable from Hitler's, and conversely Hitler's gametes produce someone whose life is indistinguishable from Nixon's. Let's assume that problems about dates in this world are not an issue, i.e. that Nixon and Hitler are exact contemporaries.[15] Could the branched model accommodate Johnston's example?

---

[14] Johnston (1977: 414).
[15] April 1889 was not the month in which Richard Nixon was born, but is noteworthy for having witnessed the birth of Adolf Hitler, Ludwig Wittgenstein, and Charlie Chaplin.

It could in this sense, that at the time of Nixon's and Hitler's birth there may indeed be a possible future branch, call it *B*, on which the person coming from Nixon's gametes leads Hitler's life, and the person coming from Hitler's gametes leads Nixon's life. But, in the branched model, both Nixon and Hitler are themselves branched individuals, and branch *B* is only one of many branches on their respective life-trees. On others, the individual coming from Nixon's gametes leads Nixon's life, and the individual coming from Hitler's gametes leads Hitler's life. The identity of the former branched individual is not affected by the fact that it has a Hitler-like branch. It can have a Hitler-like branch, i.e. be qualitatively indistinguishable from the actual Hitler on one or more branches, without *being* Hitler. That is one of the nice features of the branched theory of individuals: individuals as branched objects are large enough to accommodate many possible outcomes or careers. Nevertheless, their *identity* is not constituted by these possibilities. Their identity is constituted by their origin: in the case of humans and other organisms which reproduce sexually, by the union of their gametes.

Let us use a term that Graeme Forbes has borrowed from biology, and speak of an organism's *propagules*.[16] The term propagule covers both sexual and asexual reproduction; thus an amoeba which divides produces two similar individuals from a single propagule by *fission*, while a fertilized ovum or zygote comes from two dissimilar propagules by *fusion*. In the branched model, an organism's propagules are essential to it in the sense that whatever does not come from those propagules cannot be that organism. It is in this sense that an organism derives its identity from its propagules.

Let 'propagule essentialism' be the version of origin essentialism which maintains that an organism's propagules are essential to it. (As will be seen below, origin essentialism in the strict sense is a stronger theory than propagule essentialism, because it upholds the essentiality of other features of a thing's origin apart from its propagules.) The 'propagule essentialism' that is supported by the branched model does not differ in any important respect from Forbes's theory of essentialism, nor from McGinn's. However, although Forbes and McGinn reach the same conclusions, they disagree on how to argue for them. Forbes (1985) criticizes the fact that McGinn bases his argument for essentialism on the alleged d-continuity of zygote with adult human being. Forbes's objections are the following.

[16] Forbes (1980: 353; 1985: 133).

Forbes agrees with McGinn that *zygote*, like *child*, may be used as a phase sortal which falls under the sortal *human being*.[17] We may speak of the 'zygote phase' of a human, and when we do it will be true to say that the human is identical with the zygote he was, just as it is true to say that the human is identical with the child he was.[18] However, Forbes claims that the word 'zygote' is more normally used in another sense to mean 'zygote cell', and in this sense it is not true to say that a human is identical with a zygote. We must distinguish, Forbes says, between a human being and the cells that constitute him at various stages of his development. By Leibniz's law, he is not identical to the sum of the cells which make him up at any moment, since the human characteristically outlasts any such sum. Therefore, during the zygote phase, he is not identical to the single cell which makes him up then, for the cell ceases to exist when it divides, while the *zygote* (where the word is now used as a phase sortal) does not, any more than the *child* ceases to exist when it grows up.[19]

This argument, I believe, is based on a confusion. It is true that a human being, or more generally an organism, is not constituted by the same cells over a period of time. Organisms are constantly gaining and losing matter. But it does not follow from this that a human being or an organism is not constituted by, or identical with, the very cells which make it up at any moment. At time t_1, the rabbit which I hold in my hand is identical with *this* body; at time t_2, after a good meal, it is identical with *that* body. To suppose that a rabbit, or a human being, is identical with something other than the body which constitutes it at each moment is to make assumptions which I doubt Forbes would wish to make. Finally, to acknowledge that the body with which organisms are moment-by-moment identical *changes*, over time, is in no way to cast doubt on the identity in question. This is what transtemporal identity is all about. We are, then, perfectly free to assume with McGinn that, during the zygote phase, every human being is identical with its *zygote cell*. Being identical, it is necessarily identical, and consequently in no possible world could that same human being fail to be identical with, and hence fail to come from, that cell. That, I take it, is McGinn's argument for essentialism.

Forbes's criticisms of McGinn are the criticisms of one essentialist by another. The objections raised by M. S. Price are of a different order, for their author sets out to demonstrate that

---

[17] Strictly speaking, *human zygote* falls under *human being*, *tiger zygote* under *tiger*, etc.
[18] Forbes (1985: 137).
[19] Cf. Forbes (1985: 136–7). See also Forbes (1986: 28–9 nn. 11 and 12).

propagule essentialism is false. Price asks us to imagine the following rather desperate circumstances. George VI and his wife produce no gametes and hence no heir to the throne. Cambridge scientists are called in to manufacture a zygote from scratch, and the zygote they produce turns out (presumably by chance) to be in every respect, molecule for molecule, identical with Elizabeth II's actual zygote. The zygote is duly implanted in Elizabeth's mother's uterus, and from then on the life of the person it develops into is indistinguishable from the life of Elizabeth II. Imagine that all this happens in some possible world. In that world, Price claims, we would have no reason to deny that the person who came from the implanted zygote was identical with Elizabeth II, despite not having the same gametes. To imagine the situation is precisely to imagine Elizabeth's having been a test-tube baby.[20]

What Price has done, using this example, is to attempt to replace the criterion of *continuity* that McGinn and Forbes make use of, in establishing identity, by the criterion of *exact similarity*. McGinn argues that no human being could come from gametes with which that human being was not d-continuous. Forbes maintains that a person's identity depends on that person's coming from certain propagules, and this requirement entails spatio-temporal continuity between propagules and zygote. Price, however, destroys the continuity by envisaging a situation in which a zygote is synthesized in the laboratory. Here the continuity between the Queen's gametes and zygote is broken, and the thing that tempts us to say, nevertheless, that the resulting individual is identical with the Queen is the exact similarity of her zygote and subsequent life to that of the Queen. The question is, can exact similarity replace spatio-temporal continuity when identity is in question?

Here is an argument to show that it cannot. Imagine a situation in which scientists are not attempting to replicate anything as complex as a human zygote, but are synthesizing simple viruses consisting of strands of DNA. A scientist has under her microscope two viruses, one which is natural and came from its propagule by fission, and another which is synthetic and was manufactured from chemical components in the laboratory. The two viruses are qualitatively indistinguishable in every respect. Let VN denote the natural one, and VS the synthetic one. A visitor enters the lab, and the following dialogue takes place:

[20] Price (1982: 35).

VISITOR   This is fascinating work you are doing. Could VN have been
synthesized in the lab?

SCIENTIST   Yes, we have now perfected the technique of exactly replic-
ating viruses of VN's type.

VISITOR   I see that, but my question was not whether viruses of VN's
type could be synthesized, but whether VN *itself* could have been
synthetic?

SCIENTIST   Are you asking whether I could have mixed up VN and
VS? The answer is no—VN is on the left and VS on the right.

VISITOR   That wasn't my question either. What I meant was, is there
any possible world in which you and I are looking at VN and VS,
these very same viruses, but in which it is VN that is synthetic and
VS that is natural?

SCIENTIST   I assume that what you have in mind is not just a situation
in which the names 'VN' and 'VS' are interchanged.

VISITOR   That's right, the names are rigid designators.

SCIENTIST   Then my answer is that there is no such world. The only
thing that distinguishes VN from VS is the fact that VN is natural,
VS synthetic. This is a fact about their histories, not their present
constitution. If VN were synthetic, VN wouldn't *be* VN. Consequently,
VN couldn't have been synthesized in the lab.

This thought-experiment with viruses parallels Price's thought-
experiment, except that in the case of the viruses we have two
exactly similar individuals in the same possible world. The question 'Which one of these is VN?' cannot be answered by appealing
to similarity, but only by appealing to history, i.e. spatio-temporal
continuity with propagules. In Price's example there was only one
candidate, a person qualitatively identical with the Queen, and in
the absence of a rival candidate the ultimate test of identity,
continuity with the Queen's actual propagules, was not employed.
But in the virus example it had to be employed, and is the only
test that could provide an answer. Exact similarity, therefore,
cannot replace continuity in questions of identity.[21]

Let us focus on the viruses for another minute. In the branched
model, since reproduction by fission goes on constantly within a

---

[21] The same conclusion applies to artefacts. Kripke (1971: 86) asks whether *this lectern*,
the lectern he is presently using, could have been made out of ice instead of wood, and
the answer is no. But could this lectern have been made out of exactly the same substance
it is in fact made of, molecule for molecule, but in another workshop, by a different
carpenter, in a different country? The answer is again no; another exactly similar lectern
could have been made somewhere else, but not *this* lectern, which is a branched object of
a certain definite shape located at a certain position in the universe tree. On different
branches the lectern may be revarnished, sawn up, converted into a hat-rack or burnt, but
these events do not affect its identity, which consists not in the identity of the matter
composing it, or even in identity of both form and matter, but in identity of origin.

given species of virus, it might appear that the entire species was composed of one huge tree. This might in fact be true—every virus of a given species might be descended from a single ancestor-virus, which millions or even billions of years ago formed the root of the tree of which all other viruses of that species are parts. However, we want a sense of the word 'individual' in which there now exist in the world a zillion individual viruses of species X, not just one super-virus in the shape of a tree. Let us therefore stipulate, each time a virus or other organism replicates by fission, that the parent ceases to exist and that the two progeny viruses each constitutes a new individual coming into existence for the first time.[22] These new individuals are of course branched, in the way all individuals are, their branches reflecting the different possible fates in store for them, but this is the normal kind of branching, not the branching of a virus super-tree. What the stipulation does is break down super-trees, representing in some cases whole species of organisms, into sub-trees representing single individuals.

The stipulation that was made for organisms that reproduce by splitting will also be made for organisms that reproduce sexually, by means of gametes, but which may on occasion divide. Forbes (1980) contains a description of how identical twins are produced by an abnormal division of the zygote in which the daughter cells separate instead of remaining together.[23] By our stipulation concerning fission, the two zygotes thus produced will each be distinct newly formed individuals. As a consequence the life-tree of a single human being will not include, as sub-trees, the life-trees of identical twins, even though these twins will have come from the same gametes. No doubt of many of us it would be true to say that our zygotes could have divided in such a way as to produce identical twins. But, because of the stipulation that if and when this happens the original zygote ceases to exist, and the two new zygotes thus produced each constitute the root of a new life-tree, none of us is faced with the logical embarrassment of having to say, 'I might have been twins.'[24]

---

[22] This stipulation has the advantage of disposing of the unanswerable question 'Which of the two progeny is identical with the parent virus?' The answer is, neither.

[23] Forbes (1980: 353–4).

[24] We are not even faced with saying, 'I might have been one of a pair of identical twins', since the 'I' ceases to exist when the zygote divides abnormally. Forbes (1980: 354–5) rules out the possibility of someone x who is not an identical twin being an identical twin by means of the principle that there can be no *ungrounded identities*. For let the two twins be y and z. Then, since at the time of their formation the two twins are qualitatively identical, the identity of x with y rather than with z would be a bare metaphysical fact, not grounded in any significant difference which could make y rather than z identical with x.

To conclude this section, the difference between propagule essentialism, on the one hand, and what was referred to above as strict origin essentialism, on the other, must be discussed. Propagule essentialism is the doctrine that no organism could have failed to come from the propagules it actually came from. This holds whether the propagules in question are natural or synthetic. In the case of each organism its propagules fused to create it, or divided to create it, at a particular place and time. The organism's propagules are essential to it, but are the place and time at which it came into existence essential to it? That is, could the same organism be formed from the same propagules, but at a different place and time? Is every organism tied to the exact place-time at which it actually came into existence, or could it have come into existence at some other spatio-temporal location?

This question is raised in a colourful way by McGinn, who points out that if the exact time and place of birth of an individual such as Nixon is essential to him, then all the events surrounding his coming to be will also be essential. Not only time and place, but *everything* true of Nixon at the moment of his creation would be necessarily true of him, even that he started to be in a room containing a vase of geraniums.[25] This theme is taken up by Coburn, who asserts that if time and place of origin are essential, then if his wife came into existence in a room with purple wallpaper, she has as an essential property that she came into existence in a room with purple wallpaper.[26] This looks, at first sight, like a *reductio ad absurdum* of the essentiality of time and place of origin. We should, in McGinn's words, be alert to the dangers of any view that lets the necessity of origin attach to the *circumstances* of origin.

Moving to propagules might appear to solve the problem, in the first instance at least. If what is essential to Coburn's wife is that she originated from the gametes she actually came from, then many restrictions on time and place of origin are lifted. Her gametes could have fused in a room with purple wallpaper, or with geranium wallpaper, or with no wallpaper. We are talking, it seems, about the year 1951, and so the spatio-temporal region in which she could have started to exist was fairly limited, but today that region would be much larger. Frozen sperm and ova might be carried on a spacecraft, and the individual to which they give rise come into existence thousands of years after his or her parents have perished. A person born of gametes which at one

---

[25] McGinn (1976: 130).   [26] Coburn (1986: 179).

point were considered for such a voyage might say, quite truth-
fully and with a straight face, 'I could have been born in AD
4000'. Such is the liberating power of propagule essentialism.

On the branched model, the implications of the theory of
propagule essentialism are the following. Since the propagules
which fused to produce S.McC. could have fused at different times
and in different circumstances there will be, in the branched model
at times prior to my conception, a set of individual S.McC.'s, any
one of which I could have been. These individuals constitute what
one might call an 'equivalence class' of S.McC.'s. What unites
them is that they all originate on different branches of the model
from S.McC.'s propagules, and they serve as the semantic truth-
makers for counterfactuals like 'If my father and mother had
settled in Connecticut, I would have been born an American.'

In creating this equivalence class, propagule essentialism recognizes the existence of a large and rather messy class of non-actual
individuals who (had branch attrition worked out differently)
would have been S.McC. The advantages of such a move are that
some of the unwelcome consequences of strict origin essentialism
are avoided. If I happened to come into existence in a room with
geranium wallpaper, that fact is not essential to me. Nor, since
each of my propagules also belongs to an equivalence class, is it
essential that they came into existence at the particular place and
time they did, nor for that matter that my grandfather wore purple
socks on his third birthday, assuming that he in fact did so.[27] The
set of non-actual individuals who under different circumstances
would have been my father also constitutes an equivalence class,
and some members of it originate on branches where my grandfather wore purple socks while other members do not. In this way
propagule essentialism avoids origin essentialism's unwelcome implication, to the effect that all events lying in an individual's past
(or in its propagules' past) are essential to it.

(iii) *Natural Kinds*

In the early 1970s there occurred a revolution in the theory of
meaning, brought about by the work of Kripke and Putnam in
analysing the meaning of common nouns such as *gold, lemon,
tiger*, and *water*.[28] Before that time, the generally accepted view
had been that the meaning or intension of such words was given

[27] Cf. Coburn (1986: 179).
[28] See Putnam (1970; 1973; 1975c); Kripke (1971; 1980).

by a set of defining characteristics or properties, the truth of any sentence of the form '*x* is a lemon', or '*x* is gold', being established by *x*'s possessing all or a substantial number of these properties. Thus the *Concise Oxford Dictionary* defines *lemon* as 'pale-yellow oval acid-juiced fruit used for flavouring', and before 1970 most philosophers, myself included, took the view that to be a lemon consisted precisely in possessing this or some similar set of properties.

Kripke and Putnam changed all that. For them, the meanings of common nouns like *lemon* and *gold* are not determined by sets of defining properties or defining characteristics. Instead these words function much more like proper names; the meaning of *lemon* and *gold* consists in the fact that nouns like these name *natural kinds*. A natural kind is a naturally occurring species of thing, not an artefact, about which empirical discoveries are still being made. By no means all common nouns name natural kinds, and the meanings of those that do not fall into quite a different category from the meanings of those that do. Nouns like *triangle* and *bachelor*, for example, do not denote natural kinds, and one of the things that Kripke and Putnam showed was that between these nouns, on the one hand, and between *gold* and *tiger*, on the other, there is a great gulf fixed.

A triangle is a plane figure bounded by three straight lines, and the full and complete meaning of the word is given by these defining characteristics. To say that a triangle is a plane figure so bounded, or that a bachelor is an unmarried male, is to make an assertion that is *analytic*. This is not the case for a statement like 'Lemons are yellow', or 'Gold is the element with atomic number 79'. As Kripke points out, the fact that gold has atomic number 79 was an empirical discovery, not a truth of logic. No one could have said, at the time that atomic weights and numbers were in the process of being determined, 'If that metallic sample doesn't have atomic number 79, it can't be gold.' They could, on the other hand, have said, 'If that plane figure doesn't have three straight sides, it can't be a triangle.' So it is true, but not analytic, that gold has atomic number 79, or that water is H_2O. Moreover such propositions, once their truth has been established, are not only true but *necessary*; the word 'gold' rigidly designates the substance with atomic number 79 in all possible worlds. Natural kind terms, occurring in statements like 'Gold has atomic number 79', or 'Water is H_2O', yield examples of Kripke's newly discovered category of *necessary a posteriori* truths.

What might be called the 'central problem' of natural kinds is this: is there a sharp division or boundary between any two of them? Or do they merge into one another by imperceptible degrees? If the latter were the case rather than the former, this would constitute for some philosophers a strong reason for concluding that natural kinds do not exist. A set of natural kinds that blended smoothly into one another, without gaps or boundaries, would be for them not a *set of kinds* at all, but a continuum.[29] The 'friends of natural kinds', therefore, will be concerned with the question of whether natural kinds must be separated by natural divisions. In this matter the branched model may be of some assistance.

If colours were natural kinds, e.g. if the set of all red things constituted a natural kind, then it would not be true that natural kinds had sharp boundaries. The set of red things shades off insensibly into the set of orange things. But although colours are frequently used as examples by Quine in the first article in modern times that is explicitly devoted to natural kinds,[30] they do not constitute natural kinds in the sense understood here. A natural kind as we shall understand it is a species of substance in the Aristotelian sense, moreover a species of substance that occurs naturally, not something like a hammer or bicycle that is man-made, or something that is a mathematical entity like a triangle. To give the natural kind to which something belongs is to answer the question 'What is it?', and Aristotle insists that it is no answer to this question to give something's qualities, e.g. to say that it is red. Hence the fact that colours form a continuum constitutes no obstacle to maintaining that natural kinds are separated by sharp divisions.

A different sort of obstacle is the following. Natural kinds fall into two subgroups. In the first place there are kinds the members of which Aristotle calls *individual substances*, e.g. individual human beings, tigers, lemons, frogs, etc. These kinds are denoted by count nouns, or what Quine calls terms of divided reference.[31] The second type are what Aristotle calls *homoeomerous substances*, e.g. gold, lead, water, acid, sugar, etc., denoted by mass nouns.[32] Homoeomerous substances might be thought to constitute an exception

[29] Worse still, natural kinds might overlap at the most basic level (as distinct from overlapping hierarchically), so that a given individual might belong to two or more ground-level kinds. In the *Historia animalium* Aristotle appears to support the idea that natural kinds constitute a continuum when he says: 'Nature proceeds little by little from things lifeless to animal life in such a way that it is impossible to determine the exact line of demarcation, nor on which side thereof an intermediate form should lie' (588^b4–7). For arguments pro and con Aristotelian natural kinds as forming a continuum see Franklin (1986; 1989); Granger (1987; 1989; and other references cited therein).

[30] Quine (1969).   [31] Quine (1960: 90 ff.).

[32] *Metaphysics* Z 1028^b8–13. 'Homoeomerous' means literally 'of or having the same parts'.

to the principle that natural kinds do not blend insensibly, for we may start with a jug of water and, by adding small amounts of acid and letting it overflow, convert it by gradual degrees into a jug of acid.

For Aristotle, and for everyone up until the nineteenth century, basic differences separated individual substances from homo-eomerous substances. But today the differences seem minor and trivial. What is it that gives gold its characteristic properties, its essence? According to Locke the 'real essence' of gold, its inner constitution, is unknown to us. With the advent of the atomic theory of matter, however, the situation has changed. What gives gold its characteristic properties is the gold atom, either alone (atomic weight, atomic number, valence), or in conjunction with other gold atoms (crystalline structure, melting-point, malleab-ility). But a single gold atom is an individual substance, not a homoeomerous substance. Thanks to the atomic constitution of matter, homoeomerous substances reduce to collections of indi-vidual substances. Therefore the question of whether gold can be changed insensibly into lead, or water into acid, becomes the question of whether an atom of one kind can be changed insens-ibly into an atom of another kind.

When the question is put in this way, the answer is obviously no. A gold atom, which has 79 protons and electrons, is different in kind from a mercury atom, which has 80. Between gold and mercury there are no intermediate natural kinds, with say 79.5 protons. The natural kinds of physics and chemistry, therefore, comprising such things as *proton, electron, gold atom, water molecule, DNA*, etc., are discrete in the sense that they are separ-ated by sharp boundaries. For them, the 'central problem' of natural kinds is settled. However, there exist two other cases in which the central problem remains an issue.

In the first place, although it is true that atoms with different atomic numbers constitute distinct natural kinds, the atomic struc-ture of matter is below the threshold of perception. It is not certain that the phenomenal properties of natural kinds are such that the discrete character of chemical substances is preserved at the phenomenal level. For example, the phenomenal properties of undecane, a hydrocarbon with a carbon chain of length eleven, are to all intents and purposes indistinguishable from the phe-nomenal properties of dodecane, with a chain of length twelve. Therefore the sharp boundaries between chemical kinds which exist at the atomic level may not always be preserved at the phenomenal level, where the properties of distinct natural kinds may blend smoothly. If this were so, i.e. if precise boundaries between natural kinds were to be found only at the microscopic level, then the concept of a natural kind would lose much of its philosophical interest.

Secondly, the 'central problem' of natural kinds remains open in the case of organisms, where we face what Putnam calls the difficulty of 'abnormal members'.[33] Not all lemons are yellow. Not all tigers are four-legged. Given converging sequences of abnormal members of two different natural kinds, each of which is slightly more abnormal than its neighbour, may it not be impossible to say where one kind begins and the other leaves off? The existence of abnormal members was, according to Locke, a stumbling-block for those who wished to maintain that what constituted the difference between two natural kinds was the possession of a distinct 'real essence'. This hypothesis,

which supposes these essences as a certain number of forms or moulds, wherein all natural things that exist are cast, and do equally partake, has, I imagine, very much perplexed the knowledge of natural things. The frequent production of monsters, in all the species of animals, and of changelings, and other strange issues of human birth, carry with them difficulties, not possible to consist with this hypothesis.[34]

It is at this point that the branched model is able to contribute to the discussion. Take the problem of abnormal members. As Putnam remarks, a green lemon is still a lemon, and a three-legged tiger is still a tiger. The reason why, on the branched model, 'abnormal' forms such as these do not lead to a blurring of the boundaries separating natural kinds is that individuals are branched, and examination of the branches will reveal every abnormal form consistent with an individual's belonging to a given natural kind. Membership of the individual in the kind is determined not by its momentary state on a single branch, but by the shape of its whole life-tree. No matter how much the branches may diverge from the sortal stereotype, the individual continues to belong to its kind as long as it exists. Three-legged tigers, then, and Locke's 'changelings', do not constitute intermediate forms which bridge the differences between natural kinds, but are merely individuals of a specific kind as they appear on some branches rather than on others. A tiger which is three-legged on one branch may be perfectly normal on others. Its species is fixed by its origin, and is unaffected by whatever alterations it may suffer thereafter, even

[33] Putnam (1970), in Putnam (1975b: 140). See also Fales (1982).
[34] Locke (1689: III. iii. 17).

if on one branch a surgeon, or a geneticist, were to succeed in making it resemble a leopard.

The same applies to the phenomenal properties of the natural kinds of physics and chemistry. Although in most circumstances undecane and dodecane may be phenomenally indistinguishable, there will be some tests to which they can be put which will distinguish them. That is to say, on some branches samples of undecane and dodecane behave in observably different ways, even though on most branches they may be indistinguishable. Two substances which were indistinguishable in all circumstances, in every interaction with different measuring devices, no matter what tests were applied, would not be two substances but one. But if a sample of a pure substance *A* differs from a sample of a pure substance *B* in its reaction to one single test, i.e. in its behaviour on just one variety of branch of the model, then *A* and *B* are distinct natural kinds.

The branched model, then, can help resolve the 'central problem' of natural kinds in the following way. For Locke and others, the ever-present possibility of intermediate forms between any two natural kinds blurred the boundaries between them, and made implausible the idea that members of a kind shared a unique 'real essence'. But, on the branched model, the existence of an individual that was a genuine intermediate or hybrid, located between two natural kinds and belonging to neither or both, would be rare indeed. There are plenty of abnormal individuals that may *appear* to be intermediate, and to blur the borders between kinds. Occasionally, for example, a large wild dog may be mistaken for a member of *Canis lupus*. But on other branches that dog, brought up as a puppy in domestic surroundings, is wholly amiable, and on all branches its natural kind is *dog*, not *wolf*.

To be sure, the theory of evolution teaches that organic natural kinds are not fixed, but evolve slowly and give rise to new species over time. Consequently, the essential properties of these kinds also evolve, and new 'real essences' come into existence while old ones disappear. But evolution does not seem to be moving in the direction of filling in the gaps between species. On the contrary, the boundaries between natural kinds, once established, have been meticulously preserved over millions of years, and new kinds, with new boundaries, are constantly emerging. The branched model, housing individuals which belong to natural kinds in virtue of the permanent shape of their life-trees, encourages us to view the world's ontology as composed of these individual substances.

It should be emphasized that the theory of natural kinds which fits most easily with the branched model is not one based on Aristotelian essentialism, nor is it linked to laws of nature. In the case of biological natural kinds, Aristotelian essentialism would imply the existence of fixed essential properties necessarily possessed by every member of the kind in question. Species, however, evolve with time, losing old properties and gaining new ones. As is argued by Michael Ruse (1987), biological species should be viewed neither as Aristotelian, each possessing a real essence, nor as Lockian, reflecting merely a nominal classificatory schema, but as objective kinds, the definition of which is based upon a combination of morphological, reproductive, historical-evolutionary, and genetic factors. In the branched model, a species is pictured as a type of *natural four-dimensional branched shape*, each type being species-specific in so far as it is stamped with a characteristic dynamic pattern, yet loose enough to allow for individual variation, and with each individual's origin linked to propagules proceeding from other members of the same species. Through history, a species viewed in this four-dimensional way resembles an enormous linked and branched chain, embedded in space-time, evolving slowly, and differing in precisely specifiable ways from the chains of other species.[35] Its shape is well enough defined to be proof against the sceptical criticisms of authors such as Churchland (1985), who would insist that natural kinds be limited to those entities required by the basic laws of physics. Contrary to this view, natural kinds in the branched model are simply objective volumes of a certain characteristic shape, embedded in the four-dimensional geography of the world. Some of these shapes represent fundamental particles, but most do not.

It is worth observing, in conclusion, that the solution offered to the central problem of natural kinds by the branched model supports a notion of 'substance' that is more in keeping with the ideas of both modern physics and Aristotle than the notion we have inherited from philosophers as diverse as Descartes, Locke, and Kant.[36] In the seventeenth century, philosophers adopted a view of material substance as extended in space either homogeneously (Descartes) or in a corpuscular way (Boyle).[37] The prevailing

[35] If there are e.g. as many as 290,000 species of beetle, there will be 290,000 corresponding varieties of 4-dimensional volume in the branched model, each one a separate natural kind. Cf. Dupré (1981: 76). The individuals of each species constitute the separate links of the branched chain. See Fig. 7.2 above.

[36] In what follows I am indebted to Michael Ayers (1981) for his perceptive article 'Locke versus Aristotle on Natural Kinds'.

[37] Boyle's corpuscularian views are discussed in Alexander (1985: ch. 3).

philosophy of natural things was a mechanical one, and, in the words of Ayers, 'The mechanist's world is one in which all differences are differences of degree, and everything is in principle indefinitely mutable' (Ayers 1981: 255). Consider, as Locke did, the innumerable ways in which material substance, in the form of springs, gears, cog-wheels, and bearings, may be assembled together in the form of a watch. To the watchmaker, there may be thousands of different kinds of watches, and, between each of these kinds, thousands more that can be imagined and constructed if desired. For Locke there are no 'natural kinds' or species in the world of watches, apart from nominal essences formed from the complex ideas of watchmakers.

There are some watches that are made with four wheels, others with five; is this a specific difference to the workman? . . . It is certain that each of these hath a real difference from the rest; but whether it be an essential, a specific difference or no, relates only to the complex idea to which the name watch is given.[38]

No doubt an expert, with enough skill, could transform a watch into a radio by insensible degrees. Such is the character of mechanisms, based upon the idea that they are made out of a common material substance that is infinitely malleable. But it is precisely in this mechanistic way that Locke thought of biological species as well. A few lines below the quotation just given, he continues:

Just thus I think it is in natural things. Nobody will doubt that the wheels or springs (if I may so say) within, are different in a *rational man* and a *changeling*; no more than that there is a difference in the frame between a *drill*[39] and a *changeling*. But whether one or both these differences be essential or specifical, is only to be known to us by their agreement or disagreement with the complex idea that the name man stands for: for by that alone can it be determined whether one, or both, or neither of those be a man.

For Aristotle, as distinct from Locke, 'substance' is not what members of natural kinds are made out of. It is, instead, what they *are*. But it is what they are in the plural; *substances* not *substance*. Individual substances, in space and time, are what comprise the world. And individual substances are not transformable into one another. They are distinct, and retain the characteristics of the natural kinds to which they belong as long as they exist. In modern particle physics, it is the Aristotelian conception

---

[38] Locke (1689: III. vi. 39).   [39] An ape or baboon.

of substances, not substance, that prevails. Electrons, protons, quarks, muons, photons, and neutrinos are individual substances, some of which may pass in and out of existence but each of which belongs unequivocally to a specific kind while it does exist. The old seventeenth-century conception of a single material substance which runs through everything, and which (for Kant) survives all change, seems at least for the time being to be in retreat. In its place is the Aristotelian notion of individual substances, ranging in size from elementary particles to whales, which if represented four-dimensionally are identifiable by their four-dimensional shape, and which retain their species-specific essential properties on every branch. It is these substances which constitute the most natural ontology of the branched model.

9

Decision and Free Will

A CENTRAL feature of the branched model is *random selection*. Out of the indenumerably many branches above the first branch point, one and only one is chosen to become actual, i.e. part of the main stem of the model, and the choice is random. Nothing in the model indicates beforehand which branch will be selected. There is no 'preferred' branch. But this randomness of branch selection seems to conflict with belief in free will, and belief in our power to influence what the future will be like. In the progression of the first branch point up the model, are human beings at the mercy of indeterministic branch selection? Or is there some way in which we can so to speak 'navigate' our way up the tree, intentionally bringing it about that one type of branch is selected rather than another?

This question was raised in Chapter 3, when it was asked whether a scientist testing the conductivity of copper is forced to sit around in his lab waiting for the initial conditions he is interested in to reproduce themselves. These conditions consist in the two ends of a copper wire being joined to the terminals of a battery. Must he wait, or can he actively go out and create them? If indeed he cannot deliberately do this, but has to sit idly by until the right kind of initial conditions result from random branch selection, then scientific experimentation is a very different thing from what we believe it to be. The scientist does not sit still until the wires are joined: he joins them. Somehow or other, the model has to account for the obvious fact that human beings can decide for themselves what at least one small part of the future is going to be like, and act in such a way that it *is* like that. If any readers have doubts about this, let them ask themselves whether they are able to stop reading at this point, put the book down, and think about Robespierre or Helen of Troy instead of free will.

There is, of course, an answer to such talk of human powers and abilities that has become so firmly entrenched in our philosophical consciousness as to be second nature to many of us. This is, that we may *think* that we have the power to decide whether *A* will be the case rather than *B*, or *B* rather than *A*, but such beliefs are strictly speaking delusions. They will be delusions if the world we live in is Laplacian deterministic, so that whether it will be the case that A rather than B is determined by causes that existed before we even began to deliberate over the matter. If that were so, it would not be within our power to decide in any genuine way whether A or B; the question of A or B would have been decided long before we were born, and any so-called decision we came to would be determined by earlier states of our brain. This line of argument does not attempt to show that 'decisions' and 'powers' in some attenuated form do not exist, but on the contrary that they do exist and are compatible with universal determinism.

Whatever degree of persuasiveness this argument may have, it cannot be used here. Our problem is not to reconcile freedom of decision with determinism. Instead, the whole discussion is set within an *in*deterministic framework, witnessed by the fact that the branched model allows a single set of initial conditions to be followed by a large number of different physically possible outcomes. If determinism were true, there would be only one outcome, not many. Hence, for the branched model, the problem of seeing how determinism and freedom can be made compatible does not arise.[1]

We may have avoided the problem of how to make freedom and the power of decision compatible with determinism, but only, it will be said, by creating an even greater one. This is the problem of making freedom and the power to decide compatible with *in*determinism. If it is purely a matter of chance which of the many possible futures we are faced with becomes actual, how can human beings be said to decide anything? Instead of being caught in the grip of a rigid determinism, may we not be tossed like a cork on a sea of chance and indeterminism, equally unable to influence the course of events? This is indeed our problem, and it will be the object of this chapter to show how a viable theory of free action and decision can be built upon indeterministic foundations.

In an article published in *Mind* in 1961 that has become justly famous, J. J. C. Smart attempts to demonstrate that the so-called libertarian view of freedom, which regards freedom and determinism as incompatible and attempts to base freedom on indeterminism, is misconceived if not internally inconsistent. For Smart, as for most other philosophers, free will and chance are mutually

---

[1] Of course, as was said above in Ch. 1, the world may in fact *be* deterministic, and the real solution to the problem of freedom may *be* a compatibilistic one. In that case all bets are off, and the branched model is given a decent burial.

incompatible. To say 'This action was performed freely and intentionally by an agent', and also to say 'This action came about by pure chance', is contradictory. The libertarian, therefore, must hold that human behaviour, when free, is neither deterministic nor random, meaning that a free action is such that it cannot be explained deterministically, nor can it be attributed to chance. But it is precisely in saying this that the libertarian leaves himself open to Smart's rebuttal. For, says Smart, every event either is causally determined or, if not, is due to chance. Between determinism and chance there is no logical space. As Smart puts it, outside of 'This event happened as a result of unbroken causal continuity', and 'This event happened by pure chance', there is no third possibility.[2]

This is the libertarian's dilemma. He wishes to avoid the Scylla of determinism, and at the same time stay clear of the Charybdis of chance. But, Smart points out, to define freedom negatively, as that which is neither determined nor a matter of chance, will not do. C. A. Campbell, for example, attempts to define what he calls 'contra-causal freedom' in this way.[3] If indeed there is no third alternative to chance and determinism, then the definition will not work and the libertarian is embarking on an impossible mission. Between Scylla and Charybdis there is no room to pass.

But *is* there no third alternative? Is there no species of event the occurrence of which is neither causally determined nor a matter of chance? It will be the thesis of this chapter that there is in fact such a species of event, constituted by human choices and decisions. The way in which the argument proceeds is to show, first of all, that there are two components to the notion of chance. One of these involves the existence of a set of physically possible alternative outcomes to certain initial conditions, and the other involves the non-existence of any reason or explanation why one outcome is realized rather than another. Next it is shown that decisions resemble chance in requiring a set of alternative possible outcomes, but are unlike chance in that there exist reasons or explanations for them. Similarly, a free action resulting from a decision will also resemble a chance event in the first of these respects, and will differ from it in the second.

To sum up, an action which results from a decision will be a 'chance' event in the sense of being undetermined, but at the same

---

[2] Smart (1961: 296).
[3] Campbell (1951). Smart notes that neither 'determinism', nor 'chance', nor 'freedom' is defined at all precisely by Campbell, but in his estimation this is what 'contra-causal freedom' comes to.

time will be unlike a chance event in having an explanation why it occurs. In this way a gap will be opened up for it between the two extremes of chance and determinism.

(i) *Chance, Deliberation, and Decision*

What is a chance event? A chance event is an event such that (i) it is one of two or more physically possible outcomes of a single set of initial conditions, and (ii) there exists no reason or explanation for its occurring rather than one of the other outcomes. Since time immemorial, such things as flips of a coin or throws of a die have been considered to be examples of chance events. Relative to the initial conditions of flipping the coin or rolling the die, there are two possible outcomes for the coin and six for the die. If these are genuine instances of chance events, there will be no reason or explanation why the coin lands heads rather than tails, or why the die shows one face rather than another.

Of course, tosses of a coin or throws of a die may not be genuinely chance events. In the case of the coin the outcome may have been causally determined by the precise way the coin was tossed, the air currents it encountered, and the elasticity of the surface it landed on.[4] If so, replace the classical examples of coins and dice by quantum examples of radioactive decay, or of orbit jumps by the electron of a hydrogen atom, or of individual photons passing through a polarization analyser. The common element in all these examples is that a single set of initial conditions has two or more different physically possible outcomes, and that there is no causal explanation or reason why one outcome is selected rather than other.

On the branched model, which is an indeterministic model *par excellence*, the two components of chance are represented by (i) the many physically possible branches that occur above a single branch point or set of initial conditions, and (ii) the random selection of one and only one of these branches as actual, without any causal explanation why one branch is selected rather than another. These two components are represented by separate and distinct features of the model. Both the everyday notion of chance, therefore, and the indeterministic character of the model, appear to rest upon two distinct theses. The first thesis is existential and

---

[4] Coin tosses would be instances of objective chance if the spinning of the coin in mid-air were partly dependent upon the making and breaking of bonds with individual air molecules, which were in turn based on quantum phenomena. See Lewis (1980: 84 and 119).

affirmative, and the second is universal and negative. The existential thesis asserts the existence of two or more alternative outcomes to certain sets of initial conditions. The universal thesis denies the existence of any physical or causal explanation why one of these outcomes is realized rather than another. Representing the first in the model is multiple branching, and representing the second is random branch selection. It is the first of these features, but not the second, that is a necessary component of the human activity of deliberating and deciding.

Let us examine what is involved in the familiar process of deliberation. People can deliberate over many things: whether or not to go to graduate school, whether or not to get married, which movie to see, what to order for dessert. What is essential in all these cases is that there be at least two different courses of action open to the person who deliberates. If there is only one, then deliberation is out of place. Consider the example of a person contemplating jumping off a bridge attached to a bungee cord. This is certainly something that can be deliberated about. But if the decision is taken to jump, it is not only useless but inappropriate to deliberate whether to go up or down. We can deliberate about what is causally possible, but not about what is causally necessary or causally impossible, or at least about what we know to be necessary or impossible.

The type of deliberation in which we are interested is *practical deliberation*, deliberation about which course of action to pursue. Practical deliberation, which is discussed by Aristotle in books 3 and 6 of the *Nicomachean Ethics*, concerns the future, and differs from two other kinds of deliberation, which we may call *cognitive deliberation* and *deliberation about values*. Cognitive deliberation is not deliberation about action but deliberation about truth: deliberation about what to believe or accept as true as opposed to deliberation about what to do. The distinction between practical and cognitive deliberation underlies, for example, the division of labour in jury trials. The jury deliberates over the truth or falsehood of the proposition that the accused is guilty, while the judge, once the cognitive decision has been rendered and the prisoner found guilty, deliberates over what sentence to impose. The jury's deliberations are cognitive, the judge's practical.

The third kind of deliberation, deliberation about values, could be regarded as a subspecies of cognitive deliberation. Like the latter, it concerns what to think rather than what to do, the thinking in question being about values and ends and what ought to be rather than what is. A crude (and perhaps cynical) example relating all three types of deliberation would be the following: in Canada, the House of Commons deliberates over how the public's money *ought* to be spent, the Auditor-General deliberates over how it *was* spent, and the Cabinet deliberates over *how to spend it*.

Here, our concern is with practical deliberation, which has to do with action. As we have seen, deliberation requires a minimum of two alternative courses of action open to the agent, each one of which he can perform. Or, at least, each one of which he *believes* he can perform. As I sit at my desk, I deliberate over whether to walk home for supper, or to take the bus. But unknown to me the buses are not running. Does this mean that my deliberations are sham deliberations? Suppose that after due deliberation I decide to take the bus. In effect, I will have decided to do something that is impossible. Can one decide to do the impossible? We must proceed carefully here.

Aristotle says that we deliberate about things that are in our power and can be done (*Ethics* 1112^{a}30). But it is possible that he wasn't thinking about cases where we believe wrongly that something is in our power: if he had been, he would have amended his assertion to read that we deliberate about things that we *believe* to be in our power. The only discussion I have been able to find of this point is by Richard Taylor, who constructs the following example. Adam has dined at a friend's house and has been invited to spend the night. He weighs the pros and cons carefully before deciding. But suppose that, unknown to Adam, the last train has left and there is no other way for him to get home. Clearly, Taylor says, Adam can still deliberate about whether to remain or go, as long as he is ignorant of the non-availability of the train. Once he learns of it, he can no longer deliberate. It is therefore not the *actual* availability of the means of going, but the deliberator's *belief* about that availability that makes it possible for deliberation to take place.[5]

Taylor's example shows that what was said earlier, namely, that what is essential for deliberation is the existence of at least two different courses of action open to the deliberator, needs to be altered slightly. What is essential is that the deliberator have the power to *represent* to himself, in some way or other, the options open to him, and that these options must number at least two. Representation, of course, is a human activity that can be done well or done badly, and the deliberator may not succeed in correctly representing the alternatives that are in fact available.

[5] Taylor (1964: 76–7). I have slightly simplified Taylor's example.

However, although the correctness or incorrectness of his representation may affect his power to *bring about* what he decides to do as a result of his deliberations, it does not affect either the deliberations themselves or the choice he makes at the end of them. In ignorance of the fact that the buses are not running, I may sit in my office and deliberate about whether to walk or ride. What is necessary to the process of deliberation is that the deliberator have some representation, correct or incorrect, of the alternatives he faces. If this representation accurately portrays the alternatives, then it is a schematic picture of part of the enormous and complicated branched universe model, with its individual alternatives representing whole sets of branches on the model. If it is inaccurate and incorrect, then it mis-portrays the model. But, whether accurate or inaccurate, deliberation requires representation as its first step, and without the power to represent future courses of action deliberation is impossible.

Returning to the comparison between deliberation and chance, it now appears that deliberation does not necessarily require a *real* set of alternatives, but only a *representation* that is believed by the deliberator to be accurate. The presumption is, however, that such representations are for the most part correct. If this presumption were unfounded, i.e. if all representations of future possible courses of action presented a totally false picture of the alternatives available, then again deliberation would be impossible. A deliberator can be wrong from time to time about what is possible and what is not, but someone who was always wrong could not be a deliberator. To deliberate requires the ability to form a picture of future alternative courses of action that is, generally speaking, accurate.[6]

So far, the comparison between chance and deliberation is fairly close. What is necessary for the former is a set of alternative possible outcomes, and for the latter a reasonably accurate representation thereof. When we move from deliberation to decision, however, an important dissimilarity emerges. It is this dissimilarity

[6] An objection from David Armstrong: could a brain in a vat, whose representations of the outside world were always wrong, not deliberate? A brain in a vat could engage in *cognitive* deliberation, but not I think in *practical* deliberation, unless there were some way in which the brain's thoughts could change its environment. If the brain 'decided' (after suitable false representations of the alternatives available) to walk home rather than take the bus, then that decision would in the normal course of events have to result in ambulatory impressions, rather than bus-riding impressions, being conveyed to the brain. Without this reverse influence of brain on vat, in addition to vat on brain, such brains would not deliberate, and in the event that they did deliberate what they would be deliberating about would be the mix of chemicals or nervous impulses to be fed to them.

that constitutes the difference between a chance event and an action that is performed deliberately, as the result of a conscious decision.

What is a decision? A decision is the culmination of a process of deliberation, and consists in choosing or fixing on one of the alternatives which have been examined in deliberation. As before, the focus is on *practical decision*, where the option selected is a future course of action. Corresponding to cognitive deliberation and deliberation over values there is *cognitive decision* and *value decision*, but these, though interesting in their own right, will not be discussed here. The whole process which begins with deliberation and ends with decision can be analysed into three separate components:

1. representation of the available alternatives;
2. evaluation of the alternatives;
3. choice of one of them.

Of these three, representation has already been discussed. We turn now to evaluation and choice.

In discussing evaluation and choice, it will be useful to compare human decision-making with machine decision-making. For example, the *modus operandi* of chess-playing machines is somewhat analogous to that of people when they make practical decisions, and the similarities and differences will throw light on both. At first sight, the way in which a chess-playing machine operates looks very much like the way a human being operates when making a decision. The same components of representation, evaluation, and choice are present. Thus the machine has some method by which it symbolically represents all the possible moves it can make when faced with a given board configuration. Secondly the machine evaluates each of these possibilities, either by trying to look ahead further than its opponent ('brute force'), or by use of heuristic criteria based on board position. Finally, the machine selects one of the alternatives and plays it. So far, the process closely parallels human decision-making.

On closer examination, however, important differences emerge. Chess-playing machines fall into two categories, deterministic and indeterministic, and the differences in the case of deterministic machines are more obvious. A deterministic chess-playing machine is one which produces one and only one output for any given input, the input being (i) a distribution of chess-men on the board, combined with (ii) an internal state of the machine, and the output being a move. The machine constructs a representation of all the possible moves open to it, and then evaluates them. But following the process of evaluation, it would be incorrect to say that there were a number of alternative moves that the machine could make. In fact there is only one possible output—only one move that the machine's program allows it to make. This would normally be the move that the machine evaluates most highly, although a deterministic machine could perhaps be programmed to play its second-best move from time to time in order to throw its opponent off balance. However, such a tactic would not really accomplish much, since if the opponent were equipped with an identical deterministic machine, the opponent's machine could predict exactly where and when the original machine would play its second-best move.

The truth is that no deterministic machine is ever confronted with a set of alternative moves, each one of which is physically possible relative to the initial conditions then prevailing, and any one of which it can play. Let us call a set of such alternatives a 'choice-set'. Because each board configuration plus internal state generates a unique output, for deterministic machines there is no choice-set. That is, there is never more than one move open to such machines at any point in the game. Hence the situation of a deterministic machine is very different from that of a human chess player, who has a wide range of different possible moves she can play at any given stage.

The second type of chess-playing machine is one that functions indeterministically. The usefulness of being able to do this might lie, for example, in situations where the machine evaluated two different moves equally highly, and where those two moves came at the top of its list. Since a machine that remained paralysed if this happened would be a hopeless chess player, some way of breaking ties is needed. A deterministic machine could be programmed to play the move it happened to evaluate first, or alternatively to use some pseudo-random means of breaking the tie, such as computing the first hundred decimals of pi and playing one move if the resulting number was odd, the other if it was even.[7] The disadvantage here, as before, is that if the opponent had the use of an identical machine, using the same pseudo-random methods, this machine could faithfully predict the move before it was made. But if the original machine were a genuinely indeterministic one, with a built-in randomizer based for example

[7] An excellent discussion of pseudo-random devices, and of robots that employ randomizers in the process of deliberation, is to be found in Dennett (1984b: 115–22). What is missing in Dennett, however, is the concept of a 'choice-set', a notion with which no determinist can be comfortable.

on quantum indeterminacy, no prior prediction of its moves would be possible. Such a machine could be designed to make use of its randomizer not only when two equally good moves topped its list, but even when only small differences separated the best two or three moves it could play. Of this machine it would be true to say, as it is true to say of a human being, that it was confronted with a choice-set, namely that more than one possible course of action was open to it.

But although it might seem as if progress were being made, a great gap still exists between the operation of an indeterministic chess-playing machine and the process of human decision-making. Let us suppose that the installation of a non-pseudo-randomizer has given such a machine a range of alternative options, and has made unpredictable in principle whatever move it will play at a certain point in a chess match. Does its behaviour now resemble that of a human chess player? Not really. Its situation in having two or more possible moves open to it instead of only one is indeed like that of a human decision-maker, but a crucial difference remains. This is that, whereas it is purely a matter of chance which one of its possible moves the indeterministic machine plays, without there being any reason at all why it plays A rather than B, such is not the case with the human being. The human chess player moves queen to d2, instead of rook to d5, for a reason. Her move is not selected in some random way from among the different possible moves she could make. Which move she makes is not a matter of chance. At least it is not *normally* a matter of chance, although nothing stops her, if she wishes, from flipping a coin if she cannot make up her mind between A and B. But the case is different with the indeterministic machine. If the employment of its non-pseudo-randomizer allows each of the three following alternatives to be possible at time t_1:

(a)   playing A at time t_2,
(b)   playing B at time t_2,
(c)   playing C at time t_2,

then it is entirely a matter of chance which of (a)–(c) gets selected by the randomizer. The use of the randomizer creates a choice-set for the machine consisting of the three alternatives (a), (b), and (c), but because of the role played by the randomizer the existence of the choice-set is incompatible with there being any reason or explanation why one member of the set gets selected rather than one of the others. Randomizers create choice-sets, but at the price of destroying explanations.

The comparison between human and machine decision-making has brought out the difference between a (human) decision and a purely chance event in a clear way. Machines can indeed be built which have a choice-set—a set of possible actions open to them at any given point. Such machines resemble human deliberators, who also require a choice-set. But the difference lies in the fact that the machine selects the action it performs by a process of random selection out of the choice-set, and the random or chance nature of the selection procedure precludes there being any explanation of any kind why one action is selected rather than another. By contrast, human beings select the action they perform, from among the alternatives in the choice-set, not by chance but by deliberate choice, and in their case there *is* a reason or an explanation why one action is selected rather than another. Consequently, human actions which result from a decision *resemble* chance events in coming from a choice-set, but are *unlike* chance events in that there frequently exists a reason or explanation why they were performed.

In bringing out the differences (and also the similarities) between chance and decision, the beginnings of an answer to Smart's apparent refutation of libertarianism have been provided. Smart claimed there was no logical room for a third alternative between determinism and chance, and we have argued that this is not so. But much more remains to be said. In this section decision-making was analysed into (1) representation, (2) evaluation, and (3) choice, and in the next section we shall go more deeply into the nature of evaluation and choice, and the all-important relationship between them. We shall also examine the sense in which there exist 'explanations' for deliberate actions based on a choice-set.

(ii) *Deliberation-Reasons and Explanation-Reasons*

Consider the case of a person who is deliberating whether to do *A* or *B*. In evaluating the alternatives she finds that she has a reason to do *A*, and that she also has a reason to do *B*. Call these reasons *deliberation-reasons*, or delib-reasons for short. When we deliberate, we weigh delib-reasons. All this is part of the process of evaluation. But to decide is to do more than just deliberate. After evaluation comes choice. When we decide, there may or may not be a reason why we decide to act on the delib-reason for *A* rather than the delib-reason for *B*. This further reason, if there is one, cannot itself be a delib-reason. Let us call it an *explanation-reason.* A typical decision, therefore, will involve two different kinds of reasons. First there are the delib-reasons for the different alternatives. These reasons are duly weighed and considered. Finally, one of the alternatives is chosen. This choice will not in general be arbitrary, but will be made for a reason: an explanation-reason not a delib-reason. No adequate account of deliberation is possible unless delib-reasons and explanation-reasons are clearly distinguished.

It may be that there are cases of decisions where there exist delib-reasons but no explanation-reason. Buridan's ass, for example, starved to death when situated equidistant between two equally tempting piles of hay, because of an equal balancing of delib-reasons. This is not exactly the example we are looking for, because the unfortunate ass ended up not making a decision, and what we are looking for is a case where a decision is made, but without any explanation-reason. Such a case, however, is easy to construct. We can imagine a slightly more intelligent cousin of Buridan's ass who realizes that indecision is the road to death, and who chooses one of the piles of hay in a completely arbitrary fashion. For this intelligent cousin, the process of deliberation involves equally balanced delib-reasons, followed by the choice of one of them without any explanation-reason. This is an instance of what the scholastics called *liberum arbitrium indifferentiae*, or what has more recently been described as *criterionless choice*.[8]

A more modern version of the dilemma of Buridan's ass, based on fear rather than desire, is the 'railroad dilemma'.[9] You are hiking with your spouse in a wild horseshoe-shaped valley with steep sides through which runs a railway line, and while you are crossing the track a heavy branch falls from a tree, pinning your spouse's leg. While pulling vainly on the branch you hear the whistle of an approaching train. You could succeed in flagging down the train if you ran down the track, but the echoes in the valley make it impossible to tell which direction the train is coming from. Game theory tells us that running in either one of the two possible directions is better than staying put, but exercising our powers of *liberum arbitrium indifferentiae* in these circumstances is not easy. How many of us might not run a little distance down the track, then back again at the next whistle in the fear

---

[8] See Ofstad (1961: 46). Rescher (1959) gives an excellent account of the history of Buridan's problem. That there can be such a thing as criterionless choice, or choice without preference, has been denied by C. I. Smith (1956) and Daveney (1964), and asserted by McAdam (1965) and Hancock (1968).

[9] McCall (1987: 276-7).

that our choice of direction was wrong? Even though we know that oscillating back and forth leads to death, it takes a strong-minded person to choose one direction and run down the track without stopping until the train appears. There *are* no doubt persons who have the power to choose between these equally balanced delib-reasons, when much hangs on the choice. But most of us would in all probability be caught in hopeless and irrational vacillation.

It is not my purpose to argue that human beings have the unaided power of criterionless choice, though it seems probable that some do. Rather I want to suggest that the two components of *evaluation* and *choice* in the process of decision-making are distinct, and need not necessarily be linked. Choice may be *influenced* by evaluation, but it is not, in modern parlance, a function of it. This lack of any necessary connection between evaluation and choice is not a feature of decision-making that has figured prominently in discussions of the subject up to now. Nevertheless, more needs to be said about criterionless choice, and what is said will provide additional reasons for distinguishing between human and machine decision-making.

Faced with a criterionless choice, what is the best way of resolving the impasse? The standard method is to flip a coin. (It isn't even necessary that the coin be a fair coin; if the alternatives are truly indifferent a biased coin will do just as well.) But flipping a coin is not the end of the matter, for a decision could later be made to act *against* the fall of the coin. Someone could say 'Heads left; tails right', and then change her mind after the coin fell.[10] But it is precisely this power to revoke a choice, or to leave the choice to a randomizing device and then overrule the device, that humans have but machines don't. A chess-playing machine can certainly be programmed to use a randomizer to break ties, or to select alternatives. But what the machine cannot do is go against its own method. If it uses a randomizer to select one of two possible moves, it cannot then play the other. This, however, is what human beings *can* do. Like machines they can employ random-izers, but unlike machines they don't have to do what the ran-domizers indicate.

Let us return to the difference between delib-reasons and explanation-reasons, and to the related difference between chance and deliberative choice. At the beginning of this section the process of deliberation was characterized as a weighing of delib-reasons.

---

[10] It is said that a good way of making a difficult choice, when two options are evenly balanced, is to resolve to act tomorrow on the fall of a coin, and flip it. Whichever way the coin falls, the next day you will know what your real choice is.

If the choice-set includes three possible actions A, B, and C and if there is some reason for doing each, then there will be three separate delib-reasons. But if A is chosen, and if there is a reason why A was chosen over B and C, then this reason cannot itself be a delib-reason. It was called, for want of a better name, an explanation-reason. Not all choices and decisions have reasons for them: some like Buridan-choices are criterionless. But when they do have an explanation-reason, that explanation-reason serves as an answer to the question, 'Why did x choose A rather than B or C?'

Examples of explanation-reasons come to hand when we analyse the process of evaluation. If A is chosen over B and C, a typical explanation for this choice would be that A was evaluated more highly. A chess player might deliberate for ten minutes over three different moves, and when asked why she selected A reply that A gave her a better chance of winning. This explanation cannot be ignored or dismissed; unless we suspect her of lying we accept it as *the* reason why A was chosen over B and C. Even if the chess player's judgement was faulty, and B was in fact the best move, the correct explanation-reason has been given and the chess player really did think that A was preferable. But this entails in turn that the selection of A was not a matter of chance. There exists an explanation-reason for the decision in favour of A, and this explanation-reason is distinct from the delib-reason for A, the delib-reason for B, and the delib-reason for C. Since the chance or random selection of A would precisely imply that there was *no* reason or explanation why A was selected rather than B or C, it follows that A—the moving let us say of the queen's rook's pawn—did not occur by chance.

Here then, in somewhat greater detail, is the reply to those who claim with J. J. C. Smart that there is no logical space between chance and determinism. As is shown by the choice of A over B and C, where there exists an explanation-reason, there *is* a third alternative. The decision to opt for A was not determined, since each of the three alternatives remained causally possible up until the moment of decision. Nor was it a matter of chance. The selection and the subsequent enacting of A—in this case the moving of the pawn—show that a theory of free action can be based upon an indeterministic framework.

We shall return to the question of evaluations, choices, and explanation-reasons in section (v), where the parallels between human and machine decision-making are examined at greater length. But first a possible objection must be considered. In the example of the three possible chess moves A, B, and C, the provision of a reason why A was chosen constituted an *intentional* explanation. Such an explanation could not be derived merely by examining the branched structure of the model and comparing the proportions of sets of branches on which move A is made with those on which B or C is made. Explanation in terms of branch-proportionality is what might be called *causal-probabilistic* explanation, and is the kind of explanation that would be used in explaining why, for example, aspirin is good for the flu, or why an electron with a certain spin-orientation was measured spin-up by a Stern–Gerlach apparatus eight times out of ten. But such explanations are very different, in form at least, from the explanation of why the chess player moved the pawn instead of the knight. The objection that I have in mind goes as follows: despite this apparent difference, how can we be sure that so-called intentional explanation is not simply causal-probabilistic explanation in disguise?

This is a serious objection. If no real difference exists between intentional and causal-probabilistic explanations, then the attempt to find a third alternative between chance and determinism will fail. For consider how a causal-probabilistic explanation of the chess player's choice of A over B and C might go. Imagine in the first instance (1) that, during the period of time in which the chess player is deliberating, the three types of future branch—A-branches, B-branches, and C-branches—exist in roughly the same proportion, i.e. about one-third each. In that case there is *no* causal-probabilistic explanation of why A is chosen rather than B or C, and the door is left open for intentional explanation. But a more difficult case is this: imagine (2) that during the process of deliberation the proportion of A-branches is 98 per cent, while that of B and C-branches is only 1 per cent. In this case there *is* a perfectly good causal-probabilistic explanation for the choice of A if A is chosen, based purely on the higher proportionality of A-branches and without regard to intentional explanations of any kind. If all decision-making took place in circumstances like this, it would be difficult to argue that human beings had the power to 'navigate their way' up the universe tree, or to influence the future in any significant manner.

Difficult, perhaps, but not impossible. The flaw in causal-probabilistic explanations of decision and choices is that they convey a false picture of the decision-making process. Even if for purposes of argument we were to concede the main causal-probabilistic premiss, namely that in every case in which A is chosen over B and C the proportion of future A-branches is much higher than that of B or C, the conclusion cannot be drawn that intentional explanations must be excluded. On the contrary, intentional explanations remain essential. Even if on the physical level of branch-proportionality the probability that A will be selected is much greater than the probability that either B or C will be selected, the agent can still choose B or C. And when she does, there will be an intentional explanation of her choice.

The flaw in the causal-probabilistic picture is that no matter how probable A may be, B can still be chosen, and chosen for a reason. The deliberation process has this feature, that a choice can be made *against the probabilities*, normally in fact in complete ignorance and disregard of the probabilities. The ability to *represent* options is crucial here. What is possible can be represented by a deliberating intelligence, no matter on how few branches it may occur, and what is represented can be chosen.[11] Furthermore, when such an option is selected an intentional explanation can be given for its having been selected, though the fact that it occurs on relatively few branches precludes any causal-probabilistic explanation. That is what was meant in saying that causal-probabilism conveys a false picture of the decision-making process. Options can be chosen which are probabilistically unlikely, and, when they are selected, their selection can be accounted for by means of an intentional explanation. This explanation is plainly not a causal-probabilistic explanation.

Besides the possibility of choosing (for a reason) an unlikely option, there are other features of decision-making that fit ill with causal-probabilism. These include change of mind, the revoking of a decision once made, inability to come to a decision, and lack of commitment.[12] Let us briefly examine change of mind.

Suppose someone is deliberating over whether to move her queen's rook's pawn. Eventually she moves it, but then immediately

[11] An example of the choice of a highly improbable option may be found in Paul Brickhill's *The Great Escape*. No doubt the proportion of branches on which Allied prisoners of war succeeded in escaping undetected from their prison-camp in Germany by tunnelling under the sand was very small, but in fact that option was chosen and implemented. The choice of this mode of escape would have been impossible without the imaginative power to represent such a highly unlikely outcome.

[12] The different infirmities to which decisions fall prey are discussed in McCall (1987: 278–82), where they are given the name of *impedances*. It is to J. L. Austin that we owe the insight that as much can be learned about decisions when they go wrong as when they go right: see e.g. his paper 'A Plea for Excuses' in Austin (1961), which throws light on (normal) action and (normal) responsibility by studying actions which don't quite come off, are not fully intentional, or for which their authors endeavour to avoid being held responsible.

realizes that this is a mistake and takes it back. Changing one's mind and taking back a decision are common experiences, but difficult if not impossible to deal with in causal-probabilistic terms. Remember that a causal-probabilistic 'decision' involves a huge majority of branches containing the chosen option. What then is the revoking of a decision? A situation in which the huge majority was not there in the first place? A situation involving two successive decisions, one with a huge majority for option A, and the second with a huge majority for not-A? In cases of genuine indecision, does the universe oscillate back and forth between majorities of this kind? These questions are impossible to answer, and give the project of providing causal-probabilistic accounts of decision an air of unreality. Yet all these phenomena, of change of mind, revoking, indecision, etc., certainly take place in the real world, and are explicable at the intentional level.

What conclusion should be drawn from these considerations? The conclusion I would urge is this: that in the indeterministic world of the branched model what happens, i.e. what actually takes place, is not explicable solely in terms of random branch selection amongst branch-sets of different proportionalities. In particular, if an agent is present with sufficient intelligence to represent, evaluate, and choose amongst different options, what the future brings will not just be what is most probable in terms of branch proportionality. What is highly improbable can in fact take place, if it is selected by an intelligent chooser with a list of aims and objectives. The face of planet earth with its fields, highways, and networks of lights, in contrast to the face of Mars or the moon, bears witness to the power of choice. We may conclude, then, that human beings do indeed have the ability in some sense to 'navigate their way' from one node to another on the model, and that a third alternative exists between chance and determinism.

(iii) *The Causal Theory of Action*

We are interested, in this chapter, in situations in which an agent has two or more alternative courses of action which he can follow, and has to select one. Situations like this, although they do not cover all cases in which an agent acts, are common enough. What is interesting is that they serve to refute the most popular current theory of human action, the causal theory.

Causal theories of action have not always been popular. In the 1950s and early 1960s many philosophers, under the influence of

Gilbert Ryle and Elizabeth Anscombe, took the view that any attempt to explain the performance of actions in terms of causes, whether mental or physical, was mistaken and wrong-headed. Ryle was not averse to the idea of searching for lawlike explanations of human behaviour, but his preferred model for such explanations was one based on dispositions rather than causes. For Ryle, people have normally got motives for acting, and to appeal to a person's motive as the explanation of his action is like appealing to a glass's brittleness as the explanation of why it shattered. Motives, according to Ryle, 'are not happenings, and are therefore not of the right type to be causes'.[13] Miss Anscombe too rejects the idea that actions are, in general, susceptible of being given causal explanations; some actions are, but they are of the Jones-jumped-when-the-gun-went-off variety, and do not form a particularly interesting or important class.[14] A. I. Melden pursues this line of thinking in his book *Free Action*, and gives his objections to trying to explain actions by means of the concept of physical causality in the following passage:

For even in cases in which we have an immediate response (*e.g.* the startled jump of a person when a fire-cracker suddenly explodes behind him) . . . the question is not 'What caused the action?' but 'What caused him (or her) to do . . . (to jump, scream, or withdraw the hand)?' Here the reference to the agent is essential in the way in which it is not in the case of the reflex jerk of the leg, the twitch of a muscle or the movement of the intestines. And here we have extreme cases which shade almost imperceptibly . . . into the cases in which there is calculated and reasoned behaviour, in which the agent is getting what he wants for good and sufficient reasons. . . . In none of these cases, varied though they may be, is causation in the sense in which this term applies to physical events applicable to the actions of agents.[15]

A major change in the fortunes of the causal theory took place in 1963, with the publication of Donald Davidson's article 'Actions, Reasons, and Causes'. Davidson argues that whereas explanations of actions based on *causes* have often been contrasted unfavourably with explanations based on *reasons*, in fact these two types of explanation are one and the same. In the domain of human action reasons, or rather what Davidson calls primary

[13] Ryle (1949: 113). Gardiner, following Ryle, remarks of the assertion, *John hit you with a hammer because he is bad-tempered*, that 'It would be absurd to deny that this is an explanation: but it would be equally ludicrous to imagine that it could in some manner be "reduced" to an explanation asserting a causal relation between two events or processes, one of which is labelled "John's bad temper" ' (Gardiner 1952: 124–5).
[14] Anscombe (1957: 15–19).   [15] Melden (1961: 205–6).

reasons, *are* causes. Primary reasons, in turn, are analysed in terms of *pro attitudes* and *beliefs*. Davidson sums up his causal theory in the following two propositions.[16]

C1. *R* is a primary reason why an agent performed the action *A* under the description *d* only if *R* consists of a pro attitude of the agent towards actions with a certain property, and a belief of the agent that *A*, under the description *d*, has that property.

C2. A primary reason for an action is its cause.

The following example illustrates the complex and elegant way in which pro attitudes and beliefs combine to yield primary reasons for actions in Davidson's theory. Suppose that a naval officer presses a button on a warship, thinking that in so doing he will summon a steward to bring him a cup of tea.[17] The officer's reason for pressing the button consists of (i) a desire to summon the steward, and (ii) the belief that in pressing the button he was performing an action that would summon the steward. This reason, composed of the desire and belief in question, causes the officer to press the button. But, unknown to him, the button is connected to the torpedo tubes, and in pressing it he fires a torpedo and sinks the *Bismarck*. In so doing, Davidson maintains, the officer performs not three actions—pressing the button, firing the torpedo, and sinking the *Bismarck*—but only one. Under the first of the three descriptions the action is intentional; under the second and third, unintentional. The reason why the officer presses the button is not because he wants to sink the *Bismarck* (though in fact he may want to sink the *Bismarck*), but because he wants a cup of tea.

Davidson wishes to account for the difference between doing something *and* having a reason for it (which may not be the reason why, in this case, the agent did it), and doing something *because of* a reason.[18] In both cases there is a reason for an action, and in both cases the action ensues, but only in the second case does the reason explain the action. According to Davidson the explanation in question is a causal one; if reasons were not causes we would lack an analysis of the 'because' in 'He did it because . . .'.

---

[16] Davidson (1963), repr. in Davidson (1980: 5 and 12).

[17] Davidson (1980: 54).

[18] 'A person can have a reason for an action, and perform the action, and yet this reason not be the reason why he did it. Central to the relation between a reason and an action it explains is the idea that the agent performed the action *because* he had the reason' (Davidson 1980: 9). On p. 232 Davidson remarks that 'A man might have good reasons for killing his father, and he might do it, and yet the reasons not be his reasons in doing it (think of Oedipus).'

Even though the naval officer may have had a good reason to sink the *Bismarck*, and in fact sank it in pressing the button, he did not sink it *because* of that reason. What caused him to sink the *Bismarck* was not his desire to perform that action under the description 'sinking the *Bismarck*', but his desire to perform the very same action under the description 'pressing the button', coupled with the mistaken belief that the action in question would also satisfy the description 'summoning the steward'.

Although he maintains that primary reasons cause actions, Davidson does not wish to suggest that there exists any special domain of 'mental causation' distinct from the causation of physical science. In fact he explicitly denies this. Davidson gives the word 'cause' its strict Humean or nomological interpretation in holding that behind every actual instance of causation—every true singular causal statement—there stands a causal law.[19] Thus ' "*A* caused *B*" is true if and only if there are descriptions of *A* and *B* such that the sentence obtained by putting these descriptions for "*A*" and "*B*" in "*A* caused *B*" follows from a true causal law' (1980: 16).

Reasons cause actions, but Davidson would not have us look for any direct linking of reasons with actions via a causal law from which the sentence 'Reason *X* caused action *Y*' is deducible. The law in question which underlies the singular causal statement is not a law which links events classified as reasons with events classified as actions. It is more likely, Davidson says, to be a law which links the same events under totally different descriptions— neurological, chemical, or physical (p. 17). Since the concepts and laws of physical science promise to constitute a comprehensive, closed, exact system, in contrast to mental concepts (pp. 219 and 223–4), mental events can instantiate laws only in so far as they can be described in the language of physical science. But since mental events are identical with physical events (p. 209), they always can be so described. Hence mental events fall under causal laws, not as mental events, but as physical events. This, together with the further thesis that mental events, described as mental events, 'resist capture in the nomological net of physical theory' (p. 207) and cannot be correlated with physical events or indeed other mental events in any lawlike way, constitutes the position that Davidson describes as 'anomalous monism'.

Davidson's causal theory of action is subtle, profound, and impressive, but founders on the twin rocks of deliberation and decision.

---

[19] Davidson (1980: 16–17, 158–60, 208, 215, 231).

As was seen in the previous sections of this chapter, these activities require the existence of a choice-set comprised of two or more possible courses of action. The alternatives remain possible throughout deliberation, up until the instant one of them is chosen by the agent. Following this the chosen alternative (assuming that the agent's representation of the alternatives was accurate) is actualized by a bodily movement or series of bodily movements.

If this account of deliberation and decision is correct, it follows immediately that no causal theory can be devised which applies to actions resulting from decisions. There will exist reasons for these actions, but the reasons cannot be causes, and the explanation of why the actions are performed cannot be causal explanations, except in so far as they adduce the decision itself as cause. In the process of deliberation there may indeed be, and generally are, reasons for each of the represented alternatives. If an agent is deliberating over which of three possible courses of action A, B, and C to perform, then there will be a delib-reason for A, a delib-reason for B, and a delib-reason for C. As long as the process of deliberation continues, these delib-reasons cannot be causes; they are merely reasons. Then, at the moment of decision, one of A, B, and C is chosen. This choice will not in general be arbitrary, but will be made for a reason: what I have called an explanation-reason. But this reason cannot be a cause either, since it comes into existence only at the same time as the choice. Hence the only possible cause we are left with, for the action which results from the decision, is the decision itself. And this decision, as we have seen, has no cause. It *is* a cause, but it doesn't *have* a cause. All this, of course, on the assumption that indeterministic branched structures provide the best models for deliberation and decision.

It would appear that Davidson himself realizes that the causal theory won't work when applied to cases where the agent is confronted with a difficult choice. In his paper entitled 'How is Weakness of the Will Possible?' Davidson examines situations of moral conflict, where an agent must choose between, say, pleasure and continence. The problem for Aristotle, and for Davidson, is: how can someone know perfectly well, all things considered, that it is better to do A rather than B, and yet do B? This is the problem of *akrasia*, weakness of will. For Davidson, who sees reasons as causes, there can be no solution to the problem. The incontinent man has every reason to behave continently, yet somehow these reasons do not constitute a cause. Why, Davidson asks, would anyone ever perform an action when he thought that, everything considered, another action would be better? (1980: 42)

And yet, of course, the *akrates* does just that. There is no *causal* explanation of why the incontinent man does what he does. Davidson acknowledges as much in the last paragraph of his paper, when he says that in going against his own better judgement, the *akrates* has no reason (p. 42). Given Davidson's own theory, this would appear to be equivalent to saying, for the *akrates'* action there is no cause.

The causal theory, then, cannot cope with weakness of will. But a theory of action which takes choice or decision as its basic concept has a chance of succeeding where the causal theory fails. Either (i) weakness of will takes the form of an agent's performing an action at the same time that he believes that, all things considered, it would be better to do something else, in which case the action can be accounted for, as it is in this chapter, by the lack of any necessary connection between evaluation and decision, and by the primacy of decision. Or (ii) weakness of will takes the form of indecisiveness, vacillation, and lack of commitment, in which case it constitutes one of the infirmities to which decision is subject.[20] Either way, *akrasia* is explicable, whereas it does not appear to be explicable if we adopt the causal theory.

(iv) *Responsibility for Decisions*

An important topic that remains to be discussed is responsibility. Here the problem faced by anyone who wishes to deny determinism, and to build a viable theory of human action on indeterministic foundations, is to account for the fact that agents are normally (though not invariably) held to be accountable and responsible for what they do. The determinist can resolve this problem relatively easily, by calling attention to the fact that an agent's character, intentions, and circumstances determine his actions in roughly the same way as the molecular structure of salt determines its solubility. For a determinist, actions flow from character as water flows from the hills. But if indeterminism reigns, and if the causal links between an agent and the actions he freely performs are broken, then what becomes of responsibility? This is the indeterminist's dilemma.

At present, there exists no convincing indeterministic account of responsibility. Most of the running in this regard has been made by the determinists. Despite its somewhat baroque quality,

[20] See above, n. 12.

it would be hard to find a more eloquent statement of the determinist's position than that of 'R. E. Hobart' (Dickinson Miller):

Indeterminism maintains that we need not be impelled to action by our wishes, that our active will need not be determined by them. Motives 'incline without necessitating'. We choose amongst the ideas of action before us, but need not choose solely according to the attraction of desire, in however wide a sense that word is used. Our inmost self may rise up in its autonomy and moral dignity, independently of motives, and register its sovereign decree.

Now, *in so far* as this 'interposition of the self' is undetermined, the act is not *its* act, it does not issue from any concrete continuing self; it is born at the moment, of nothing, hence it expresses no quality; it bursts into being from no source. The self does not register *its* decree, for the decree is not the product of just that '*it*'. The self does not rise up in *its* moral dignity, for dignity is the quality of an enduring being, influencing its actions, and therefore expressed by them, and that would be determination. *In proportion* as an act of volition starts of itself without cause it is exactly, so far as the freedom of the individual is concerned, as if it had been thrown into his mind from without—'suggested' to him—by a freakish demon. . . . *In proportion* as it is undetermined, it is just as if his legs should suddenly spring up and carry him off where he did not prefer to go. Far from constituting freedom, that would mean, in the exact measure in which it took place, the loss of freedom. It would be an interference, and an utterly uncontrollable interference, with his power of acting as he prefers. In fine, then, *just so far* as the volition is undetermined, the self can neither be praised nor blamed for it, since it is not the act of the self.[21]

If the events which take place within a person's brain are conceived of as spread out four-dimensionally in a single Minkowski manifold, then it is easy to understand R. E. Hobart's misgivings about indeterminism. If an act of volition 'starts of itself', then there will be neural events[22] which cause actions but which themselves have no cause. Of these events we might as well say, Hobart maintains, that they are 'thrown into the mind' from without. It is as if they were placed there by a 'freakish demon'.[23] The agent does not cause them, does not produce them. Hence the actions which result from them cannot be said to be the agent's own actions; he is not accountable for them, and cannot be praised or blamed for them.

[21] Hobart (1934: 6–7).

[22] Or mental events. Nothing in what is said here hangs on whether materialism or dualism is true.

[23] Van Inwagen (1983: 130–5), makes some delightful dialectical capital out of Hobart's freakish demon.

But let us change the paradigm. In place of a single Minkowski manifold, with chance events which pop up unannounced, conceive of the branched model. According to the branched model there occur in the brain from time to time sets of initial conditions (neurological states) which have, not unique successor states, but two or more different types of successor states on different sets of branches. Furthermore the brain can represent, in a schematic way, these sets of branches. The representations are simplified pictures of courses of action the agent can perform, or events he can bring about by moving his body.

Now the final step: the agent chooses one of these possible courses of action. Again, nothing pops into his mind from without. No freakish demon suggests things to him. It is not at all as if the agent's legs suddenly spring up and carry him off somewhere. The agent simply chooses. Nor would it be correct to say that he causes his choice.[24] The natural thing would be to say that he makes his choice, or that the choice is his choice. When things are conceived in this way, Hobart's charge that indeterminism destroys responsibility is answered. Events are not 'thrown into' the agent's mind from outside; instead the agent deliberately and intentionally decides which possible events will be actual. Normally he decides for a reason, not arbitrarily. The decision is *his* decision, and he is responsible for whatever results from it. The possessive adjective does the philosophical work in this instance— if it were not his decision, the agent would not be responsible.

To sum up, it is incorrect to assert that responsibility requires determinism. Using branched alternatives, and the concept of decision, an indeterministic account of responsibility can be constructed which accords with our intuitions as well as, or better than, any other account. The metaphysics of the branched model provides a framework for conceiving of human beings, or indeed of any agent with similar powers of deliberation and decision, as free and responsible individuals.

(v) *Indeterministic Mechanisms which Select their own Future States*

It is time to pull together the themes of indeterminism, decision, and action, and to pose the question of what it is about human

[24] This might be the preferred way of putting the matter for those who support agent causation. See Taylor (1958 and 1966: ch. 9); Chisholm (1966); and Thorp (1980: ch. 6). The present chapter supports agent causation only to the extent of arguing that a free agent is one that selects its own future states. See section (v).

beings that gives them the power of *prohairesis*, as Aristotle called it: deliberative choice. In this section I shall propose a very materialistic answer to this question, namely that what gives humans the power of deliberative choice is their *brain*.

In section (i) above, decision was analysed into (1) representation, (2) evaluation, and (3) choice. How could a brain, a physical mechanism of great complexity but still part of the material world, engage in the intentional activities of representation, evaluation, and choice? In what follows I shall attempt to show that such activities are indeed possible for a complex indeterministic mechanism. In particular it will not be necessary to invoke the existence of a non-material mind or agent, above and beyond the brain itself, to explain their operation.

As before, it will be useful to compare the functioning of the brain with the functioning of a chess-playing machine. This machine, like a human being, represents and evaluates its possible moves, and then selects one of them which it plays. Being familiar with chess and its rules, it is natural for us to think of the set of playable moves at each stage in the game as an objectively defined set, which exists in its own right and which the machine may or may not represent to itself correctly. But in fact this is not how things stand with chess-playing machines. If the machine is deterministic, it is true that it may be able to construct an internal representation of the set of all permissible moves at each stage, but still there is only one move that it can play. A deterministic machine, of necessity, always plays the move it evaluates most highly. Even an indeterministic machine would employ a randomizer only to break ties between options, or perhaps to throw off the opponent by occasionally and unpredictably playing its second-best move. In that case there would be only two or perhaps three moves it would be physically possible for the machine to play, and whichever of these was in fact played would be a matter of pure chance.

The point to be emphasized is that indeterminism *creates* possible moves, i.e. possible courses of action. With the help of a randomizer, a chess-playing machine can increase the number of possible moves it can make from one to two or three. But there are limits. No one would design a machine equipped with a randomizer which gave it a chance of exposing its king to a simple checkmate in two moves. A well-designed machine would be *incapable* of making such a move, not in the sense that a good human chess player would be incapable of committing a simple blunder, but in the sense that the move would be *physically impossible* for the machine to perform. On the branched model, there would be no future branch on which the machine moved in that way. But (and this is the crucial difference) there are always *some* future branches on which human chess players, including Grand Masters, make stupid moves, even though it is true that most players move their muscles in such a way that the 'stupid' branches drop off.

What creates such choices? Why is it that the choice-set of a human chess player may comprise sixty to a hundred different moves, most of them admittedly bad, while the choice-set of the machine contains at most two or three? The difference, I claim, lies in the brain, which functions indeterministically and in so doing makes it possible for any of the sixty to a hundred moves to be played. Indeterminism creates choice-sets, and one of the brain's roles is to do just that. But simply to create choices is not enough. A chess-playing machine which used a randomizer in order to make it possible for it to lose in two moves would be a poorly designed machine. The trick is, to fashion a wide choice-set, but at the same time also create a mechanism of evaluation and choice so that none of the physically possible but foolish and undesirable options actually gets selected.

We have come to the crux of the matter. How can a brain be a mechanism which is indeterministic enough to create choices, but not so indeterministic that the selection of the acted-upon alternative is by pure chance? Let's start with the first part.

It has been maintained by some philosophers that the brain cannot function in a genuinely indeterministic fashion, because to do so would require its behaviour to be based upon quantum effects. The brain, being a macroscopic object, obeys classical not quantum laws. But it is not true that the brain, although of macroscopic size, cannot be influenced by events at the microscopic or quantum level. Some macroscopic bodies, such as photomultipliers or Stern–Gerlach apparatuses, serve precisely to amplify quantum effects. The brain is not a Stern–Gerlach apparatus, but there are dozens of possible ways in which it might be sensitive to events at the quantum level. Here is one of them.

The brain is the locus of complex patterns of electrical activity, which can be measured by suitably placed electrodes. Much of this activity is periodic and rhythmical, and some small-scale rhythmic activity may be based on mini-circuits of interconnected neurons. If so, the precise pattern of electrical activity will depend upon the exact time that each electrical impulse takes to complete a circuit. If two or more impulses arrive at a given neuron at the same instant they may create an effect (e.g. cause the neuron to fire), whereas if one arrives before the other the effect may fail. But the speed at which these impulses travel down the axon of a neuron, or the rate at which they cross a synapse, is subject to minute variations, so that events at the quantum level could make a difference. For instance, the transmission of an impulse down an axon depends on the diffusion of individual sodium and potassium ions through the walls of the cell, and the exact velocity of such diffusion could be subject to quantum variability. The simultaneous arrival of the impulses at the terminal neuron would consequently be unpredictable in principle, based upon quantum indeterminism.

It might be objected that the exact length of time it takes for an impulse to complete one round of a single mini-circuit in the brain may be subject to quantum indeterminism, but that does not prove that the overall functioning of the brain is indeterministic. Small-scale indeterminism in the timing of individual circuits may be washed out by being averaged over billions and billions of circuits, so that large-scale functioning is predictable within very narrow limits. This is true, but on the other hand it is also true that the brain *could* operate in such a way that small-scale indeterminism at the level of individual mini-circuits was amplified into large-scale indeterminism. Suppose that a mechanism existed which monitored a collection of thousands of mini-circuits and which registered 'off' at all times except when the principal neurons of each circuit fired simultaneously. If and when that happened, the monitoring mechanism would register 'on', remain 'on' for a moment, and then revert to 'off'. The situation would be not unlike that of the old 'Bell-Fruit' slot machines we used to see in bars, where a pull on the lever would rotate the wheels and (one hoped) produce an alignment of three Bell-Fruits and a jackpot of cash.

I am not saying that the brain operates, at some intermediate indeterministic level, like a gigantic Bell-Fruit machine. But it could. The analogy is perhaps not an entirely happy one, but if the brain worked something like this it would amplify quantum indeterminism and create action-options which a deterministic mechanism lacked.[25] The fact of one monitoring mechanism being

---

[25] Daniel Dennett in his book *Elbow Room* argues that contrary to what has just been said, indeterminism does not of itself create options or opportunities for an agent. Dennett's Mark I Deterministic Deliberator, for example, makes as much of its opportunities, and survives as well if not better, in a hostile environment as the Mark II Random Deliberator (1984b: 115–19). Both types of deliberating robot, according to Dennett, are confronted

'on' would signify that option *A* was open, the fact of another mechanism being 'on' would signify the availability of *B*, and so forth. Each option might become available several times per second. In this way, the brain could use quantum indeterminism to create options, i.e., create a choice-set. However, as was noted earlier, the creation of a choice-set is only half the story. An intelligent indeterministic mechanism must not only create its own set of possible future courses of action, it must also exercise *prohairesis*. This is more delicate.

As we saw above, an indeterministic chess-playing machine could employ a randomizer to create options, but if so the selection of one of these options to be acted upon was by sheer chance. This is not good enough. The brain not only needs to create options, using something like the indeterministic intermediate-level mechanism just described, but it also needs to evaluate its options and arrive at an order of preference. It then selects its best option and acts on it. Now the process of evaluation and the selection of the best option need not be indeterministic. One would hope in fact that they were not: that the establishment of a ranked list of options was in accordance with set aims and goals, and that the brain could use a rational procedure like the hedonistic or utilitarian calculus, or the principle of universalizability and the categorical imperative, or for that matter a rigid and complete set of rules which in casuistic fashion covered every conceivable alternative, to determine its best course of action. The upshot would be that the initial indeterminism that created the choice-set was subordinated to a higher selection process which could be entirely deterministic.

Let us take stock. I am suggesting that the brain be regarded as possessing two levels of functioning, a lower indeterministic level that creates choice-sets, and a higher more organized and structured level, deterministic or 99 per cent deterministic, that embodies rational choice. Because of its basic lower-level indeterminism the overall functioning of the brain is not deterministic: identical inputs into identical states of the brain do not necessarily yield identical outputs. The option that is unerringly selected by the higher-level process may be the option that is probabilistically the least likely, measured by branch proportionality. The

with opportunities, which in a deterministic world are purely *epistemic* possibilities, things which are possible-for-all-one-knows (p. 113). In such a world one deliberates over courses of action which may, for all one knows, be possible, or which may not be. But in the branched model each option in a choice-set is *objectively* possible, and an indeterministic deliberator has a choice-set with at least two alternatives. Cf. Prior (1968: 48–9).

indeterministic factor is essential, since if it were not there the number of objectively possible courses of action open to us at any given moment would be limited to one. Although the range of choice is generated by chance, the acted-upon alternative is not selected by chance. It is selected in accordance with a rational procedure, based on goals, objectives, rules, likes, dislikes, memories, passions, etc. Whatever the eventual chosen course of action, there will in general be an intentional explanation for it. In this way the brain embodies a process of rational choice, built upon an indeterministic option-generating foundation.

Many years ago, Karl Popper wrote a long article in which he argued that the idea of a being that could predict its own future states involved a contradiction.[26] The purpose of the article was to demonstrate that universal Laplacian determinism was impossible in a universe containing sufficiently powerful predictors. In this chapter we have been concerned not with beings that can predict their states, but with beings that can select their own states. Our emphasis has been on action, not knowledge. But the same general conclusion seems to apply: beings that can do this can exist only in an indeterministic universe, such as the universe of the branched model. In fact they do it in virtue of being themselves indeterministic, thereby being capable of generating choice-sets, and then exercising *prohairesis*. Such entities, of which our brains are a prime example, may correctly be described as indeterministic mechanisms which select their own future states, in accordance with a goal-directed or rule-governed decision procedure.

(vi) *Summing-up*

The question at issue is this: can a theory of free and responsible action be built upon indeterministic assumptions? In the context of the branched model, how is it possible consciously and deliberately to guide the first branch point into one particular area of the model, avoiding others? I hope that by now an answer to these questions will have emerged, in outline at least.

A philosopher sits at a table writing, and a deliberative question occurs to him. Should he continue writing, or should he break off, put the kettle on, and make himself a cup of coffee? These two alternatives are located on different sets of future branches, in different areas of the universe tree. It seems presumptuous to

---

[26] Popper (1950); see also Ackermann (1976: ch. 7).

think that the philosopher, by his own bare choice, should have the power to direct the first node of the model into one region rather than another. Yet this is so. In the first place, the philosopher is not like a deterministic Turing machine. The quantum-based indeterministic functioning of his brain guarantees the existence of a choice-set containing at least two options, both of which are physically possible. Secondly the philosopher is not like a chess-playing machine equipped with a randomizer: the selection of one of these options will not be by chance. The philosopher deliberates, and makes up his mind to forget about coffee and continue working for another hour. The explanation-reason for this decision is that he wants to finish the book he is writing.

The philosopher, an intelligent deliberator, has exercised his ability to represent alternative courses of action, evaluate them, and choose one. In the example just given there were two deliberation-reasons and one explanation-reason. It is not necessary to believe that the eventual choice was the result of anything but a deterministic or quasi-deterministic rational decison procedure. The philosopher's family could have predicted his decision with a high degree of accuracy. But the explanation of his choice is still a reason, not a cause. It cannot be a deterministic cause because of the antecedent existence of the choice-set, and it cannot be a probabilistic cause because of the phenomenon of change of mind, and because an alternative of low probability can still be chosen.

In the branched model, indeterminism plays two different roles. First, the model is indeterministic because of its branched structure. A deterministic model would be unbranched. Secondly, the selection of one branch out of indenumerably many above a node is itself random and indeterministic. That is to say, it is random and indeterministic unless an intelligent deliberator is present. Such deliberators, who have the power of *prohairesis*, constrain the indeterminism of the branched model by providing intentional explanations for branch attrition. Their powers are admittedly limited, but they do account for some tiny fraction of the global phenomenon of branch attrition. However, they do nothing to touch the branching character of the structure itself, without which there would be no intelligent choice and no explanations. Only within the context of indeterminism do deliberation, decision, and freedom assume their true shape, guided by rational or quasi-rational procedures and undistorted by deterministic fetters.

10

Conclusion

THE time has come to pull together the threads of previous chapters. As was stated at the beginning, the book is written as a long 'argument to the best explanation'. A model has been constructed, and the claim has been made that this model illuminates and resolves certain important philosophical problems. The extent to which the model is successful in doing this must now be judged, and the inference considered, whether the model truly represents the overall structure of space-time.

The questions to be asked are these. If the universe had structure S, would this explain X, Y, and Z? Would anything else explain X, Y, and Z equally well, or better? If the answer to the first question is yes, and if the answer to the second question is no, then an inference to the best explanation may lead us to conclude that the universe has structure S. To be sure readers must draw this conclusion for themselves, or alternatively decide that the conclusion is unwarranted.

A good explanation is one which accounts for a wide range of diverse phenomena in an economical and elegant way. In this respect the hypothesis of the branched model is not unlike the theory of evolution. Darwin was confronted with a vast accumulation of facts about living creatures which, for an intelligent and observant naturalist, stood in need of explanation. Why in the Galapagos Islands do we find various species of finches occupying ecological niches reserved for different birds in other parts of the world? Why do the embryos of land vertebrates have gill slits? Evolutionary theory is able to account for these and other facts, in a way that is so natural and compelling that Huxley, when he first read *The Origin of Species*, is said to have exclaimed: 'How stupid not to have thought of that!' The theory of evolution, when it appeared in 1858-9, was a brilliant example of what Whewell describes as the colligation of facts by means of a true and appropriate conception.

So it is, I submit, with the branched model. It also colligates, perhaps not facts, but philosophical problems, some old and some new. It sheds light on them, by bringing them all within the scope of a very general hypothesis about the physical world. This hypothesis, I contend, is confirmed in the same way as Darwin's hypothesis was confirmed, by an argument to the best explanation. The fact that the model is capable of dealing with such a wide variety of problems constitutes strong evidence that the world really has a branching space-time structure.

If we examine the overall logical structure of *The Origin of Species*, which is similar to that of the present work, we find a

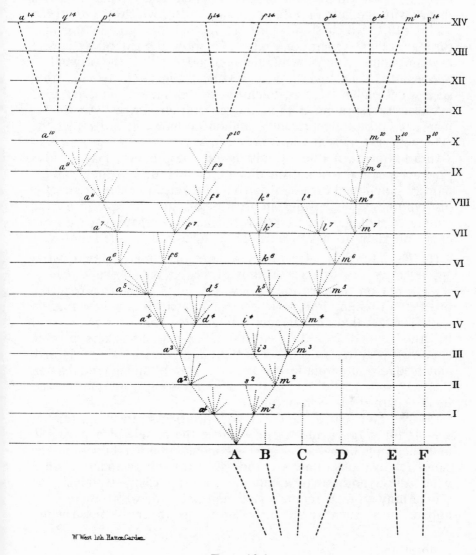

FIG. 10.1

hypothesis being put forward which, Darwin argues, is superior in explanatory power to its main rival, the view adopted by almost all educated people in 1859. Darwin's hypothesis is that organic species are not fixed but change by slow degrees over long periods of time, in such a way that any two species, no matter how remote, trace their origin to a common progenitor. The orthodoxy of 1859, on the other hand, held that species were fixed and immutable, each species having been separately created. Darwin illustrates his theory with the help of a diagram, one half of which is reproduced as Fig. 10.1, which shows how six species A–F gradually evolve through 14,000 generations into nine species, eight of which are new and come from A, while the remainder, F^{14}, is the original F. Species B, C, D, and E have become extinct, as well as other species descended from A. Each horizontal line on the diagram is intended to represent a thousand generations or, as Darwin remarks, could even represent a million or more, in order that the extremely slow rate at which evolution proceeds may be depicted.[1]

In addition to his hypothesis *that* species change, Darwin also presents a causal mechanism which accounts for *how* they change, namely individual variation and natural selection. He argues for his hypothesis by showing, in great detail, how it explains a wide variety of biological facts which on the rival creationist view are either puzzling or inexplicable. These facts include the following.[2]

(1) The fact that hard and fast lines of demarcation between species often fail to exist. This is explicable on Darwin's theory, though not on the creationist view, if species are normally in the process of dividing into varieties, which later become new species.

(2) The fact that animals in a habitat often annex the role played by other animals in a different habitat, e.g. woodpeckers in some countries prey on insects on the ground, thrush-like birds dive and feed on sub-aquatic insects, etc. This is understandable on the view that species tend to diversify and occupy any unoccupied or ill-occupied place in nature.

(3) The fact that a human hand, formed for grasping, that of a mole for digging, the leg of a horse, the paddle of a porpoise, and the wing of a bat, should all include similar bones, in the same relative positions. Creationists can only wonder at this, which the hypothesis of a common ancestor easily explains.

(4) Many animals retain organs and parts for which there is no apparent use, such as the shrivelled wings under the fused wing-

---

[1] Darwin (1876: 95–101).
[2] Darwin gives a summary of what his theory explains (1876: 430–8).

covers of many beetles, the hind leg bones of whales, etc. Such disused body parts constitute an embarrassment for the creationist, but lend support to Darwin's theory.

(5) The amazing similarity of the embryos of animals as different as the dog, the bat, the rabbit, and *Homo sapiens*.

I have cited these facts, which Darwin's theory explains and its rival does not, to illustrate the broad range of questions which evolution answers. Darwin's theory possesses a high degree of *consilience*, to use Whewell's word. But how good, from a logical point of view, is the method of arguing to the best explanation? As Nancy Cartwright points out, the inference from '*P* explains *Q*' and '*Q* is true' to '*P* is true' scarcely wears its validity on its sleeve.[3] Larry Laudan, in an influential article attacking the belief that the sequence of more and more successful theories in science supports scientific realism in the sense that these theories are best understood as progressively 'more true' of an independently existing real world, also points out, with the help of numerous examples, that we cannot argue from the *success* of a scientific theory to its *truth*, or even to its *approximate truth*. Of the chemical and physical theories of the late nineteenth century, for example, which explicitly assumed that matter was neither created nor destroyed, Laudan says that he is aware of 'no sense of approximate truth (available to the realist) according to which such highly successful, but evidently false, theoretical assumptions could be regarded as "truthlike" '.[4]

Considerations of this kind should give us pause. Whatever we as inferers to the best explanation may say, there will still be a gap between success and truth. No matter how successfully a theory explains a wide range of salient phenomena, the theory in question may still be false. However, the gap may be small, and the likelihood of the theory's being true may be great. The possibility that Darwin's evolutionary hypothesis should turn out to be false, for example, is very slim.[5] In the same way, I shall argue, the chance that the branched model should successfully explain the many things that it does, and yet turn out not to represent the actual structure of space-time, is also very slim.

Laudan asserts the principle that *a realist would never want to say that a theory was approximately true if its central theoretical terms failed to refer,*[6] and certainly the branched model could not

---

[3] Cartwright (1983: 89).   [4] Laudan (1981*b*: 35).

[5] 'I must freely confess, the difficulties and objections are terrific; but I cannot believe that a false theory would explain, as it seems to me it does explain, so many classes of fact.' Letter of 17 Dec. 1859 to Hugh Falconer, in Darwin (1903: 455).

[6] (1981*b*: 33).

be a true or approximately true picture of the world if 'set of space-time branches' or 'space-time structure branching along the time-axis' failed to refer. The explanatory power of the model, whatever that power may be, lies precisely in its branching structure, and to assert both that the model explains something, and that the term 'branching structure' fails to refer to anything, is self-defeating. Nor can 'branching structure' refer merely to something abstract, to part of a purely mathematical or conceptual model. If the branched model successfully explains something real, such as time flow or non-locality, or the probability value of a quantum event, then the explanation works only on the assumption that the real world exhibits the branching structure in question. In this respect the branched model differs from Laudan's examples of past theories which have proved to be explanatorily successful but not even approximately true, such as Ptolemaic astronomy, 'catastrophist' geology, the caloric theory of heat, etc. The difference is that the hypothesis of the branched model could not be explanatory without being true, at least to the extent that the term 'spatio-temporal branching structure' refers to something real.

A possibility which at present seems remote, but which should not be excluded, is the possibility of constructing an *empirical* test of the branched hypothesis. As was said earlier, the structure of the branched model is a matter of empirical fact, not logically necessary or a priori. If this is so, a test should in principle be able to be devised to examine the empirical consequences of branching. For example, in Appendix 1 two different kinds of space-time forking are discussed, namely upper cut and lower cut. Could an experiment be found which yielded one result if branching were upper cut, and a different result if branching were lower cut? If this were possible, more direct means than the argument to the best explanation would exist of examining the truth or falsehood of the branched hypothesis.

To sum up, inference to the best explanation conforms to what C. S. Peirce calls the method of abduction: '[Induction] never can originate any idea whatever. No more can deduction. All the ideas of science come to it by way of Abduction. Abduction consists in studying facts and devising a theory to explain them. Its only justification is that if we are ever to understand things at all, it must be in that way.'[7] But abduction and inference to the best explanation should not be expected to satisfy the canons of deductive validity. Not to put too fine a point on it, the inference schema:

---

[7] Peirce (1931–5: v. 5.145)

P explains *Q*.
Q is true.
Therefore, *P* is true.

is formally invalid. However, the form of argument considered in this chapter is a different one:

P explains *Q*, *R*, *S*, *T*, . . .
Q, *R*, *S*, *T*, . . . are diverse, seemingly unconnected phenomena, no other or no better explanation of which is presently known.
Therefore, there is a good probability that *P* is true.[8]

Put in this way, inference to the best explanation seems sensible and uncontroversial. To infer to the best explanation is part of what it is to be rational.[9] The major burden, therefore, would appear to lie in showing that the branched model really does succeed in explaining certain things that philosophy has hitherto left unexplained.

In the case of time, directionality is built into the tree-like branching structure, thus obviating the necessity of accounting for it in terms of asymmetric physical processes like entropy increase. Time flow, which at present is entirely excluded from scientific metaphysics and is generally regarded as a subjective phenomenon related to consciousness, has in the branched model an objective correlate in the progressive loss of branches. The set of branches represents the set of all physically possible or potential futures, and the transition from potentiality to actuality consists in the selection at each instant of one and only one branch as part of the 'actual' history of the world.

Recent discussions of 'antirealism' in science have denied the existence of any objectively real world which the laws of nature describe, the success of the scientific enterprise being measured by the degree to which these laws approximate to a correct and complete description. The branched model, on the other hand, serves as the objective truth-maker for a complete or more precisely completable set of natural laws, to which our present laws are at best an approximation. Since exact probability values are built into the branched model through the notion of proportionality of sets of branches, the set of laws grounded in the model includes probabilistic laws. As such the model is particularly appropriate to an inherently probabilistic science such as quantum mechanics.

[8] The word 'probability' used here denotes an epistemic probability, not one of the objective, *de re* probabilities that are the subject of the book.
[9] Cf. Armstrong (1983: 59).

An important and much-discussed concept in recent years is that of non-locality, a feature which appears to characterize the behaviour of two 'entangled' particles which are spatially separate. The difficulty lies in providing any kind of physical interpretation of or basis for the property of non-locality, and here the branched model finds a possible application. In the model the branches divide along three-dimensional spacelike hyperplanes, and the distant correlations revealed in the EPR experiment may be explained by the joint presence or absence of properties of the two particles on sets of branches with a fixed proportionality measure. Since one and only one of these branches is selected at each instant as the 'actual' branch, a measurement performed on one of the particles can appear instantaneously to affect the other by determining the outcome of a similar measurement performed some distance away. The effect is produced, in the model, solely by branch selection, and since no information travels from one particle to the other, Einstein's prohibition of faster-than-light signal transmission is not violated.

A recurrent problem in philosophy is the problem of defining the concept of probability. The 'logical' definition was proposed by Keynes and Carnap, the 'relative frequency' definition by Venn, von Mises, and Reichenbach, the 'subjective' definition by Ramsey, de Finetti, and Savage, and the 'propensity' theory by Popper. From the branched model a fifth theory of probability may be derived, different from the others, which assigns to every future event a precise, objective, single-case probability value which is grounded in the branching structure. These values may be unknown, or alternatively they may be known with a high degree of certainty as in some sciences, but whether known or unknown they exist and can be shown to obey classical probability laws. The existence of objective single-case probability values provides the basis for a new semantics for counterfactuals and other conditionals.

On the subject of identity, the branched model provides criteria of both transtemporal and transworld identity which are strong enough to allow for the 'full-blooded' existence of the same individual in different future possible branches or worlds. Such individuals, which derive their identity by tracing back to a common origin, are identical in the strict sense when viewed as three-dimensional beings persisting through time, and are not merely counterparts of one another.

It is argued that the two descriptions of a thing (i) as a three-dimensional object which endures through time, or alternatively (ii) as a four-dimensional object with temporal extension, are mutually intertranslatable and equivalent. If this is so, then the 'branched' form of an object in four dimensions can be viewed as a space-time picture of that object's potentialities or powers, with branch proportionality supplying the probability that any given potentiality will be realized.

An essential property of an individual is one that it possesses at all times on all branches from the moment of its origin, while accidental properties are those that characterize it on some branches but not on others. For example a three-legged tiger is still a tiger: it is accidentally three-legged and essentially a tiger. The branched model supports (without strictly implying) an ontology of natural kinds, based on collections of essential properties, which are either fixed and unchanging as in the physical sciences or flexible enough to permit of evolution as in the biological sciences. The idea of 'substance' inherent in this conception is Aristotelian, based on individual substances, rather than Lockian, based on a single material substance which is malleable and transformable.

Finally, the branched model provides the foundations for a new analysis of deliberation, decision, and free will. The process of deliberation typically involves (i) the representation of alternative future courses of action, each one realizable in the sense of being physically possible for the agent to bring about, (ii) the evaluation of these, and finally (iii) the choice of one of them which is the decision. The branched model permits a 'choice-set' of possible actions, and it is the intentional selection of one of these, rather than a purely random or chance selection, that distinguishes a world containing intelligent agents from a lifeless world. The traditional problem of free will is to find some way of allowing free will to coexist with determinism. In the case of the branched model, the problem is to find a way that free will, i.e. intelligent choice, can coexist with indeterminism. The book concludes with a discussion of how individuals capable of what Aristotle calls *prohairesis*—deliberative choice—can be defined (without recourse to Cartesian dualism) as beings capable of selecting their own future states.

The argument to the best explanation is complete. If the world really does possess the dynamic treelike structure of the branched model, then accounts become available of temporal direction and temporal flow; of the ontological basis of laws of nature; of the interpretation of quantum mechanics including quantum non-locality, state vector reduction, and measurement; of the definition of probability; semantics for counterfactuals and other conditionals; transworld identity of individuals; essential properties; and finally deliberation, decision, and free will. In the same way as Darwin's evolutionary hypothesis explains biological facts and sheds light on the world of living things, so I claim that the branched hypothesis explains and sheds light on a number of important philosophical and scientific issues. No other hypothesis is successful, over so wide a range. If the argument to the best explanation is accepted, the extent to which the branched model is a true picture of the world is directly proportional to its explanatory power. It is therefore appropriate to end the book by putting the question with which we started. Is the explanatory power of the model sufficiently strong to allow us to infer that in all likelihood the four-dimensional space-time structure of the world is branched? If so, the author's objective has been achieved.

Appendix 1

The Topology of Branched Spaces

The branched space-time model presented in Chapter 1 has a rich and complex topological structure, which this appendix and the next will attempt to sketch in greater detail. As will be seen there are many unanswered questions about branched topology.

The best way to get clear about the nature of a branched topological space is to begin with one-dimensional examples. Of these, the simplest is a one-dimensional point set in the shape of the letter 'Y'. This space is not homeomorphic with the real line, meaning that there exists no continuous mapping, with a continuous inverse, of one on to the other. Like a closed loop, a 'Y' is inherently different from a line and must be treated as a distinct variety of topological space.

Even at the one-dimensional level, a fundamental distinction exists between different types of branched space. A 'Y' is a tree with two branches, and may be regarded as a topological space consisting of three disjoint sets of points, the trunk T and the two branches A and B. In topology, a *neighbourhood* N_x of a point x is defined as an open set that contains x, and a *limit point* of a set S is defined as a point y such that every neighbourhood N_y of y contains at least one point in S which is distinct from y. (A limit point of S may or may not be in S.) With the aid of this terminology, two quite different relationships between trunk and branches of a tree may be defined.

(1) In the first case, the trunk T contains its own limit point and the two branches A and B do not. We say that the trunk is 'closed at the split', while the branches are 'open at the split'. Branching of this kind has been named 'lower cut' by Nuel Belnap,[1] and has the property that for every such branching there is a unique *branch point* which serves as a common limit point for trunk and branches.

(2) In the second case, the trunk is 'open at the split' (does not contain any of its limit points), while each of the branches is closed at the split. The trunk is in effect an open set with two distinct limit points, each of which belongs to one of the branches. Belnap calls branching of this kind 'upper cut'. It is unlike lower cut in lacking any unique branch point.

One-dimensional examples of lower cut and upper cut branched spaces are depicted in Fig. A1.1, which indicates the common limit point and the separate limit points of the branches in the two cases. Because

---

[1] See Belnap (1991*a*; 1992; 1993), three papers which no student of branching can afford not to read. The difference between lower and upper cut branching is discussed in McCall (1990).

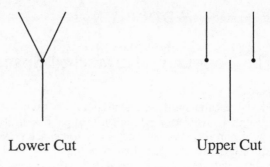

Lower Cut   Upper Cut

FIG. A1.1

an upper cut space cannot be embedded in the Euclidean plane, the example given of it does not appear to be connected, although for a one-dimensional creature living in the space it is both connected and path-connected.

I am indebted to Roy Douglas of UBC for a precise topological description of the upper cut example (Douglas 1994). In McCall (1990) a different picture of upper cut branching is given, which is misleading in that it depicts the greatest lower bounds of the two branches as distinct points arbitrarily close together in the plane, whereas in fact (because no non-Hausdorff space can be embedded in any Hausdorff space) they lie in no common plane. See Belnap (1991a: 147) for a different picture of upper and lower cut spaces.

The exact description of Douglas's example involves defining a *topological space* U consisting of a set Z of points together with the set of open subsets of Z. In the diagram, imagine that the two limit points of the branches lie on the x-axis, that the trunk lies on the y-axis and consists of points $(0, y)$, where $y < 0$, while the two branches consist of points $(-1, y)$ and $(1, y)$ respectively, where $y \geqslant 0$. This yields Z. The open sets of Z are defined by first defining an *open interval* to be one of the following five types of subset of Z. (In these five, α and β are arbitrary real numbers, which are constants for any one specific interval.)

$W_1(\alpha, \beta) = \{(0, y) \mid \alpha < y < \beta \leqslant 0\}$
  (i.e. the set of all points $(0, y)$ such that $\alpha < y < \beta \leqslant 0$)
$W_2(\alpha, \beta) = \{(-1, y) \mid 0 \leqslant \alpha < y < \beta\}$
$W_3(\alpha, \beta) = \{(0, y) \mid \alpha < y < 0\} \cup \{(-1, y) \mid 0 \leqslant y < \beta\}$
$W_4(\alpha, \beta) = \{(1, y) \mid 0 \leqslant \alpha < y < \beta\}$
$W_5(\alpha, \beta) = \{(0, y) \mid \alpha < y < 0\} \cup \{(1, y) \mid 0 \leqslant y < \beta\}$

The *open sets* of the branched topological space U then consist of those subsets of Z which are unions of open intervals.

The upper cut space U having been defined, we shall see how it differs from a lower cut space. One of the principal differences between lower and upper cut spaces is that the latter are not Hausdorff spaces. Hausdorff spaces satisfy the *Hausdorff separation axiom*, which states that for any two distinct points x and y of a topological space X, there exist disjoint neighbourhoods N_x and N_y of x and y. Satisfaction of this axiom is considered to be a relatively mild extra condition to impose on a topological space, and most of the spaces that topologists study are Hausdorff spaces. But, although lower cut branched spaces are Hausdorff spaces, upper cut spaces are not, as may easily be seen.

Let a and b be two distinct limit points of the branches of an upper cut branched space (Fig. A1.2). Any open set which contains a cannot 'end' at a, but must also contain points of the trunk. The same goes for any open set containing b. Therefore no two neighbourhoods N_a of a and N_b of b will be disjoint, and the Hausdorff axiom is violated. If on the other hand the branching were lower cut, the trunk and each of the branches would share a unique limit point and the Hausdorff axiom would hold.

Although upper cut branched spaces are not Hausdorff spaces, there is a weaker separation axiom due to Fréchet, known as the T_1 axiom, which they do satisfy. The T_1 axiom states that for any two distinct points x and y, there exists a neighbourhood N_x of x which excludes y. Upper cut spaces conform to the T_1 axiom and are T_1 spaces, with the consequence that it can be shown that in them all finite point sets are closed.[2]

A second difference between upper and lower cut spaces concerns the property of path connectedness. Given a topological space X, a *path* in X from x to y is a continuous function $f: [a, b] \to X$ of some closed interval in the real line into X, such that $f(a) = x$ and $f(b) = y$. A space X is said to be *path-connected* if every pair of points of X can be joined by a path in X. Now upper and lower cut spaces are both path-connected, but with a difference. Some paths in upper cut spaces double back on themselves, forcing those who follow them to retrace their steps. If we name such paths 'doubled paths', then in one-dimensional upper

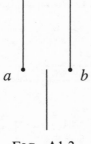

FIG. A1.2

---

[2] See e.g. Munkres (1975: 98) for this consequence of the T_1 axiom.

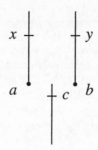

FIG. A1.3

cut spaces two points on different branches can be connected only by a doubled path, whereas this is not true of a lower cut space. Let x and y be points on different branches of an upper cut space (Fig. A1.3). A path from x to y cannot go directly from a to b without intermediate points, which means that the path must go a little way down into the trunk and then up again, traversing some points c twice.[3] This problem does not arise in the case of lower cut spaces, since any path from one branch to another goes directly through the branch point.

In two and higher dimensions, a point x on one branch of an upper cut space can be connected with a point y on another branch by an undoubled path that enters the trunk at one limit point, travels through it, and then exits into the other branch at a different limit point, as in Fig. A1.4.

Closely related to the unorthodox path-connectedness of one-dimensional upper cut spaces is the fact that no distortion can make the union of two branches homeomorphic to the real line. Similarly in two dimensions the union of two branches cannot be deformed into the Euclidean plane, and so on for higher dimensions. This is not true of lower cut space, which can be modelled in three dimensions by folding a piece of

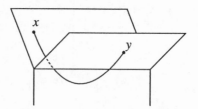

FIG. A1.4

---

[3] A proof of this may be found in McCall (1990). A 'doubled path' is a continuous function of the real line, the inverse of which is not a function, while an 'undoubled path' is a continuous one–one function.

paper and taping a third piece along the convex edge of the fold. But trying to flatten out the two branches of an upper cut space is like trying to press flowers in a book: even if the petals are infinitely thin, any attempt to flatten out two adjacent leaves of an upper cut space always results in an overlap or a ridge, which no amount of cutting or shrinking can eliminate.

Lower cut spaces, then, are Hausdorff and the union of any two branches is Euclidean; upper cut spaces are non-Hausdorff and the union of any two branches is not Euclidean. A third difference is that lower cut spaces are not *locally* Euclidean, whereas upper cut spaces are. A space is said to be 'locally Euclidean' if it is homeomorphic to Euclidean space for sufficiently small neighbourhoods around any point. Lower cut space is not Euclidean for any neighbourhood around a branch point, no matter how small, but upper cut space, despite the fact that it is not globally Euclidean, is locally Euclidean throughout. For example, Douglas's upper cut space U is locally Euclidean because every open interval around any point of U is homeomorphic with the real line. The table summarizes the differences between the two spaces that are so far known: no doubt other differences will emerge with further study.

| *Lower cut space* | Hausdorff | Union of any two branches Euclidean | Not locally Euclidean |
| *Upper cut space* | T_1 | Union of any two branches not Euclidean | Locally Euclidean |

I do not know of any attempt to give a topological description of all the different varieties of branched space. In McCall (1990) I attempt an inductive definition of n-dimensional lower cut spaces, based on the

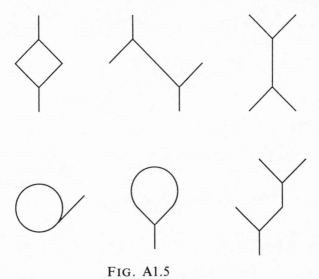

FIG. A1.5

notion of a branch point or branch surface, and perhaps some who read this appendix will be stimulated to try their hand at listing and describing the different kinds. Fig. A1.5 shows a few types that occurred to me; no doubt there are others. Only the last is a tree, which is of course the structure of principal interest in this book.

Appendix 2

Branched Four-Dimensional Space-Time

In identifying the frame-invariant features of branched four-dimensional space-time, our point of departure will be that of Belnap (1992). Belnap defines 'Our World' as a huge set of *point-events*, i.e. space-time points each of which is characterized by attributes such as mass, charge, curvature, and various field properties. Point-events are not empty space-time locations, but are individual punctiform occurrences. We may, if we wish, speak of two point-events which share the same properties as instantianting the same *event-types*, the latter being universals while point-events are particulars.

As it is conceived in this appendix, the world is a *topological space* with a binary relation—the so-called causal relation of special relativity—defined on it. More precisely, the world is an ordered triple $[\mathbf{W}, O, \leqslant]$ where $\mathbf{W}$ is the set of all point-events, O is the set of open subsets of $\mathbf{W}$, and $\leqslant$ is the causal relation which defines the light-cone structure of $\mathbf{W}$. The relation $\leqslant$ is reflexive, transitive, and antisymmetric, so that $\leqslant$ is a partial ordering of $\mathbf{W}$.

Two essential notions are those of *chain* and *slice*. A *chain* is a subset of $\mathbf{W}$, any two members of which are comparable by $\leqslant$. A maximal chain is *lightlike* or *null* if for any two points x and y on it, every point z which is between x and y, i.e. which is such that $x < z$ and $z < y$, is already a member of the chain. A *non-null timelike path* is a maximal chain with the property that for any two points on it, some point between them is not on the path. The surface of the double light cone associated with every point x of $\mathbf{W}$ is the set of all points connected by a null track to x.

In order to define the sense in which $\mathbf{W}$ is branched, the concept of a slice is required. Two points x and y are *incomparable* if $x \nleqslant y$ and $y \nleqslant x$. A slice is then defined to be a maximal connected set of pairwise incomparable points. A slice is an infinitely thin hypersurface—only one instant thick. The requirement of connectedness is needed in order that a 'slice' should not consist of the union of two disjoint open sets, each belonging to a different branch, where 'branch' is defined as follows.

The relation '$x \leqslant y$', from which the strict relation '$x < y$' ('x precedes y') can be derived, can also be read 'y is an upper bound of x'. An *upward directed set* is a set, any two of whose members have a common upper bound in the set. A *branch* of $\mathbf{W}$ can then be defined as a maximal upward directed set, i.e. one which is properly contained in no upward directed superset. $\mathbf{W}$ as a whole is a downward directed set, and to assert that the overall shape of $\mathbf{W}$ is tree-like, i.e. that $\mathbf{W}$ (i) is branched and (ii) branches only towards the future, requires the notion of in-accessibility.

Two points are defined to be *mutually inaccessible* if (i) they are incomparable, and (ii) no slice contains them both. We can now say what it is for **W** to be a *branched* topological space. If **W** contains three points x, y, and z such that $x < y$, $x < z$ and such that y and z are mutually inaccessible, then **W** is branched. To rule out branching towards the past, we stipulate that if two points are mutually inaccessible they have no common upper bound.

In this appendix I shall assume that all branching in **W** is lower cut branching (see Appendix 1). I believe that what I say holds also if upper cut branching is permitted, but the argumentation is more easily carried through if all branching is taken to be lower cut.

A point x of **W** is said to be a *choice point* if x lies on at least two distinct maximal chains C_1 and C_2, and if the portion of C_1 which is later than x and the portion of C_2 which is later than x are mutually inaccessible. The *prism structure* of **W** rests on the fact that above any choice point x, and within the same prism, there stands a denumerable infinity of other choice points. The choice points on every chain above x form a discrete set, with an accumulation point which is the base node of the next prism. Choice points which stand at the base of prisms are *basic choice points*, and above them stand *subsidiary* choice points of level 1, level 2, ... level n, ..., before the next basic choice point (see Chapter 4, p. 89). The order type of the choice points is that of the ordinal numbers, so that every choice point has an immediate successor but not necessarily an immediate predecessor.

Up to now, nothing that has been said about **W** hangs upon the notion of a reference frame. However we now move from a frame-invariant description of **W** to a description which is associated with a frame of reference, where the particular frame can be arbitrarily chosen (within certain limits). The move is from a description of **W** as it existed at the beginning of time to a description of **W** at a later time, and the latter description will necessarily be associated with a frame of reference. So described, **W** assumes the form of one of the familiar frame-dependent branched models that have been featured throughout the book.

At the time of the big bang, the universe consisted of the full set **W**. But now, 15 billion years later, the size of **W** has been considerably reduced. The vast majority of the physically possible alternatives available at earlier times were not selected, and the corresponding unac-tualized point-events no longer form part of **W**. Let S be any spacelike hypersurface or slice passing through here-now which contains no subsidiary choice points, and which is such that it contains a basic choice point for every subsidiary choice point that is later than it (i.e. that is later than some member of S). Thus S contains the base node of every prism that it intersects. Each such S determines a subset $\mathbf{W_S}$ of **W**, namely S plus all members of **W** which are either earlier or later than S. The set $\mathbf{W_S}$ depends on the choice of S, which is arbitrary provided it includes the point here-now. With each S is associated a reference frame, determined by the simultaneity-classes of events which S defines.

The composition of the subset $\mathbf{W}_S$ is in part arbitrary, depending on the choice of S, and in part non-arbitrary, depending on which of the original branches of $\mathbf{W}$ has turned out to contain the point here-now. What remains to be shown is that $\mathbf{W}_S$ branches along parallel hypersurfaces.[1]

We define first the relation of *proper precedence* for slices:

$$A << B =_{\mathrm{Df}} (x)(x \varepsilon A \supset (\exists y)(y \varepsilon B \& x < y)).$$

If $A << B$, then A and B do not intersect and we shall speak of them as being parallel. We construct in succession the set $\{S_{12}\}$ of all slices which are parallel to S and which contain all subsidiary choice points one level above the basic choice points of S, is followed by the set $\{S_{13}\}$ of slices parallel to S and containing subsidiary choice points two levels above S, followed by . . ., followed by the set $\{S_{1n}\}$ . . ., followed by the set $\{S_{21}\}$ of slices containing the basic choice points of the next tier of prisms, . . . etc. This family $\{S_{ij}\}$ of parallel slices specifies the framework of the branching structure of $\mathbf{W}_S$.

The family $\{S_{ij}\}$ contains all the *branch points* of $\mathbf{W}_S$, i.e. the slices or hypersurfaces along which $\mathbf{W}_S$ branches, but it does not contain all members of $\mathbf{W}_S$. To fill in the gaps we require a principle which states that if T and U are slices such that $T << U$, then there exists a set of pairwise parallel slices $\{V_i\}$ such that every point between T and U belongs to one and only one member of $\{V_i\}$. That is, $\{V_i\}$ partitions the space between T and U. Given this principle, the family $\{S_{ij}\}$ of branch points above S is extended to a family of parallel hypersurfaces which partitions $\mathbf{W}_S$. The relation $<<$ is a strict partial ordering of this family.

We have arrived at the conclusion we sought, namely that the set $\mathbf{W}_S$ of points of $\mathbf{W}$ which is defined by S branches along hypersurfaces parallel to S, and is partitioned by a maximal family of non-intersecting hypersurfaces ordered by $<<$. $\mathbf{W}_S$ is therefore one of the branched models depicted earlier in the book. Furthermore, choice of a different hypersurface S' leads to a different model $\mathbf{W}_{S'}$. Underlying and common to all such models is the branched topological space comprising the set of point-events $\mathbf{W}$. What this appendix has provided is a timeless, frame-invariant description of $\mathbf{W}$, together with various frame-dependent descriptions of $\mathbf{W}$-at-a-time, i.e. the different subsets $\mathbf{W}_S$ of $\mathbf{W}$. These constitute the branched models introduced in Chapter 1.

---

[1] [Added in proof.] It may be preferable to take the surfaces along which $\mathbf{W}_S$ branches, linking spacelike separated basic choice points, to be not slices or hyperplanes but surfaces with alternating peaks and valleys, the slope of which is always that of a light ray. The advantage of such surfaces is that they are not imposed arbitrarily on $\mathbf{W}$, but are objective features of $\mathbf{W}$ as structured by the causal relation $\leqslant$.

Appendix 3

A Branched Picture of Mermin's EPR Experiment

To test the ability of the branched model to explain the distant correlations observed in the EPR experiment, this appendix contains an analysis of a particularly striking thought-experiment devised by David Mermin.[1] Mermin's statistics illustrate the impossibility of providing a 'local hidden variable' explanation of EPR phenomena, and strongly support the adoption of a non-local model.

A 'Mermin device' consists of a source S and two detectors A and B the internal workings of which are unimportant. No wires or connections of any kind link the three parts of the apparatus, except that from time to time S emits a pair of 'particles', the arrival of which is registered at the detectors by the flashing of a red or green light. Each detector has a switch with three settings, labelled 1, 2, and 3, and the setting of the two switches is varied randomly throughout the experiment (see Fig. A3.1). A single observation or trial consists of four elements: the setting of the switches at A and B, and the colours flashed by A and B. For example, the observation '32RG' records that A's switch was set to 3, B's to 2, that A flashed red, and B green. An experimental run consists of a sequence of trials, e.g. 12RG, 31RR, 23GR, . . .

Mermin's statistics have the following properties:

(1) If the two switches have the same setting 11, 22, or 33, the lights always flash the same colours.

(2) In any sufficiently long run, the frequency of trials in which the two colours are the same approaches, to any desired approximation, the frequency of trials in which they are different; i.e. half

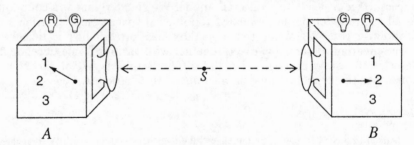

FIG. A3.1

[1] Mermin (1981; 1985).

the time the lights flash the same colour, and half the time different colours.

These statistical results demonstrate, as will be seen, that the operation of the Mermin device cannot be explained by any local realist or hidden variable theory. Suppose, for example, someone attempted to explain things by postulating that each particle, as it left the source, was equipped with a set of instructions telling the detector how to flash for each of the three possible switch settings. The instruction set RGR, for instance, would tell the detector to flash 'red' if the switch were set to 1, 'green' if it were set to 2, and 'red' if it were set to 3. If the two particles of any pair emitted by the source were both equipped with the same instruction sets, then this would ensure that, if the two switches had the same setting, the same colours would be flashed. Moreover, no other mechanism which was local, i.e. which did not involve some form of instantaneous communication between the two detectors, could ensure this. The particles, when they leave the source, cannot 'know' how the switches on the detectors will be set, and therefore they must carry instruction sets to cover all possible settings. Again, neither detector can 'know' how the switch on the other detector is set. Their only contact with the other parts of the apparatus, assuming locality, is via each arriving particle. Therefore, to ensure that if the switches are set the same way the detectors flash the same way, pairs of particles must carry identical instruction sets.

The postulation of local hidden variables, in the form of instruction sets for the two particles, can explain property (1) of Mermin's statistics, but it cannot explain property (2). This is that, on the average, half the time the detectors flash the same colour, and half the time they flash different colours. Since the setting of the switches is random, each of the nine possible combinations 11, 12, 13, 21, 22, 23, 31, 32, 33 will be equiprobable. But if both particles carry identical instruction sets, as is required by property (1), then, Mermin argues, the two detectors should flash the same colours not half the time, but at least $\frac{5}{9}$ of the time.

To see this, suppose that a given pair of particles carries the instruction set RRG. In that case, the detectors will flash the same colour when the switches are set to 11, 22, 33, 12, and 21. If the switches are set to 13, 31, 23, or 32, different colours will be flashed. Therefore, since each of these settings is equiprobable, the detectors should flash the same colour $\frac{5}{9}$ of the time, and different colours $\frac{4}{9}$ of the time. By parity of reasoning, the same result holds for any of the instruction sets GRR, RGG, GGR, RGR, and GRG in which one colour appears twice and the other once. In each of these cases, the detectors will flash the same colour $\frac{5}{9}$ of the time. But this leaves only the sets RRR and GGG, and these result in the same colour flashing *all* of the time. Therefore, Mermin concludes, if instruction sets exist, the same colours will flash in at least $\frac{5}{9}$ of all trials, no matter how the instruction sets are distributed and no matter how the switches are set.

This consequence of the hypothesis of local hidden variables, in the form of instruction sets for particles, constitutes Bell's inequality for the Mermin device, and is violated by Mermin's statistics. Properties (1) and (2) of Mermin's statistics, however, are precisely in accordance with what quantum theory would predict, if Mermin's simple device were replaced by a sophisticated EPR apparatus using electrons and Stern–Gerlach magnets, or photons and polarization analysers. We are left, therefore, without any local hidden variable account of the statistical correlations observed in the EPR experiment. Failing such an account, to what are the correlations to be attributed?

Mermin gives no answer to this question, except to say that the EPR experiment is as close to magic as any physical phenomenon he knows of, and to provide us with a marvellous quotation from Richard Feyn-

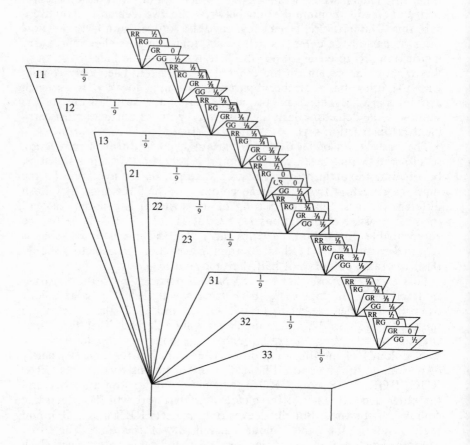

FIG. A3.2

man.[2] The branched interpretation, however, provides a non-local account which is able to explain Mermin's statistics. Fig. A3.2 shows the branched space-time model corresponding to Mermin's experiment.

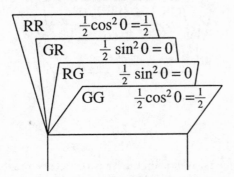

FIG. A3.3

In the model, each of the nine possible settings of the two switches has probability $\frac{1}{9}$. The three different settings of the switch on each detector correspond (in a real version of Mermin's *Gedankenexperiment* using photons) to three different polarization analysers, each one oriented at an angle of 120° to the others. For example, setting 1 on each detector corresponds to an HV-analyser, setting 2 to a 120°-analyser, and setting 3 to a 240°-analyser. In that case, if the settings are the same on the two detectors, the relative angle ϕ between the analysers is zero, and the probabilities for the joint results RR, GR, RG, and GG are as shown in Fig. A3.3.[3] But if the settings are different, the relative

---

[2]

We have always had a great deal of difficulty
understanding the world view
that quantum mechanics represents.
At least I do,
because I'm an old enough man
that I haven't got to the point
that this stuff is obvious to me.

Okay, I still get nervous with it . . .

You know how it always is,
every new idea,
it takes a generation or two
until it becomes obvious
that there's no real problem.

I cannot define the real problem,
therefore I suspect there's no real problem,
but I'm not sure
there's no real problem.

Feynman (1982), rendered in verse form by Mermin (1985: 47).

[3] In Mermin's version of the experiment the two polarizers flash the same colour when

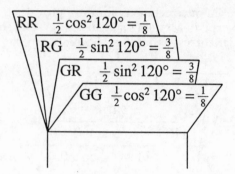

$$RR \quad \tfrac{1}{2}\cos^2 120° = \tfrac{1}{8}$$
$$RG \quad \tfrac{1}{2}\sin^2 120° = \tfrac{3}{8}$$
$$GR \quad \tfrac{1}{2}\sin^2 120° = \tfrac{3}{8}$$
$$GG \quad \tfrac{1}{2}\cos^2 120° = \tfrac{1}{8}$$

FIG. A3.4

angle ϕ between the analysers in the two detectors must be 120°, and the probabilities are as shown in Fig. A3.4.

All this information is contained in Fig. A3.2, which explains the two features (1) and (2) of Mermin's statistics in the following way. First, the diagram explains why it is that, if the two switches have the same setting, the lights always flash the same colour. The reason is that all 11, 22, and 33 branches are either RR or GG, never RG or GR. (That is, the probability for the outcomes RG and GR on those branches is zero.) Secondly, the diagram explains why, in any given trial, the probability of the lights flashing the same colour is exactly $\tfrac{1}{2}$. If we sum the probabilities of the outcomes RR and GG on each of the nine main branches we obtain:

$$p(RR \lor GG) = \frac{2}{18} + \frac{2}{72} + \frac{2}{72} + \frac{2}{72} + \frac{2}{18} + \frac{2}{72} + \frac{2}{72} + \frac{2}{72} + \frac{2}{18} = \frac{1}{2}$$

The conclusion is that Mermin's statistics are explained by proportionalities of sets of branches in the branched model.

set in parallel. This would require (as he notes) that if his experiment were performed with real photons emitted by positronium decay, rather than by an atomic cascade, one of the two detectors would have to be wired in reverse.

Appendix 4

Probabilities of Conditionals Defined as Conditional Probabilities

Since the year 1976, those who wish to define the probability of a conditional $p(A \rightarrow C)$ as the conditional probability $p(C \mid A)$ have had to reckon with a well-known trivialization result, due to David Lewis. In his paper (1976) Lewis shows that in a language which satisfies the postulates of the classical probability calculus, and which permits the definition

(1) $p(A \rightarrow C) = p(C \mid A)$ if $p(A)$ is positive,

where p belongs to a class of probability functions closed under conditionalization, there can be expressed at most two sentences, the probabilities x and y of which are such that $0 < x, y < 1$. In this appendix it will be shown that this trivialization result can be avoided, and that $p(A \rightarrow C)$ can be defined as $p(C \mid A)$ in the branched model consistently with the existence of unlimited numbers of sentences whose probability lies between zero and one.

Lewis's proof, which makes use of the principle

(2) $p(A \rightarrow C \mid B) = p(C \mid AB)$ if $p(AB)$ is positive,

proceeds as follows. Let X and Y be two propositions which are probabilistically non-independent. E.g. in the context of rolling a die let X be that an even number comes up, and let Y be that a six comes up. The following is a classical theorem:

(3) $p(A) = [p(A \mid B) \cdot p(B)] + [p(A \mid \bar{B}) \cdot p(\bar{B})]$.

Taking A as $X \rightarrow Y$ and B as Y we obtain:

(4) $p(X \rightarrow Y) = [p(X \rightarrow Y \mid Y) \cdot p(Y)] + [p(X \rightarrow Y \mid \bar{Y}) \cdot p(\bar{Y})]$.

Now by (2) $p(X \rightarrow Y \mid Y) = p(Y \mid XY) = 1$, and $p(X \rightarrow Y \mid \bar{Y}) = p(Y \mid X\bar{Y}) = 0$.

Putting these values into (4) we obtain:

(5) $p(X \rightarrow Y) = p(Y)$.

The definition of probabilities of conditionals as conditional probabilities therefore yields

(6) $p(Y \mid X) = p(Y)$,

which asserts the independence of X and Y, contrary to hypothesis.

Lewis's conclusion is that the definition (1) leads to inconsistency, at least in non-trivial languages permitting the formulation of more than two propositions with probability values greater than 0 and less than 1 (from which a pair of non-independent propositions analogous to X and Y can always be constructed).

In the branched model, the principle which fails to go through is (2). Where probabilities are objective, (2) is intuitively false. Looking at a barn at Lachute in 1993 I say, truly, that if a wind of 80 km./hr. hits it in 1997, the probability that the barn will be blown down is 0.75. That is, $p(B \mid W) = p(W \to B) = 0.75$. Someone now says, 'Assume that the barn *is* blown down in 1997. What is the probability today, conditional upon its being blown down in 1997, that if the barn is hit by a 80 km./hr. wind in 1997, it will be blown down in 1997?' The answer is, that the probability today is still the same, that $p(W \to B \mid B) = 0.75$, not unity. Nor is $p(W \to B \mid \bar{B})$ zero. The probability of a conditional today is not affected by whether the conditional's consequent is realized next year. In general, $p(A \to C \mid B) \neq p(C \mid AB)$. Note that this immunity to change in probability value in the light of subsequent events is confined to conditionals and conditional probabilities. It may well be that $p(B) = .001$, even though $p(B \mid W) = 0.75$.

In letters to the author in the summer of 1989 Lewis gives an argument to show that adopting the definition (1) leads, even in the branched model, to a contradiction. In his argument Lewis makes use of the assumption that, in the branched model, the probability functions at different levels are closed under conditionalizing on the propositions in a certain finite partition. He demonstrates the truth of this assumption, but only on the further assumption that the probability of a conditional at a node on the model is given by the proportion of branches above the node on which the conditional is *true*. However, it is precisely this assumption that was seen in Chapter 6 to be incorrect. Many type A conditionals have probability values in the model without being true (or false) on any branch. Hence the attempt to derive an inconsistency from definition (1) fails, given the theory of conditionals of Chapter 6.

Fig. A4.1 illustrates Lewis's argument, which uses inverse probabilities. It will suffice to consider a case in which there are only two

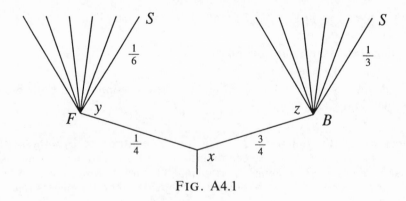

FIG. A4.1

alternatives, for example the selection of a fair die (F) or a biased die (B), each with a certain probability. Let $p(F) = \frac{1}{4}$, $p(B) = \frac{3}{4}$. If the fair die is rolled, the probability of six (S) is $\frac{1}{6}$; if the biased die is rolled the probability of six is $\frac{1}{3}$.

For the three nodes x, y, and z we have three probability functions p, p', and p'' such that $p(S) = (\frac{1}{4} \times \frac{1}{6}) + (\frac{3}{4} \times \frac{1}{3}) = \frac{7}{24}$, $p'(S) = \frac{1}{6}$, and $p''(S) = \frac{1}{3}$. Consider the backwards conditional $S \to F$: 'If it's six, the die is fair.' Since $p(FB) = 0$, standard probability theory gives us that

(7) $p(S \to F) = [p(S \to F \mid F) \cdot p(F)] + [p(S \to F \mid B) \cdot p(B)]$.

Suppose for *reductio*, Lewis says, that definition (1) holds for the three probability functions p, p', and p''. We then obtain:

(8) $p(S \to F) = p(F \mid S) = p(FS)/p(S) = \frac{1}{24} \times \frac{24}{7} = \frac{1}{7}$.

(9) $p'(S \to F) = p'(F \mid S) = p'(FS)/p'(S) = 1$.

(10) $p''(S \to F) = p''(F \mid S) = p''(FS)/p''(S) = 0$.

Since the probability functions p, p', and p'' are closed under conditionalizing on the propositions F and B (which constitute a partition of the required sort), we have that

(11) $p'(S \to F) = p(S \to F \mid F)$.

(12) $p''(S \to F) = p(S \to F \mid B)$.

Putting the values derived from (9), (10), (11), and (12) into (7) we obtain that $p(S \to F) = 1 \times \frac{1}{4} + 0 \times \frac{3}{4} = \frac{1}{4}$, which contradicts (8). Hence, on the branched model, definition (1) leads to an inconsistency.

Lewis's argument requires steps (11) and (12), namely that p' and p'' come from p by conditionalizing. But, in the case of probabilities of conditionals in the branched model, this is not so. To prove the general principle on which (11) and (12) are based requires assuming that conditionals have truth-values on every branch, and this they do not have. The general principle that Lewis needs for appropriately positioned A, B, and C is:

(13) $p'(A \to C) = p(A \to C \mid B)$.

For the proof of (13), let the function p hold at node u, the function p' at node v, let B be true only at v, and let $b_1, \ldots b_n$ be a fan of branches above v (Fig. A4.2). (It will simplify the calculations, without

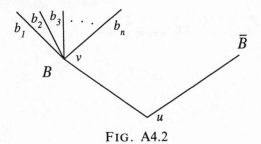

FIG. A4.2

affecting the generality of the conclusions, to assume that n is finite.) The probability of $A \to C$ at u is $p(A \to C)$, and at v it is $p'(A \to C)$. In accordance with Lewis's view that probability is always probability of *truth*, $p'(A \to C)$ must be given by summing the probabilities of the branches above v on which $A \to C$ is true. The proof of (13) then proceeds as follows:

$$p'(A \to C) = \sum_{i=1}^{n} \left[p'(b_i) \cdot \begin{cases} 1 \text{ if } A \to C \text{ is true on } b_i \\ 0 \text{ if } A \to C \text{ is false on } b_i \end{cases} \right]$$

$$= \sum_{i=1}^{n} \left[p(b_i \mid B) \cdot \begin{cases} 1 \text{ if } \ldots \\ 0 \text{ if } \ldots \end{cases} \right]$$

$$= \sum_{i=1}^{n} \left[p(b_i \& B) \big/ p(B) \cdot \begin{cases} 1 \text{ if } \ldots \\ 0 \text{ if } \ldots \end{cases} \right]$$

$$= \sum_{i=1}^{n} \left[p(b_i \& B) \cdot \begin{cases} 1 \text{ if } \ldots \\ 0 \text{ if } \ldots \end{cases} \right] \big/ p(B)$$

$$= p(A \to C \& B)/p(B)$$
$$= p(A \to C \mid B)$$

In this proof, the assumption that $A \to C$ must be either true or false on each of the branches $b_1, \ldots b_n$ is made explicitly, and without this assumption the proof does not go through. Hence probabilities of conditionals are not closed under conditionalizing in the branched model, and $p(A \to C)$ can be defined as $p(C \mid A)$ without trivialization and without inconsistency.

Finally, the fact that conditionals can have well-defined probability values and yet in some cases not have truth-values does not mean that conditionals are not propositions. In all important senses of the word, conditionals *are* propositions. They can be asserted, denied, refuted, and confirmed; they imply other propositions and are implied by them; they can be, and frequently are, either true or false as distinct from merely being assertible or justifiably believable. But they do not *always* have truth-values. As we have seen, type **A** conditionals with false antecedents are neither true nor false, unless their probability values equal zero or one. Like the propositions in Łukasiewicz's many-valued logics, they violate the principle of bivalence.

References

Ackermann, Robert (1976), *The Philosophy of Karl Popper* (Amherst, Mass.).

Ackrill, J. L. (1963), *Aristotle's Categories and De Interpretatione* (Oxford).

Adams, Ernest (1965), 'The Logic of Conditionals', *Inquiry*, 8: 166–97.

—— (1970), 'Subjunctive and Indicative Conditionals', *Foundations of Language*, 6: 89–94.

—— (1975), *The Logic of Conditionals* (Dordrecht).

Adams, Robert M. (1974), 'Theories of Actuality', *Nous*, 8: 211–31; reprinted in Loux (1979).

Akhundov, Murad D. (1986), *Conceptions of Space and Time* (Cambridge, Mass.).

Albert, David, and Loewer, Barry (1988), 'Interpreting the Many-Worlds Interpretation', *Synthese*, 77: 195–213.

—————— (1990), 'Wanted Dead or Alive: Two Attempts to Solve Schrödinger's Paradox', in A. Fine, M. Forbes, and L. Wessels (eds.), *PSA 1990* [*Proceedings* of the Philosophy of Science Association] (East Lansing, Mich.), i. 277–85.

Alexander, Peter (1985), *Ideas, Qualities and Corpuscles* (Cambridge).

Angel, Roger (1980), *Relativity: The Theory and its Philosophy* (New York).

Anscombe, G. E. M. (1957), *Intention* (Oxford).

—— (1971), *Causality and Determination* (Cambridge), reprinted in Sosa (1975).

Armstrong, David M. (1978), *Universals and Scientific Realism*, 2 vols. (Cambridge).

—— (1980), 'Identity through Time', in van Inwagen (1980).

—— (1983), *What is a Law of Nature?* (Cambridge).

Austin, J. L. (1961), *Philosophical Papers* (Oxford).

Ayer, A. J. (1972), *Probability and Evidence* (London).

Ayers, Michael (1981), 'Locke versus Aristotle on Natural Kinds', *Journal of Philosophy*, 78: 247–72.

Ballentine, L. E. (1970), 'The Statistical Interpretation of Quantum Mechanics', *Reviews of Modern Physics*, 42: 358–81.

—— (1990). *Quantum Mechanics* (Englewood Cliffs, NJ).

—— and Jarrett, Jon P. (1987), 'Bell's Theorem: Does Quantum Mechanics Contradict Relativity?' *American Journal of Physics*, 55: 696–701.

Barrow, J. D., and Tipler, F. J. (1986), *The Anthropic Cosmological Principle* (Oxford).

Bell, J. S. (1975), 'On Wave Packet Reduction in the Coleman–Hepp Model', *Helvetica Physica Acta*, 48: 93–8; reprinted in Bell (1987).
—— (1981), 'Bertlmann's Socks and the Nature of Reality', *Journal de Physique*, Coll. C2 42 (3): 41–61; reprinted in Bell (1987).
—— (1987), *Speakable and Unspeakable in Quantum Mechanics* (Cambridge).
—— (1989), 'Towards an exact Quantum Mechanics', in S. Deser and R. J. Finkelstein (eds.), *Themes in Contemporary Physics II* (Singapore).
—— (1990), 'Against "Measurement" ', in A. I. Miller (ed.), *Sixty-two Years of Uncertainty* (New York).
Belnap, Nuel (1991*a*), 'Before Refraining: Concepts for Agency', *Erkenntnis*, 34: 137–69.
—— (1991*b*), 'Backwards and Forwards in the Modal Logic of Agency', *Philosophy and Phenomenological Research*, 51: 777–807.
—— (1992), 'Branching Space-Time', *Synthese*, 92: 385–434.
—— (1993), 'Agents in Branching Time', typescript, University of Pittsburgh.
Bennett, Jonathan (1984), 'Counterfactuals and Temporal Direction', *Philosophical Review*, 93: 57–91.
—— (1988), 'Farewell to the Phlogiston Theory of Conditionals', *Mind*, 97: 509–27.
Black, Max (1952), 'The Identity of Indiscernibles', *Mind*, 61: 152–64.
—— (1955), 'Why Cannot an Effect Precede its Cause?' *Analysis*, 16: 49–58.
Bradley, F. H. (1883), *The Principles of Logic* (Oxford).
Brand, Myles (1979), 'Causality', in P. D. Asquith and H. Kyburg (eds.), *Current Research in Philosophy of Science* (East Lansing, Mich.), 252–81.
—— (1980), 'Simultaneous Causation', in van Inwagen (1980).
Broad, C. D. (1923), *Scientific Thought* (London).
—— (1938), *Examination of McTaggart's Philosophy*, 2 vols. (Cambridge).
—— (1959), 'A Reply to my Critics', in P. A. Schilpp (ed.), *The Philosophy of C. D. Broad* (New York).
Bub, Jeffrey (1968), 'The Daneri–Loinger–Prosperi Quantum Theory of Measurement', *Il nuovo cimento*, 57B: 503–20.
—— (1988), 'From Micro to Macro: A Solution to the Measurement Problem of Quantum Mechanics', in A. Fine and J. Leplin (eds.), *PSA 1988* [*Proceedings* of the Philosophy of Science Association] (East Lansing, Mich.), ii. 134–44.
Burgess, John P. (1978), 'The Unreal Future', *Theoria*, 44: 157–79.
—— (1980), 'Decidability for Branching Time', *Studia Logica*, 39: 203–18.
Butler, Joseph (1736), 'Of Personal Identity', appendix to *The Analogy of Religion* (Dublin).
Cahn, Steven (1967), *Fate, Logic and Time* (New Haven, Conn.).
Campbell, C. A. (1951), 'Is "freewill" a pseudo-problem?', *Mind*, 60: 441–65.

Carnap, Rudolf (1945), 'The Two Concepts of Probability', *Philosophy and Phenomenological Research*, 5: 513–32, reprinted in H. Feigl and W. Sellars (eds.), *Readings in Philosophical Analysis* (New York, 1949).

—— (1958), *Introduction to Symbolic Logic* (New York).

—— (1962). *Logical Foundations of Probability*, 2nd ed. (Chicago).

Carter, Brandon (1974), 'Large Number Coincidences and the Anthropic Principle in Cosmology', in M. S. Longair (ed.), *Confrontation of Cosmological Theories with Observational Data* (Dordrecht).

Cartwright, Nancy (1983), *How the Laws of Physics Lie* (Oxford).

Chapman, Tobias (1982), *Time: A Philosophical Analysis* (Dordrecht).

Chisholm, Roderick M. (1946), 'The Contrary-to-Fact Conditional', *Mind*, 55: 289–307.

—— (1966), 'Freedom and Action', in K. Lehrer (ed.), *Freedom and Determinism* (New York).

—— (1976), *Person and Object* (London).

Churchland, Paul (1985), 'Conceptual Progress and Word/World Relations: In Search of the Essence of Natural Kinds', *Canadian Journal of Philosophy*, 15: 1–17.

Clauser, J. F., and Shimony, A. (1978), 'Bell's Theorem: Experimental Tests and Implications', *Reports on Progress in Physics*, 44: 1881–927.

Clifton, R. K., Redhead, M. L. G., and Butterfield, J. N. (1990), 'Generalization of the Greenberger–Horne–Zeilinger Algebraic Proof of Nonlocality', *Foundations of Physics*, 21: 149–84.

Coburn, Robert C. (1986), 'Individual Essences and Possible Worlds', *Midwest Studies in Philosophy*, 11: 165–83.

Colodny, Robert G. (1972) (ed.), *Paradigms and Paradoxes* (Pittsburgh).

Costa de Beauregard, Olivier (1987), *Time: The Physical Magnitude* (Dordrecht).

Cushing, James T., and McMullin, Ernan (1989), *Philosophical Consequences of Quantum Theory* (Notre Dame, Ind.).

Daneri, A., Loinger, A., and Prosperi G. M. (1962), 'Quantum Theory of Measurement and Ergodicity Conditions', *Nuclear Physics*, 33: 297–313; reprinted in Wheeler and Zurek (1983).

Darwin, Charles (1876), *The Origin of Species*, 6th edn., vol. xvi of *The Works of Charles Darwin* (London, 1988).

Darwin, Francis (1903) (ed.), *More Letters of Charles Darwin*, i (New York).

Dau, Paolo (1986), 'Part-Time Objects', *Midwest Studies in Philosophy*, 11: 459–74.

Daveney, T. F. (1964), 'Choosing', *Mind*, 73: 515–26.

Davidson, Donald (1963), 'Actions, Reasons, and Causes', *Journal of Philosophy*, 60; reprinted in Davidson (1980), 3–19.

—— (1980), *Essays on Actions and Events* (Oxford).

Davies, P. C. W. (1974), *The Physics of Time Asymmetry* (London).

—— (1977), *Space and Time in the Modern Universe* (Cambridge).

—— (1982), *The Accidental Universe* (Cambridge).

Davies, P. C. W. (1983), *God and the New Physics* (London).

de Finetti, Bruno (1937, 1964). 'Foresight: Its Logical laws, its Subjective Sources', translated in Kyburg and Smokler (1964).

—— (1974), *Theory of Probability*, 2 vols. (London).

Denbigh, K. G. (1975), *An Inventive Universe* (London).

—— (1981), *Three Concepts of Time* (Berlin).

Dennett, Daniel (1984*a*), 'Cognitive Wheels: The Frame Problem of AI', in C. Hookway (ed.), *Minds, Machines and Evolution* (Cambridge); reprinted in M. A. Boden (ed.), *The Philosophy of Artificial Intelligence* (Oxford, 1990).

—— (1984*b*), *Elbow Room* (Cambridge, Mass.).

d'Espagnat, Bernard (1976), *Conceptual Foundations of Quantum Mechanics*, 2nd edn. (Reading, Mass.).

—— (1979), 'The Quantum Theory and Reality', *Scientific American*, Nov.: 158–81 (European edn.: 128–40).

DeWitt, Bryce S. (1970), 'Quantum Mechanics and Reality', *Physics Today*, 23; reprinted in DeWitt and Graham (1973).

—— (1971), 'The Many-Universes Interpretation of Quantum Mechanics', in *Foundations of Quantum Mechanics* (New York); reprinted in DeWitt and Graham (1973).

—— and Graham, Neill (1973), *The Many-Worlds Interpretation of Quantum Mechanics* (Princeton, NJ).

Dicke, R. H. (1961), 'Dirac's Cosmology and Mach's Principle', *Nature*, 192 (4 Nov.): 440–1.

Dieks, D. (1988), 'Special Relativity and the Flow of Time', *Philosophy of Science*, 55: 456–60.

Douglas, Roy (1994), 'Stochastically-Branching Space-Time Topology' in S. Savitt (ed.), *Time's Arrows Today: Recent Physical and Philosophical Work on the Direction of Time* (Cambridge).

Dretske, Fred. I. (1977), 'Laws of Nature', *Philosophy of Science*, 44: 248–68.

—— and Snyder, A. (1972), 'Causal Irregularity', *Philosophy of Science*, 39: 69–71.

Dudman, V. H. (1984), 'Parsing 'If'-Sentences', *Analysis*, 44: 145–53.

Dummett, Michael (1954), 'Can an Effect Precede its Cause?', *Proceedings of the Aristotelian Society*, Supp. vol. 28: 27–44.

—— (1960), 'A Defence of McTaggart's Proof of the Unreality of Time', *Philosophical Review*, 69: 497–504; reprinted in Dummett (1978).

—— (1964), 'Bringing about the Past', *Philosophical Review*, 73: 338–59.

—— (1969), 'The Reality of the Past', *Proceedings of The Aristotelian Society*, 69: 239–58.

—— (1973), *Frege: Philosophy of Language* (London).

—— (1978), *Truth and Other Enigmas* (London).

Dupré, John (1981), 'Natural Kinds and Biological Taxa', *Philosophical Review*, 90: 66–90.

Earman, John (1972), 'Some Aspects of General Relativity and Geometrodynamics', *Journal of Philosophy*, 69: 634–47.

—— (1974), 'An Attempt to Add a Little Direction to "The problem of the direction of time" ', *Philosophy of Science*, 41: 15–47.
—— (1984), 'Laws of Nature: The Empiricist Challenge', in R. J. Bogdan (ed.), *D. M. Armstrong* (Dordrecht).
—— (1986), *A Primer on Determinism* (Dordrecht).
—— (1989), *World Enough and Space-Time* (Cambridge, Mass.).
Eddington, Arthur S. (1920), *Space, Time and Gravitation* (Cambridge).
—— (1928), *The Nature of the Physical World* (Cambridge).
Einstein, A. (1936), 'Physics and Reality', *Journal of the Franklin Institute*, 221: 349–82.
—— Podolsky, B., and Rosen, N. (1935), 'Can Quantum Mechanical Description of Physical Reality be Considered Complete?', *Physical Review*, 47: 138–41; reprinted in Wheeler and Zurek (1983).
Ellis, Brian (1984), 'Two Theories of Indicative Conditionals', *Australasian Journal of Philosophy*, 62: 50–66.
Everett, Hugh (1957), ' "Relative state" formulation of quantum mechanics', *Reviews of Modern Physics*, 29; reprinted in DeWitt and Graham (1973).
Fales, Evan (1982), 'Natural Kinds and Freaks of Nature', *Philosophy of Science*, 49: 67–90.
Fetzer, James (1971), 'Dispositional Probabilities', in R. Buck and R. S. Cohen (eds.), *Boston Studies in the Philosophy of Science*, viii (Dordrecht).
—— (1981), *Scientific Knowledge* (Dordrecht).
Feynman, Richard P. (1982), 'Simulating Physics with Computers', *International Journal of Theoretical Physics*, 21: 467–88.
Fine, Arthur (1980), 'Correlations and Physical Locality', in P. D. Asquith and R. N. Giere (eds.), *PSA 1980* [*Proceedings* of the Philosophy of Science Association] (East Lansing, Mich.), ii. 535–62.
—— (1986), *The Shaky Game* (Chicago).
—— (1989), 'Do Correlations Need to be Explained?', in Cushing and McMullin (1989), 175–94.
Fitzgerald, Paul (1969), 'The Truth about Tomorrow's Sea Fight', *Journal of Philosophy*, 66: 307–29.
Fleming, Gordon (1988), 'Lorentz Invariant State Reduction, and Localization', in A. Fine and J. Leplin (eds.), *PSA 1988* [*Proceedings* of the Philosophy of Science Association], (East Lansing, Mich.), ii. 112–26.
Flew, Antony (1954), 'Can an Effect Precede its Cause?', *Proceedings of the Aristotelian Society*, Supp. vol. 28: 45–62.
Forbes, Graeme (1980), 'Origin and Identity', *Philosophical Studies*, 37: 353–62.
—— (1985), *The Metaphysics of Modality* (Oxford).
—— (1986), 'In Defense of Absolute Essentialism', *Midwest Studies in Philosophy*, 11: 3–31.
Franklin, James (1986), 'Aristotle on Species Variation', *Philosophy*, 61: 245–52.

Franklin, James (1989), 'Species in Aristotle', *Philosophy,* 64: 107–8.

Fraser, J. T. (1987), *Time, the Familiar Stranger* (Amherst, Mass).

Freeman, Eugene, and Sellars, Wilfrid (1971) (eds.), *Basic Issues in the Philosophy of Time* (La Salle, Ill.).

Friedman, Michael (1983), *Foundations of Space-Time Theories* (Princeton, NJ).

Gabbay, D., and Guenther, F. (1983–9) (eds.), *Handbook of Philosophical Logic*, 4 vols. (Dordrecht).

Gale, Richard (1965), 'Why a Cause Cannot be Later than its Effect', *Review of Metaphysics*, 19: 209–34.

—— (1968), *The Language of Time* (London).

Gardiner, P. L. (1952), *The Nature of Historical Explanation* (Oxford).

Geach, P. T. (1962), *Reference and Generality* (Ithaca, NY).

—— (1965), 'Some Problems about Time', *Proceedings of the British Academy*, 51: 321–36; reprinted in Geach (1972).

—— (1968), 'What Actually Exists?', *Proceedings of the Aristotelian Society*, Supp. vol. 42: 7–16.

—— (1972), *Logic Matters* (Oxford).

—— (1979), *Truth, Love and Immortality* (London).

Ghirardi, G. C., Grassi, R., and Pearle, P. (1990), 'Relativistic Dynamical Reduction Models: General Framework and Examples', *Foundations of Physics*, 20: 1271–1316.

—— Pearle, P., and Rimini, A. (1990), 'Markov Processes in Hilbert Space and Continuous Spontaneous Localization of Systems of Identical Particles', *Physical Review*, A 42: 78–89.

—— Rimini, A., and Weber, T. (1980), 'A General Argument against Superluminal Transmission through the Quantum Mechanical Measurement Process', *Lettere al Nuovo Cimento*, 27: 293–8.

—————— (1986), 'Unified Dynamics for Microscopic and Macroscopic Systems', *Physical Review*, D 34: 470–91.

—————— (1988), 'The Puzzling Entanglement of Schrödinger's Wave Function', *Foundations of Physics*, 18: 1–27.

Gibbard, Allan (1981), 'Two Recent Theories of Conditionals', in Harper *et al.* (1981), 211–47.

Giere, R. N. (1973), 'Objective Single-Case Probabilities and the Foundations of Statistics', in *Logic, Methodology and Philosophy of Science IV* (Amsterdam).

Gillies, D. A. (1973), *An Objective Theory of Probability* (London).

Gödel, Kurt (1949), 'A Remark about the Relationship between Relativity Theory and Idealistic Philosophy', in P. A. Schilpp (ed.), *Albert Einstein: Philosopher-Scientist* (Evanston, Ill.), 557–562.

Goldblatt, Robert (1980), 'Diodorean Modality in Minkowski Space-time', *Studia Logica*, 39: 219–36.

Goodman, Nelson (1947), 'The Problem of Counterfactual Conditionals', *Journal of Philosophy*, 44: 113–28; reprinted as ch. 1 of Goodman (1955).

—— (1951), *The Structure of Appearance* (Cambridge, Mass.).

—— (1955), *Fact, Fiction, and Forecast* (London).

Gorovitz, Samuel (1964), 'Leaving the Past Alone', *Philosophical Review*, 73: 360–71.

Granger, Herbert (1987), 'Aristotle and the Finitude of Natural Kinds', *Philosophy*, 62: 523–6.

—— (1989), 'Aristotle's Natural Kinds', *Philosophy*, 64: 245–7.

Greenberger, D. M., Horne, M., and Zeilinger, A. (1989), 'Going beyond Bell's Theorem', in M. Kafatos (ed.), *Bell's Theorem, Quantum Theory, and Conceptions of the Universe* (Dordrecht), 7–76.

———— Shimony, A., and Zeilinger, A. (1990), 'Bell's Theorem without Inequalities', *American Journal of Physics*, 58: 1131–43.

Grünbaum, Adolf (1967a), 'The Anisotropy of Time' [with discussion], in T. Gold (ed.), *The Nature of Time* (Ithaca, NY), 149–86.

—— (1967b), *Modern Science and Zeno's Paradoxes* (Middletown, Conn.).

—— (1969), 'The Meaning of Time', in Freeman and Sellars (1971), 195–228.

—— (1973), 'Karl Popper's Views on the Arrow of Time', in P. A. Schilpp (ed.), *The Philosophy of Karl Popper* (La Salle, Ill.), 775–97.

—— (1974), *Philosophical Problems of Space and Time*, 2nd edn. (Dordrecht).

Gupta, Anil (1980), *The Logic of Common Nouns* (New Haven, Conn.).

Hacking, Ian (1965), *Logic of Statistical Inference* (Cambridge).

—— (1975), *The Emergence of Probability* (Cambridge).

—— (1987), 'Coincidences: Mundane and Cosmological', in J. M. Robson (ed.), in *Origin and Evolution of the Universe: Evidence for Design?*, (Kingston, Ont.).

Hancock, Roger (1968), 'Choosing as Doing', *Mind*, 77: 575–6.

Harman, Gilbert (1965), 'The Inference to the Best Explanation', *Philosophical Review*, 74: 88–95.

—— (1968), 'Knowledge, Inference, and Explanation', *American Philosophical Quarterly*, 5: 164–73.

Harper, William L., et al. (1981) (eds.), *Ifs* (Dordrecht).

Harris, Erroll E. (1988), *The Reality of Time* (Albany, NY).

Hawking, Stephen W. (1988), *A Brief History of Time* (New York).

Healey, Richard A. (1984), 'How Many Worlds?', *Nous*, 18: 591–616.

—— (1989), *The Philosophy of Quantum Mechanics* (Cambridge).

Heisenberg, Werner (1958), *Physics and Philosophy* (New York).

Hellman, Geoffrey (1984), 'Introduction', *Nous*, 18: 557–67.

Hinckfuss, Ian (1975), *The Existence of Space and Time* (Oxford).

Hobart, R. E. (1934), 'Free Will as Involving Determination and Inconceivable without It', *Mind*, 43: 1–27.

Horwich, Paul (1987), *Asymmetries in Time* (Cambridge, Mass.).

Hughes, R. I. G. (1989), *The Structure and Interpretation of Quantum Mechanics* (Cambridge, Mass.).

Humphreys, Paul (1985), 'Why Propensities Cannot be Probabilities', *Philosophical Review*, 94: 557–70.

James, William (1897), 'The Dilemma of Determinism', in *The Will to Believe* (New York).

Jaques, Elliott (1982), *The Form of Time* (New York).

Jarrett, Jon P. (1984), 'On the Physical Significance of the Locality Conditions in the Bell Arguments', *Nous*, 18: 509–89.

Jauch, J. M. (1968), *Foundation of Quantum Mechanics* (Reading, Mass.).

Johnston, Patricia (1977), 'Origin and Necessity', *Philosophical Studies*, 32: 413–18.

Kaplan, David (1979), 'Transworld Heir Lines', in Loux (1979), 88–109.

Kemp Smith, Norman (1933) (trans.), *Critique of Pure Reason* by Immanuel Kant (London).

Keynes, John Maynard (1921), *A Treatise on Probability* (London).

Kim, Jaegwon (1984), 'Concepts of Supervenience', *Philosophy and Phenomenological Research*, 45: 153–76.

Kolb, E. W., and Turner, M. S. (1990), *The Early Universe* (New York).

Kripke, Saul A. (1971), 'Identity and Necessity', in M. K. Munitz (ed.), *Identity and Individuation* (New York); reprinted in Schwartz (1977), 66–101.

—— (1980), *Naming and Necessity*, 2nd edn. (Cambridge, Mass.).

Kroes, Peter (1984), 'Objective versus Mind-Dependent Theories of Time Flow', *Synthese*, 61: 423–46. Reprinted with minor changes in Kroes (1985).

—— (1985), *Time: Its Structure and Role in Physical Theories* (Dordrecht).

Kvart, Igal (1986), *A Theory of Counterfactuals* (Indianapolis).

Kyburg, Henry (1974a), *The Logical Foundations of Statistical Inference* (Dordrecht).

—— (1974b), 'Propensities and Probabilities', *British Journal for the Philosophy of Science*, 25: 358–75.

—— and Smokler, H. E. (1964), *Studies in Subjective Probability* (New York).

Laplace, Marquis de (1820), *Traité analytique des probabilités* (Paris).

Laudan, Larry (1981a), *Science and Hypothesis* (Dordrecht).

—— (1981b), 'A Confutation of Convergent Realism', *Philosophy of Science*, 48: 19–49.

Leggett, A. J. (1987), *The Problems of Physics* (Oxford).

Leibniz, Gottfried (1902), *Discourse on Metaphysics, Correspondence with Arnauld and Monadology*, trans. G. R. Montgomery, (Chicago).

Lewis, David (1968), 'Counterpart Theory and Quantified Modal Logic', *Journal of Philosophy*, 65: 113–26; reprinted in Loux (1979).

—— (1973), *Counterfactuals* (Oxford).

—— (1976), 'Probabilities of Conditionals and Conditional Probabilities', *Philosophical Review*, 85; reprinted in Lewis (1986a).

—— (1980), 'A Subjectivist's Guide to Objective Chance', in R. C. Jeffrey (ed.), *Studies in Inductive Logic and Probability*, ii; reprinted in Lewis (1986a), 83–132, with postscripts.

—— (1983), *Philosophical Papers*, i (Oxford).

—— (1986a). *Philosophical Papers*, ii (New York).

—— (1986b) *On the Plurality of Worlds* (Oxford).

Locke, John (1689), *An Essay Concerning Human Understanding* (London). Quotations from the text edited by A. C. Fraser (Oxford, 1894).

Loizou, Andros (1986), *The Reality of Time* (Aldershot, Hampshire).

Lombard, Lawrence Brian (1986), *Events* (London).

London, Fritz, and Bauer, Edmond (1939, 1983), 'The Theory of Observation in Quantum Mechanics', English translation of 'La théorie de l'observation en mécanique quantique' in Wheeler and Zurek (1983), 217–59.

Loux, Michael (1979) (ed.), *The Possible and the Actual* (Ithaca, NY).

Lucas, John R. (1973), *A Treatise on Time and Space* (London).

—— (1989), *The Future* (Oxford).

Łukasiewicz, Jan (1922), 'On Determinism', in S. McCall (ed.), *Polish Logic 1920–1939* (Oxford, 1967), 19–39.

—— (1930), 'Philosophical Remarks on Many-Valued Systems of Propositional Logic', in S. McCall (ed.), *Polish Logic 1920–1939* (Oxford 1967), 40–65.

McAdam, James I. (1965), 'Choosing Flippantly or Non-Rational Choice', *Analysis*, 25: 132–6.

McArthur, Robert P. (1974), 'Factuality and Modality in the Future Tense', *Nous*, 8: 283–8.

—— (1976), *Tense Logic* (Dordrecht).

McCall, Storrs (1965), 'Abstract Individuals', *Dialogue*, 4: 217–31; reprinted in Karel Lambert (ed.), *Philosophical Applications of Free Logic* (New York, 1991).

—— (1966), 'Temporal Flux', *American Philosophical Quarterly*, 3: 270–81.

—— (1968), 'On What it Means to be Future' [abstract], *Journal of Symbolic Logic*, 33: 640.

—— (1969), 'Time and the Physical Modalities', *Monist*, 53: 426–46; reprinted in Freeman and Sellars (1971).

—— (1970), 'The Cardinality of Possible Futures' [abstract], *Journal of Symbolic Logic*, 35: 363.

—— (1976), 'Objective Time Flow', *Philosophy of Science*, 43: 337–62.

—— (1979), 'The Strong Future Tense', *Notre Dame Journal of Formal Logic*, 20: 489–504.

—— (1983), 'If, Since and Because: A Study in Conditional Connection', *Logique et Analyse*, 103–4: 309–21.

—— (1984a), 'Counterfactuals Based on Real Possible Worlds', *Nous*, 18: 463–77.

—— (1984b), 'A Dynamic Model of Temporal Becoming', *Analysis*, 44: 172–6.

—— (1984c), 'Freedom Defined as the Power to Decide', *American Philosophical Quarterly*, 21: 329–38.

—— (1985), 'Incline without Necessitating', *Dialogue*, 24: 589–96.

—— (1987), 'Decision', *Canadian Journal of Philosophy*, 17: 261–87.

—— (1990), 'Choice Trees', in J. M. Dunn and A. Gupta (eds.), *Truth or Consequences: Essays in Honour of Nuel Belnap* (Dordrecht), 231–44.

McDowell, John (1978), 'On "The Reality of the Past" ', in C. Hookway and P. Pettit (eds.), *Action and Interpretation* (Cambridge).

McGinn, Colin (1976), 'On the Necessity of Origin', *Journal of Philosophy*, 73: 127–35.

Mackie, John L. (1974*a*). *The Cement of the Universe* (Oxford).

—— (1974*b*). '*De* what *re* is *de re* Modality?', *Journal of Philosophy*, 71: 551–61.

McTaggart, J. M. E. (1927), *The Nature of Existence* (Cambridge).

Maxwell, Nicholas (1985), 'Are Probabilism and Special Relativity Incompatible?', *Philosophy of Science*, 52: 23–43.

—— (1988), 'Are Probabilism and Special Relativity Compatible?', *Philosophy of Science*, 55: 640–5.

Melden, A. I. (1961), *Free Action* (London).

Mellor, D. H. (1971), *The Matter of Chance* (Cambridge).

—— (1981*a*), *Real Time* (Cambridge).

—— (1981*b*). 'McTaggart, Fixity, and Coming True', in R. Healey (ed.), *Reduction, Time and Reality* (Cambridge), 79–97.

Mermin, N. David (1968), *Space and Time in Special Relativity* (New York).

—— (1981), 'Quantum Mysteries for Anyone', *Journal of Philosophy*, 78: 397–408; reprinted in Cushing and McMullin (1989).

—— (1985), 'Is the Moon There When Nobody Looks? Reality and the Quantum Theory', *Physics Today* 38 (4): 38–47.

—— (1990*a*), 'What's Wrong with these Elements of Reality?', *Physics Today*, June: 9–11.

—— (1990*b*), 'Quantum Mysteries Revisited', *American Journal of Physics*, 58: 731–4.

Mill, John Stuart (1843, 1973), *A System of Logic*, vol. vii of J. F. Robson (ed.), *Collected Works* (Toronto).

Montague, Richard (1962), 'Deterministic Theories'; reprinted in his *Formal Philosophy* (New Haven, Conn., 1974), 303–59.

Morris, Richard (1985), *Time's Arrows* (New York).

Munkres, J. R. (1975), *Topology* (Englewood Cliffs, NJ).

Nagel, Ernest (1961), *The Structure of Science* (New York).

Newton-Smith, William (1980), *The Structure of Time* (London).

Noonan, Harold W. (1988), 'Substance, Identity and Time', *Proceedings of the Aristotelian Society*, Supp. vol. 62: 79–100.

Oaklander, Nathan (1984), *Temporal Relations and Temporal Becoming* (Lanham, Md.).

Ofstad, Harald (1961), *An Inquiry into the Freedom of Decision* (Oslo).

Otte, Richard (1987), 'Critical Discussion: McCall and counterfactuals', *Nous*, 21: 421–5.

Parfit, Derek (1984), *Reasons and Persons* (Oxford).

Pargetter, Robert (1984), 'Laws and Modal Realism', *Philosophical Studies*, 46: 335–47.

Park, David (1980), *The Image of Eternity* (Amherst, Mass.).

Pearle, Philip (1989), 'Combining Stochastic Dynamical State-Vector Reduction with Spontaneous Localization', *Physical Review* A 39: 2277–89.

—— (1990), 'Toward a Relativistic Theory of State Vector Reduction' in A. I. Miller (ed.), *Sixty-two Years of Uncertainty* (New York).

—— and Soucek, J. (1989), 'Path Integrals for the Continuous Spontaneous Localization Theory', *Foundations of Physics Letters*, 2: 287–96.

Pears, David (1956), 'The Priority of Causes', *Analysis*, 17: 54–63.

Peirce, C. S. (1931–5), *Collected Papers*, iv and v (Cambridge, Mass.).

Penrose, Roger (1989), *The Emperor's New Mind* (Oxford).

Perry, John (1975) (ed.), *Personal Identity* (Berkeley, Calif.).

Plantinga, Alvin (1973), 'Transworld Identity or Worldbound Individuals?', in M. K. Munitz (ed.), *Logic and Ontology* (New York); reprinted in Loux (1979).

—— (1974), *The Nature of Necessity* (Oxford).

—— (1985), 'Self-Profile', in J. E. Tomberlin and P. van Inwagen (eds.), *Alvin Plantinga* (Dordrecht).

Popper, Karl (1950), 'Indeterminism in Quantum Physics and in Classical Physics', *British Journal for the Philosophy of Science*, 1: 117–33, 173–95.

—— (1957), 'The Propensity Interpretation of the Calculus of Probability, and the Quantum Theory', in S. Körner (ed.), *Observation and Interpretation* (London), 65–70.

—— (1959), 'The Propensity Interpretation of Probability', *British Journal for the Philosophy of Science*, 10: 25–42.

Price, M. S. (1982), 'On the Non-Necessity of Origin', *Canadian Journal of Philosophy*, 12: 33–45.

Prior, Arthur N. (1957), *Time and Modality* (Oxford).

—— (1958), 'Time after Time', *Mind*, 67: 244–6.

—— (1962), *Changes in Events and Changes in Things*, lecture, University of Kansas; reprinted in Prior (1968), 1–14.

—— (1967), *Past, Present and Future* (Oxford).

—— (1968), *Papers on Time and Tense* (Oxford).

Putnam, Hilary (1965), 'A Philosopher Looks at Quantum Mechanics; reprinted in Putnam (1975a), 130–158.

—— (1967), 'Time and Physical Geometry', *Journal of Philosophy*, 64: 240–7; reprinted in Putnam (1975a).

—— (1970), 'Is Semantics Possible?' in H. E. Kiefer and M. K. Munitz (eds.), *Language, Belief and Metaphysics* (New York); reprinted in Putnam (1975b), 139–52, and Schwartz (1977).

—— (1973), 'Meaning and Reference', *Journal of Philosophy*, 70: 699–711; reprinted in Schwartz (1977).

—— (1975a), *Philosophical Papers*, i (Cambridge).

—— (1975b), *Philosophical Papers*, ii (Cambridge).

—— (1975c), 'The Meaning of "Meaning" ', in K. Gunderson (ed.), *Language, Mind and Knowledge*; reprinted in Putnam (1975b), 215–71.

Putnam, Hilary (1981), 'Quantum Mechanics and the Observer', *Erkenntnis*, 16: 193–219; reprinted in Putnam (1983).

—— (1983), *Philosophical Papers*, iii (Cambridge).

—— (1984), 'Is the Causal Structure of the Physical Itself Something Physical?', *Midwest Studies in Philosophy*, 9: 3–16.

Quine, W. V. (1953), *From a Logical Point of View* (Cambridge, Mass.).

—— (1960), *Word and Object* (Cambridge, Mass.).

—— (1969), 'Natural Kinds', in *Ontological Relativity and Other Essays* (New York), 114–38; reprinted in Schwartz (1977).

—— (1976a), 'Worlds away', *Journal of Philosophy*, 73: 859–63.

—— (1976b), 'Whither Physical Objects?', in R. S. Cohen (ed.), *Essays in Memory of Imre Lakatos* (Dordrecht).

—— (1981), *Theories and Things* (Cambridge, Mass.).

—— (1982), *Methods of Logic*, 4th edn. (Cambridge, Mass.).

Quinton, A. M. (1962), 'The Soul', *Journal of Philosophy*, 59: 393–409.

Rae, Alastair (1986), *Quantum Physics: Illusion or Reality?* (Cambridge).

Ramsey, F. P. (1931), *The Foundations of Mathematics and Other Logical Essays* (London).

—— (1990), *Philosophical Papers*, ed. D. H. Mellor (Cambridge).

Redhead, Michael (1983), 'Relativity, Causality, and the Einstein–Podolsky–Rosen Paradox', in Swinburne (1983).

—— (1987), *Incompleteness, Nonlocality, and Realism* (Oxford).

Reichenbach, Hans (1949), *The Theory of Probability* (Berkeley, Calif.).

—— (1957), *The Philosophy of Space and Time* (New York); originally published in German in 1928.

Rescher, Nicholas (1959), 'Choice without Preference', *Kant-Studien*, 51: 142–75.

—— and Urquhart, Alasdair (1971), *Temporal Logic* (New York).

Rietdijk, C. W. (1966), 'A Rigorous Proof of Determinism Derived from the Special Theory of Relativity', *Philosophy of Science*, 33: 341–4.

Robinson, Abraham (1966), *Non-Standard Analysis* (Amsterdam).

Ruse, Michael (1987), 'Biological Species: Natural Kinds, Individuals, or What?', *British Journal for the Philosophy of Science*, 38: 225–42.

Russell, Bertrand (1903), *The Principles of Mathematics* (London).

—— (1913), 'On the Notion of Cause', *Proceedings of the Aristotelian Society*, 13; reprinted in his *Mysticism and Logic* (London, 1917).

—— (1915), 'On the Experience of Time', *Monist*, 25: 212–33.

—— (1940), *An Inquiry into Meaning and Truth* (London).

—— (1948), *Human Knowledge: Its Scope and Limits* (London).

Ryle, Gilbert (1949), *The Concept of Mind* (London).

—— (1950), ' "If", "So", and "Because" ' in Max Black (ed.), *Philosophical Analysis* (Ithaca, NY), 302–18.

Salmon, Wesley (1967), *The Foundations of Scientific Inference* (Pittsburgh).

—— (1971), *Statistical Explanation and Statistical Relevance* (Pittsburgh).

—— (1975), *Space, Time and Motion* (Ann Arbor, Mich.).

—— (1979), 'Propensities: A Discussion Review', *Erkenntnis*, 14: 183–216.

—— (1984), *Scientific Explanation and the Causal Structure of the World* (Princeton, NJ).

—— (1988), 'Dynamic Rationality: Propensity, Probability, and Credence', in J. Fetzer (ed.), *Probability and Causality* (Dordrecht).

Saunders, John Turk (1965), 'Fatalism and Ordinary Language', *Journal of Philosophy*, 62: 211–22.

Savage, L. J. (1954), *The Foundations of Statistics* (New York).

Schlesinger, George (1980), *Aspects of Time* (Indianapolis).

Schwartz, Stephen P. (1977) (ed.), *Naming, Necessity, and Natural Kinds* (Ithaca, NY).

Seddon, Keith (1987), *Time: A Philosophical Treatment* (London).

Selby-Bigge, L.A. (1888) (ed.), *A Treatise on Human Nature* by David Hume (Oxford).

—— (1894) (ed.), *An Enquiry Concerning Human Understanding* by David Hume (Oxford).

Shimony, Abner (1978), 'Metaphysical Problems in the Foundations of Quantum Mechanics', *International Philosophical Quarterly*, 18: 3–17.

—— (1984), 'Controllable and Uncontrollable Non-Locality', in *Foundations of Quantum Mechanics in the Light of New Technology* (Tokyo), 225–30.

—— (1986), 'Events and Processes in the Quantum World', in R. Penrose and C. J. Isham (eds.), *Quantum Concepts in Space and Time* (Oxford).

—— (1988), 'The Reality of the Quantum World', *Scientific American*, Jan: 46–53.

—— (1989*a*), 'Search for a Worldview which Can Accommodate our Knowledge of Microphysics', in Cushing and McMullin (1989), 25–37.

—— (1989*b*), 'Conceptual Foundations of Quantum Mechanics', in Paul Davies (ed.), *The New Physics* (Cambridge), 373–95.

Shoemaker, Sydney (1963), *Self-Knowledge and Self-Identity* (Ithaca, NY).

—— (1979), 'Identity, Properties, and Causality', *Midwest Studies in Philosophy*, 4: 321–42.

—— and Swinburne, Richard (1984), *Personal Identity* (Oxford).

Sklar, Lawrence (1974), *Space, Time, and Spacetime* (Berkeley, Calif.).

Skyrms, Brian (1980), *Causal Necessity* (New Haven, Conn.).

Slote, Michael (1978), 'Time in Counterfactuals', *Philosophical Review*, 87: 3–27.

Smart, J. J. C. (1949), 'The River of Time', *Mind*, 58: 483–94.

—— (1961), 'Free-will, Praise and Blame', *Mind*, 70: 291–306.

—— (1967), Entry on 'Time' in Paul Edwards (ed.), *The Encyclopedia of Philosophy* (New York).

—— (1980), 'Time and Becoming', in van Inwagen (1980).

Smith, Constance I. (1956), 'A Note on Choice and Virtue', *Analysis*, 17: 21–3.

Smith, Quentin (1985), 'The Mind-Independence of Temporal Becoming', *Philosophical Studies*, 47: 109–19.

Sober, Elliott (1985), 'Two Concepts of Cause', in *PSA 1984* [*Proceedings* of the Philosophy of Science Association] (East Lansing, Mich.), ii.

Sorabji, Richard (1983), *Time, Creation and the Continuum* (London).

Sosa, Ernest (1975) (ed.), *Causation and Conditionals* (Oxford).

Squires, Euan (1986), *The Mystery of the Quantum World* (London).

Stalnaker, Robert (1968), 'A Theory of Conditionals', in N. Rescher (ed.), *Studies in Logical Theory*, APQ Monograph Series (Oxford); reprinted in Sosa (1975).

—— (1975), 'Indicative Conditionals', *Philosophia*, 5; reprinted in Harper *et al.* (1981), 193–210.

—— (1986), 'Counterparts and Identity', *Midwest Studies in Philosophy*, 11: 121–40.

Stapp, Henry P. (1989), 'Quantum Non-Locality and the Description of Nature', in Cushing and McMullin (1989), 154–74.

—— (1993), 'Significance of an Experiment of the Greenberger–Horne–Zeilinger Kind', *Physical Review*, A47: 847–53.

Stein, Howard (1968), 'On Einstein–Minkowski space-time', *Journal of Philosophy*, 65: 5–23.

—— (1970), 'Is There a Problem of Interpreting Quantum Mechanics?', *Nous*, 4: 93–103.

—— (1982), 'On the Present State of the Philosophy of Quantum Mechanics', in P. D. Asquith and T. Nickles (eds.), *PSA 1982* [*Proceedings* of the Philosophy of Science Association] (East Lansing, Mich.), ii. 563–81.

—— (1991), 'On Relativity Theory and Openness of the Future', *Philosophy of Science*, 58: 147–67.

Sudbery, Anthony (1986), *Quantum Mechanics and the Particles of Nature* (Cambridge).

Suppes, Patrick (1973), 'New Foundations of Objective Probability: Axioms for Propensities', in *Logic, Methodology and Philosophy of Science IV* (Amsterdam).

—— (1974), 'Popper's Analysis of Probability in Quantum Mechanics', in P. A. Schilpp (ed.), *The Philosophy of Karl Popper*, ii (La Salle, Ill.).

Swinburne, Richard (1981), *Space and Time*, 2nd edn. (London).

—— (1983) (ed.), *Space, Time and Causality* (Dordrecht).

Szamosi, Geza (1986), *The Twin Dimensions* (New York).

Taylor, Richard (1958), 'Determinism and the Theory of Agency', in Sydney Hook (ed.), *Determinism and Freedom in the Age of Modern Science* (New York).

—— (1962), 'Fatalism', *Philosophical Review*, 71: 56–66.

—— (1964), 'Deliberation and Foreknowledge', *American Philosophical Quarterly*, 1: 73–80.

—— (1966), *Action and Purpose* (Englewood Cliffs, NJ).

Thomason, Richmond H. (1970), 'Indeterminist Time and Truth-Value Gaps', *Theoria*, 36: 264–81.

—— (1984), 'Combinations of Tense and Modality', in Gabbay and Guenther (1983–9), ii. 135–65.

—— and Gupta, Anil (1980), 'A Theory of Conditionals in the Context of Branching Time', *Philosophical Review*, 89: 65–90.

Thorp, John (1980), *Free Will* (London).

Tooley, Michael (1977), 'The Nature of Laws', *Canadian Journal of Philosophy*, 7: 667–98.

—— (1987), *Causation* (Oxford).

Torretti, Roberto (1983), *Relativity and Geometry* (Oxford).

Vallentyne, Peter (1988), 'Explicating Lawhood', *Philosophy of Science*, 55: 598–613.

van Benthem, Johan (1982), *The Logic of Time* (Dordrecht).

van Fraassen, Bas (1970), *An Introduction to the Philosophy of Time and Space* (New York).

—— (1972), 'A Formal Approach to the Philosophy of Science', in Colodny (1972), 303–66.

—— (1980), *The Scientific Image* (Oxford).

—— (1982), 'The Charybdis of Realism: Epistemological Implications of Bell's Inequality', *Synthese*, 52: 25–38; reprinted with additions in Cushing and McMullin (1989).

—— (1985), 'EPR: When is a Correlation not a Mystery?', in P. J. Lahti and P. Mittelstaedt (eds.), *Symposium on the Foundations of Modern Physics* (Singapore), 113–28.

—— (1989), *Laws and Symmetry* (Oxford).

—— (1991), *Quantum Mechanics: An Empiricist View* (Oxford).

van Inwagen, Peter (1980) (ed.), *Time and Cause: Essays Presented to Richard Taylor* (Dordrecht).

—— (1983). *An Essay on Free Will* (Oxford).

—— (1985), 'Plantinga on Transworld Identity', in J. E. Tomberlin and P. van Inwagen (eds.), *Alvin Plantinga* (Dordrecht).

Venn, John (1866), *The Logic of Chance* (London).

von Mises, Richard (1939), *Probability, Statistics and Truth* (London). (2nd edn., London, 1957.)

von Neumann, J. (1932, 1955), *Mathematical Foundations of Quantum Mechanics* (Princeton, NJ); English translation of *Mathematische Grundlagen der Quantenmechanik*, 1932.

Weyl, Hermann (1949), *Philosophy of Mathematics and Natural Sciences* (Princeton, NJ).

Wheeler, J. A. and Zurek, W. H. (1983) (eds.), *Quantum Theory and Measurement* (Princeton, NJ).

Whitrow, G. J. (1972), *What is Time?* (London).

—— (1980), *The Natural Philosophy of Time*, 2nd edn. (Oxford).

Wiggins, David (1967), *Identity and Spatio-Temporal Continuity* (Oxford).

—— (1980), *Sameness and Substance* (Oxford).

Wigner, Eugene P. (1961), 'Remarks on the Mind–Body Question', in I. J. Good (ed.), *The Scientist Speculates* (London); reprinted in Wheeler and Zurek (1983), 168–81.

—— (1963), 'The Problem of Measurement', *American Journal of Physics*, 31: 6–15; reprinted in Wheeler and Zurek (1983), 324–41.

Wigner, Eugene P. (1970), 'On Hidden Variables and Quantum Mechanical Probabilities', *American Journal of Physics* 38: 1005–9.

Williams, Bernard (1970), 'The Self and the Future', *Philosophical Review*, 79: 161–80.

Williams, Donald C. (1951), 'The Myth of Passage', *Journal of Philosophy*, 48: 457–72.

Wippel, J. F., and Wolter, A. B. (1969) (eds.), *Medieval Philosophy* (New York).

Zwart, P. J. (1972), 'The Flow of Time', *Synthese*, 24: 133–58.

—— (1976), *About Time* (Amsterdam).

Index